MODERN MATERIALS AND MANUFACTURING PROCESSES

Second Edition

R. Gregg Bruce
Purdue University School of Technology

Mileta M. Tomovic
Purdue University School of Technology

John E. Neely
Instructor, Machine Technology, Ret.
Lane Community College, Eugene, Oregon

Richard R. Kibbe
Managing Director, Machining and Tooling Development
R.R.K. Machine and Manufacturing

Prentice Hall
Upper Saddle River, New Jersey   Columbus, Ohio

Library of Congress Cataloging-in-Publication Data
Modern materials and manufacturing processes / Gregg Bruce ... [et al.].—2nd ed.
    p. cm.
    Includes bibliographical references and index.
    ISBN 0-13-186859-4
    1. Manufacturing processes.   2. Materials.   3. Metal-work.
I. Bruce, Gregg (R. Gregg Bruce)
TS183.M617   1998   97-5993
670—dc21   CIP

Cover photo: Michael Rosenfeld/Tony Stone Images
Editor: Stephen Helba
Production Editor: Mary M. Irvin
Design Coordinator: Karrie M. Converse
Text Designer: Linda M. Robertson
Cover Designer: Russ Maselli
Production Manager: Pamela A. Bennett
Marketing Manger: Frank Mortimer, Jr.

This book was set in Times Roman by The Clarinda Company and was printed and bound by Courier Kendalville. The cover was printed by Phoenix Color Corp.

© 1998 by Prentice-Hall, Inc.
Simon & Schuster/A Viacom Company
Upper Saddle River, New Jersey 07458

Printed in the United States of America

10 9 8 7 6 5 4 3 2 1

ISBN: 0-13-186859-4

Prentice-Hall International (UK) Limited, *London*
Prentice-Hall of Australia Pty. Limited, *Sydney*
Prentice-Hall of Canada, Inc., *Toronto*
Prentice-Hall Hispanoamericana, S. A., *Mexico*
Prentice-Hall of India Private Limited, *New Delhi*
Prentice-Hall of Japan, Inc., *Tokyo*
Simon & Schuster Asia Pte. Ltd., *Singapore*
Editora Prentice-Hall do Brasil, Ltda., *Rio de Janeiro*

Preface

This textbook presents state-of-the art technology on materials and manufacturing processes in a style that can be understood by the typical technician or technology student. It is intended for students in technology, technicians, and vocational education students.

The book is organized to present clear, sequential course material for any teaching system, whether competency based or traditional. Objectives at the beginning of each chapter clarify the goals of each chapter. Case studies and case problems are included where appropriate to promote thought and discussion on the part of the students. Questions and case problems may be used for review or examination purposes. Nearly 1000 illustrations and photographs complement the text.

This book is intended primarily for survey courses in materials and manufacturing processes. Rapidly growing technologies such as those found in the plastics and aerospace industries are emphasized. We not only discuss the specific materials and processes of manufacturing normally included in a survey course, but also, in a unique approach, show how those materials and processes are integrated into today's functioning manufacturing industry. However, this book does not provide a hands-on instructional system for the operation of machinery or the use of hand tools.

Part I, on materials and their applications, discusses frequently used materials such as metals, plastics, and rubbers, which are identified and classified. Their extraction from raw materials and their processing are presented in sufficient detail to enable the student to proceed to the study of product manufacturing from these materials.

Part II introduces measurement and quality assurance with particular emphasis on tolerances, measurement, and calibration.

Part III covers specific manufacturing processes. Conventional manufacturing processes including casting, cold and hot rolling, forming, forging, plastics molding, and joining of materials are discussed in detail. Newer processes and methods are presented where they apply.

Part IV discusses the design, tooling, and production aspects of manufacturing. This part also introduces modern automation methods including computer numerical control (CNC) computer-aided design (CAD) as a designer's tool, computer-aided manufacturing (CAM), industrial robotics, and the integration of these technologies, computer integrated manufacturing (CIM), and flexible manufacturing systems (FMS), precursors to the totally automated factory of today and tomorrow. In Part IV we also discuss how the manufacturing industry functions, what job titles and general responsibilities are, and how production manufacturing is accomplished.

An extensive glossary appears at the end of the book.

Acknowledgments

The authors express their appreciation to Patricia J. Olesak, Assistant Professor, Purdue University School of Technology, for lending her expert knowledge of materials to the second edition. The authors are most grateful to the companies and individuals who have contributed materials and help in preparing this text. Special thanks go to Eugene Aluminum & Brass Foundry, Hergert's Industry, Lane Community College, and Lockheed-California Company who granted access to their facilities for photography and for supplying illustrations and technical information.

Ajax Magnothermic Corporation, Warren, OH
Alewijnse Electrical Engineering, Environment System, Nijmegen, Holland
Allis Chalmers, Milwaukee, WI
Aluminum Company of America, Pittsburgh, PA
American Die Casting Institute, Des Plaines, IL
American Iron & Steel Institute, Washington, DC
American Motors, Southfield, MI
American Society for Metals, Metals Park, OH
Apex Broach & Machine Company, Detroit, MI
Armco Inc., Specialty Steels Division, Middletown, OH
Automated Finishing Incorporated, Menomonee Falls, WI

Baird Corporation, Bedford, MA
Barrington Automation, Fox River Grove, IL
Bethlehem Steel Corporation, Bethlehem, PA

Beuhler LTD, Lake Bluff, IL
Bird-Johnson Company, Walpole, MA
Boeing Commercial Airplane Company, Seattle, WA
Bracker Corporation, Pittsburgh, PA
Buck Chuck Company, Kalamazoo, MI
Buffalo Forge Company, Buffalo, NY
Bureau of Mines, Albany, OR
Burlington Industries, Greensboro, NC
Buss-Condux, Inc., Elk Grove Village, IL

CHR Industries, Inc., New Haven, CT
Chrysler Corporation, Detroit, MI
Cincinnati Milacron, Cincinnati, OH
Cincinnati Milacron Plastics Machinery Division, Batavia, OH
Clausing Machine Tools, Kalamazoo, MI
Coherent, Palo Alto, CA
Colt Industries, Crucible Specialty Metal Division, Syracuse, NY
Consolidated Metco, Inc., Clackamas, OR
Consulate General of the Netherlands, Economic Section, NY

Del Manufacturing, Oxnard, CA
Disogrin Industries Corporation, Manchester, NH
Dixon Industries Corporation, Bristol, RI
DoALL Company, Des Plaines, IL
Dover Publications, New York, NY

Ductr Mfg., Eugene, OR
DuPont Company, Engineering Polymers Division, Wilmington, DE

ESAB North America Incorporated, ESAB Robotic Welding Division, Fort Collins, CO
Ethan Allen, Danbury, CT
Eugene Aluminum & Brass Foundry, Eugene, OR
Exxon Company, Houston, TX

Federal Products, Providence, RI
Fellows Corporation, Springfield, VT
Ford Motor Company, Dearborn, MI

Gates Learjet, Phoenix, AZ
General Motors Corporation, Detroit, MI
Giddings and Lewis, Fond du Lac, WI
Gleason Machine Division, Rochester, NY
Globe Amerada Glass Company, Elk Grove Village, IL
Goodyear Tire and Rubber Company, Akron, OH

Hamilton Caster & Mfg. Co., Hamilton, OH
Heck Industries, Redford, MI
Hergert's Industries, Drain, OR

Illinois Gear, Chicago, IL
Intergraph Corporation, Huntsville, AL

Japax Incorporated, Kanagawa, Japan
John Wiley & Sons, Inc., New York, NY

Kearney and Trecker, Milwaukee, WI
Kennametal Inc., Raleigh, NC
Kevex Corporation, King of Prussia, PA

L. S. Starrett Company, Athol, MA
Landis Tool, Waynesboro, PA
Lane Community College, Eugene, OR
Laserdyne, Eden Prairie, MN
Laserfare, Johnston, RI
Leland Balber Associates, Pittsburgh, PA
Libby-Owens-Ford, Toledo, OH
Lidköping Mekaniska Verkstads AB, Sweden
Lockheed-California Company, Division of Lockheed Aircraft Corporation, Burbank, CA
Louis Levin & Son, Inc., Culver City, CA
LTV Steel Company, Cleveland, OH

McGraw-Edison Company, Power Systems Division, Pittsburgh, PA
Magnaflux Corporation, Chicago, IL
Martin Marietta Corporation, Bethesda, MD

The Minster Machine Company, Minster, OH
Miller Electric Manufacturing Company, Appleton, WI
Mittler Machine Corporation, Providence, RI
M. L. Sheldon Plastics Corporation, New York, NY
MTI Corporation, Paramus, NJ
Mueller-Phipps International Corporation, Poughkeepsie, NY

National Machinery Company, Tiffin, OH
Niagara Machine & Tool Works, Buffalo, NY
Nikon Incorporated, Instrument Division, Garden City, NY
Northrop Corporation, Hawthorne, CA

Oregon Portland Cement Company, Portland, OR

Pacific Machinery & Tool Steel Company, Portland, OR
Parker Hannifin Corporation, Cleveland, OH
Pennsylvania Pressed Metals, Inc., Emporium, PA
Pierce Corporation, Eugene, OR
Plastics Machinery Incorporated, Williamston, MI
Porter, Westbrook, ME
Positech Corporation, Laurens, IA
Pottstown Machine Company, Pottstown, PA

Republic Steel Corporation, Cleveland, OH
Rhino Robots, Champaign, IL

Sandusky Foundry & Machine Co., Sandusky, OH
Saunders Equipment, Inc., Cold Spring, NY
Siempelkamp Corporation, Buffalo Grove, IL
Siempelkamp Corporation, Marietta, GA
The Stanley Works, New Britain, CT
Storage Technology, Louisville, CO

Teledyne Wah Chang Albany, Albany, OR
Testing Machines Inc., Amityville, NY
Tinius Olsen Testing Machine Co., Inc., Willow Grove, PA
TRW Cutting Tools Division, Augusta, GA
TVT Die Casting & Mfg. Inc., Portland, OR

United States Steel Corporation, Pittsburgh, PA
Unitron Instruments, Inc., Plainview, NY

Warner & Swasey, Sheffield Measurement Division, Dayton, OH
Wellcraft Marine Corporation, Sarasota, FL
Wheaton Industries, Wheaton Glass Division, Millville, NJ
Whirlpool Corporation, Benton Harbor, MI

Contents

Part IV   Manufacturing Systems   397

Introduction to Materials and Manufacturing

Few human enterprises represent a more aggressive and dedicated use of time, energy, money, and material resources than the activities of extracting or otherwise making raw materials and then processing these materials into the endless line of products that make the modern lifestyle possible. Since the first appearance of humankind on this planet, manufacturing, the processes and methods of converting materials into different forms and products, has gone hand in hand with human development. Many examples of toolmaking by ancient humans have been discovered. With tools, ancient peoples were able to shape natural and ultimately synthetic materials for weapons, shelters, and other products that could be traded or sold, thus generating both commerce as well as demand for more, different, and improved types of materials and manufactured products. The desire and demand for exotic materials and products stimulated the explorer and the trader to travel to the far corners of the earth.

These exploits created needs for higher technology in manufactured products. Navigational instruments and accurate maps are examples. These items were essential in order to make these voyages of discovery safer and more reliable, and to better ensure that the trader would return safely with the materials that were in such demand. Such factors stimulated the maximum efforts in design and materials application available at the time.

As ancient peoples gathered together to avail themselves of products, services and materials, manufacturing societies began to develop, and the structure for villages, towns, and ultimately cities was established. The increasing concentrations of people created needs for much improved manufacturing methods to supply demands. Specialty trades and crafts developed rapidly. This in turn created a need for faster methods of production, met by harnessing the power of animals, water, steam, and finally nuclear energy to power manufacturing machinery.

The lure of employment in manufacturing industries and the availability of manufactured products that would make life easier and more pleasant contributed greatly to the industrialization and urbanization of much of the world. Great manufacturing centers developed, usually located close to fundamental sources of raw materials such as coal, oil, and iron ore, prime ingredients on which industrialization is built. In today's world, a vast international manufacturing society exists that aggressively makes use of all newly developed technology both in materials development and application, as well as in manufacturing processes.

In today's world of exotic materials and high technology, modern manufacturing technology is constantly changing and growing. New methods of accomplishing production and applying new materials have become available almost daily and become obsolete rapidly as new technology comes online. Although most of the well-established processes for manufacturing products will always be with us, the differences will come in how these processes are used. New applications of standard processes will, at the same time, create new processes. New materials will also create new processing methods. The manufacturing technologist of today and tomorrow must be well versed in materials application and in the standard processes but will always be looking forward toward new applications of these materials and processes as well as toward the development of entirely new processes. Manufacturing of the future will become more specialized, making it more difficult for the technologist to keep informed of this rapidly changing and expanding technology.

The central purpose of this textbook is to present an overview of materials science, a survey of traditional as well as high-technology manufacturing processes, and a look at manufacturing systems in the computer high-technology age.

THE MANUFACTURING ECONOMY— PAST, PRESENT, AND FUTURE

By the middle of the nineteenth century a truly remarkable revolution, the Industrial Revolution, which had its beginning two centuries before, was well established. Goods formerly made by hand were made by machine and the era of large-scale mass production had begun.

The effects of the Industrial Revolution have been far-reaching. Cheap production of goods in large quantities, now the order of the day, has raised the standard of living for many people in industrialized societies. The Industrial Revolution has formed the basis of the modern lifestyle that most of us take for granted, and general industrialization has provided expanded employment.

In the latter half of the twentieth century a new industrial revolution, one of computers and space-age high technology, is taking place. Like that of the past century, this new technology will have far-reaching effects. The effects of the current industrial revolution will be much more far-reaching than anything even dreamed of in the past centuries.

Modern industrial technology has already given us the power and the tools to be far more productive with much less need for people to do the manufacturing work. It is not to be denied that modern technology has made the lifestyle of the modern industrial society even more easy, convenient, and filled with incredible products, most of which are within the purchasing power of the average person living in that society. However it may have temporary negative effects as well—for those who are not prepared to work in a high-technology, computerized, and automated manufacturing world. As lower skill level jobs are automated, the production worker of today must look to the future and acquire the job skills necessary to fill the higher skilled positions that are created as technology advances.

In addition, the gradual industrialization of less technologically oriented societies throughout the world is moving ahead. We of the established industrial world could, for much of the past fifty years, take little notice of this advance, as our technology was always well in front of that in less developed countries. As the wage demands of the old established industrial societies rose over time, manufacturers sought to take advantage of the lower wage structures generally prevalent in emerging countries. Thus, the great migration of manufacturing from the established industrial societies to the emerging countries began and continues today.

What have been the results of this migration? It is true that manufacturing industries that have moved abroad have for some years been able to operate under more favorable wage structures. However, the very industries that these manufacturers sought to staff with cheaper labor have, by a natural evolution, raised the standard of living for workers in the countries where the industries were located. More jobs have been created, and with jobs have come higher pay and the ability of industrial workers to buy not just necessities but goods that make life easier and more pleasant. This factor plus the often diligent attitude of foreign workers toward their jobs and resulting high level of productivity and quality, have brought industrial revolutions similar to our own in emerging countries. This process is continuing unabated and now presents intense competition to our domestic employment and economy. The effects have been and will continue to be far-reaching. In those countries where industrialization occurred first, global competition has placed the technical leadership of the established industrial old world in question.

For example, the influx of high quality goods manufactured abroad and imported into the United States for sale at an often much cheaper price has continued unabated for many years. The results are only too evident. Jobs in the domestic economy have been lost, and manufacturing industry has not always countered by keeping up with modern trends in automation, productivity, quality control, and worker efficiency. Foreign manufacturing, often operating in a very favorable unrestricted business environment, and often with the total support of local governments, has been able to take better advantage of modern technology and is engaged in an effective program of operating modern, efficiently run industries.

If the industrial structure of the old industrial societies is to remain intact and the standard of living is to remain at a high level, it is of paramount importance that the entire industrial and manufacturing community adopt new attitudes toward improving manufacturing efficiency and competitiveness. The worker, who is in the end responsible for industrial output, must be treated fairly, but quality work and high productivity must be maintained. Manufacturing industries must put to use all of the modern tools necessary to accomplish efficient quality production in the high-technology age.

It is a sure bet that product manufacturing in the future will be done on a more automated level of high technology and automation. Many of the products that will soon be commonplace have not even been designed yet, but as they are, more and more high-technology manufacturing systems and methods will have to be created to produce them.

Automation trends, already at a high level in manufacturing, will become ever more the preferred method of production. Flexible manufacturing systems (FMS)

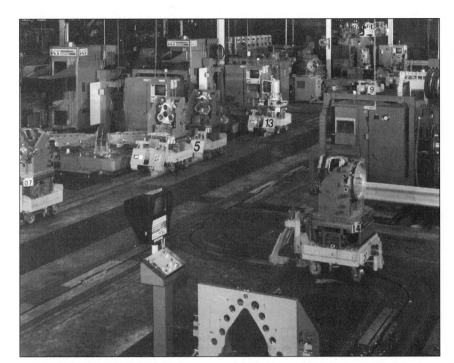

FIGURE 1
The flexible manufacturing system (FMS) with automated workpiece handling and transfer (Kearney and Trecker).

(Figure 1), the forerunners of completely automated factories already online in many industries, will become commonplace as computer-controlled automatic machines take over most of the mundane work heretofore done by fairly low-paid manual labor. The need for operating personnel will be minimized in this system, as robot-operated transporters move workpieces automatically between manufacturing stations. Industrial robots (Figure 2) will perform the routine tasks of loading and off-loading parts into computer numerically controlled (CNC) machining centers.

High technology is really not new. It has been with us since the first recorded manufacturing efforts, as manufacturing industries continually have applied the latest available materials, tools, and methods to manufacturing requirements. Until now, however, most manufacturing industries still needed people to operate machines to turn out product units. The computer, with its considerable ability to make decisions in process control applications, is already replacing people in the manufacturing process. As computer technology has become more sophisticated its decision-making powers have increased proportionally. Artificial intelligence (AI) technology is presently under rapid development, giving computers even more decision-making power, making them even more versatile and in more complete control of routine as well as sophisticated production operations.

THE COMPUTER INFLUENCE IN MANUFACTURING

The history of technological development encompasses many contributions. First there was metallurgy, making available a wide selection of metals for toolmaking and manufactured products. The Industrial Revolution of the eighteenth and nineteenth centuries brought the age of machine-made mass produced goods. Electronics, especially solid state and integrated circuit (ICs or chips) technology, has made instant communication available to practically everyone on earth. Thus the message of what is available in manufactured goods has reached ever widening markets. These technological advances have transformed industry and society as a whole. However, the effect of the computer has been much more extensive than any technological device yet developed. The concept of computer control of processes is extremely significant in that it has changed and will continue to profoundly and forever change the way manufacturing is accomplished.

Does this mean that computers will control everything including the actions of people? The answer to this question is a definite no. However, the computer will aid those people who direct the actions of people not only in manufacturing but in all phases of life. The exact impact of computers and computerized automation trends in man-

FIGURE 2
FMS work cell for CNC
turning. The robot (center)
loads and off-loads parts for
both machine tools
(Cincinnati Milacron).

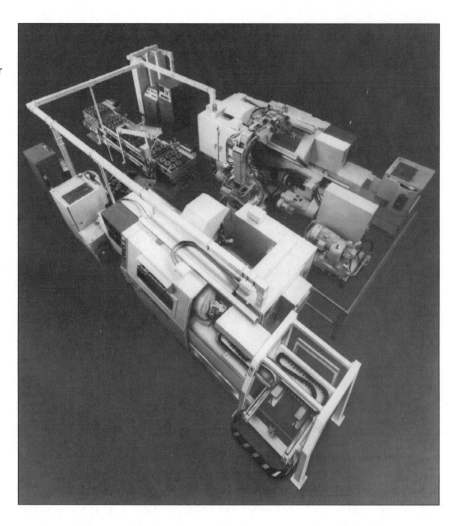

ufacturing on the future worker is uncertain. However, it is certain that the factory worker of the future will have to become more of an electromechanical technician instead of being the machine tender of today's factory.

The technician will have to learn broader skills, acquiring a more complete knowledge of electrical/electronic and hydraulic systems, transactional mechanisms, general mechanical skills, and very likely a working knowledge of computer systems and computer programming. For those not willing to learn new skills or accept retraining, future employment prospects at a suitable pay level may be limited. However, the skills and abilities of people with an open mind and a natural desire to understand how things work and how to operate, maintain, and repair complex electronic/mechanical machines are likely to be in demand.

MANUFACTURING AND YOU

For those who afford themselves of proper education and training, opportunities in manufacturing will abound. However, the future employee at any level in manufacturing will have to undergo broader and more complex training than at any time in the past.

Trade-Level Opportunities

At the trade level, many of the repetitive production jobs that now are done by minimum wage employees will be eliminated as automation such as computer numerical control (CNC) and industrial robots take over. Robots will become more and more sophisticated and capable, and they will move further into the areas of the skilled trades

FIGURE 3
Robot welder welds a gear case mounted on one of two workpiece positioners (ESAB Robotic Welding Division).

FIGURE 4
Production technician assembles laser computer disks in an ultraclean room environment (Storage Technology).

such as welding (Figure 3). The robot welder can, in many situations, outperform its human counterpart in both quantity and quality on many welding tasks.

The higher level skilled trades, such as tool and die making, will continue to be in demand, but preparation for these jobs will require several years of diligent training. The general machinist who can operate highly computerized rapid prototyping systems will be needed.

The production technician, an individual who has broad training in electric/electronics, mechanical/hydraulic systems, and computers will probably have the brightest future in trade-level manufacturing jobs. The capability of modern and future automated computerized manufacturing systems will be truly amazing, and the equipment and control systems will be complex and will require skilled maintenance. The production technician will have to be able to diagnose problems quickly and to effect repairs rapidly and efficiently so that complex equipment can be placed back into production as soon as possible. High-technology production technicians will be specialists at their work. They will assemble and test delicate high-precision equipment, often in controlled environments. In Figure 4 the technicians are assembling large capacity

laser computer disks which must be handled in an ultra-clean room environment.

Middle Management Levels

The future in manufacturing for the college-trained engineer and engineering technologist will also be bright. As automated manufacturing systems are put in place, the engineer and engineering technologist will be needed for plant facilities, tool, jig, and fixture design (Figure 5), robot and numerical control programming, and implementation of all aspects of the manufacturing system (Figure 6). As with the production technician, the engineer and engineering technologist will need to have broad training in materials applications, computers, and manufacturing processes. Such individuals will also be heavily involved with all aspects of quality control and assurance as the

FIGURE 5
Automatic wing spar
assembly tooling. New
methods of tooling will
challenge the industrial and
manufacturing engineer
(Boeing).

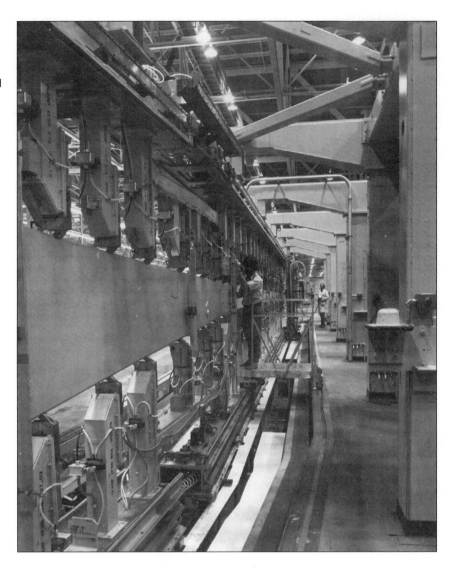

manufacture of higher technology products becomes more commonplace.

Design Engineering

Creative product designers will always be needed as the capacity of high-technology manufacturing makes possible production of a vast array of incredible new products. Tomorrow's designer will need to be thoroughly knowledgeable of the modern manufacturing methods and their capabilities. The designer of the future using the computer aided design (CAD) system will smoothly integrate new design ideas with new material and new processes as they become available. Preparation for jobs in this area will require college level education and on-the-job industrial training.

ONGOING MANUFACTURING TECHNOLOGY

It is certain that modern manufacturing technology will always be in a continuing state of development. As new materials, methods, and computer aids come into play, the manufacturing engineer and technologist will apply these to solve production problems.

Materials

The product designer will continually seek new materials that are lighter and easier and cheaper to process while meeting design requirements for strength and durability. Plastic and composite materials will play a more important role in future products, thus permitting additional

FIGURE 6
Integrated automation, machining, assembly, and testing in sequential automation (Giddings and Lewis).

energy savings both in manufacturing processes and in powering road, air, and space vehicles (Figure 7).

Design

The computer will play an ever increasing role in initial product design and design evaluation (Figures 8 and 9). Computer aided design is an absolutely essential design tool. Control programs for automated manufacturing equipment will be generated from the CAD database, thus forming integrated computer aided design/computer aided manufacturing (CAD/CAM) and computer integrated manufacturing (CIM) systems. The day is fast approaching when design specifications entered at the CAD station will shortly appear as finished products at the shipping dock.

FIGURE 7
High-technology designs and materials for tomorrow's mach five high altitude aircraft. The leading edges of the airframe will glow red hot at a temperature of 1000°F (Lockheed Co.).

FIGURE 8
The computer aided design (CAD) engineering workstation is fast becoming the preferred method of designing (Intergraph Corporation).

FIGURE 9
CAD used in vehicle designing (Ford Motor Company).

Computer aided engineering (CAE) is becoming a frequently chosen method of evaluating design, eliminating expensive prototyping. In Figure 10 complex compound contour shapes of today's aerodynamic automobile bodies are studied and evaluated using CAE methods.

Proven designs of the past will be reevaluated with an eye toward energy- and fuel-consumption efficiency. Figure 11 shows a high-technology application of the airscrew. Driven by the efficient turboprop jet engine, the new propeller design can propel aircraft with the same speed and altitude capability of modern jets, but with a large reduction in fuel consumption.

FIGURE 10
CAE for design evaluation of automotive aerodynamics utilizing a CAD model (Ford Motor Company).

FIGURE 11
A new look at old and proven technology—high technology turbojet propellers (Lockheed Co.).

Manufacturing Processes

Manufacturers in all types of industries will search for and implement new manufacturing processes. Many of these will take on some unique and interesting forms. In Figure 12 a high-pressure water jet containing an abrasive material is used as a cutting tool for titanium aircraft parts. Since almost no heat is developed in this process, distortion of the part is eliminated. Computer control is becoming the order of the day. Automated process control is found in nearly every aspect of manufacturing from electronic component insertion (Figure 13), robotic spot welding (Figures 14 and 15), and precision measurement and inspection (Figure 16). Other high-technology applications of available technology include the further use of lasers for many applications (Figure 17).

FIGURE 12
High-pressure water containing an abrasive is used to cut titanium aircraft parts (Lockheed Co.).

FIGURE 13
Electronic components are automatically inserted onto circuit boards (Whirlpool Corporation).

THE PLACE OF TRAINING AND EDUCATION IN THE HIGH-TECHNOLOGY WORLD

Whether one has a working knowledge of the newest skills in the world of high technology will determine to a great extent who will be employable at adequate pay

FIGURE 15
Underbody and two side panels are secured in a large fixture for production spot welding (Chrysler Corporation).

levels. Progressive manufacturing industries, schools, colleges, and universities will be responsible for a continuing supply of adequately trained technicians, technologists, and engineers to meet the needs of the high-technology age. Also, in-service training programs in new methods will be a high priority for all manufacturers. Students of materials and manufacturing processes in the

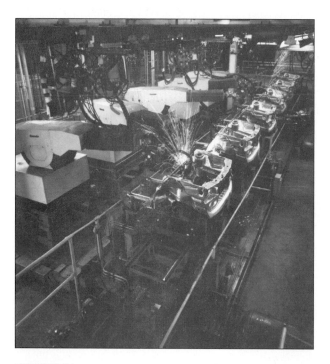

FIGURE 14
Computer-driven spot welding robots spot weld front body assemblies with great speed and accuracy (Ford Motor Company).

FIGURE 16
Computerization in precision measurement (MTI Corporation).

FIGURE 17
Laser used in precision measurement (Giddings and Lewis).

high-technology world will need hands-on experience in the new systems and in their operations and applications.

For those who are properly prepared, the world of high-technology manufacturing presents many opportunities. The technology of tomorrow will require many trained personnel to repair, service, and implement the manufacturing systems of an ongoing high-technology industrial revolution. New product designs and the manufacturing systems required to produce them will be an ever expanding field for those who acquire the proper training. The production technician, technologist, engineer, and designer will contribute greatly to maintaining a nation's international technological position. Will you be ready to meet these challenges? The following chapters in this book will help you to answer this important question.

PART I

MATERIALS OF MANUFACTURE

At the dawn of history, the only materials used by humans were taken directly from natural sources. Stones were used to erect crude shelters or structures; harder stones with cutting edges were used for arrowheads and spear points and for tools for cutting wood, bone, and leather. Clay was shaped into containers and building bricks. The discovery that firing these clay items made them stronger and water resistant was probably the first step in a long procession of the adaptation of natural materials to the needs and uses of humanity.

Today, literally thousands of materials are derived from nature. In addition, many more synthetic materials do not come directly from natural sources. All of these amazing developments have come about because of increasing knowledge and understanding of the properties of materials. This has been especially true in the last century, when we began to understand the arrangement of atoms and crystal structures and their behavior in various conditions. Plastics, elastomers, metals, ceramics, and composites are some of the important materials used today that will be studied in Part I.

The Atomic and Crystalline Structure of Materials

Objectives

This chapter will enable you to:

1. Explain a simplified model of the atom.

2. Understand several bonding arrangements and explain the role of valence in bonding.

3. Describe the crystalline structure of metals and how this affects their behavior.

4. Give an account of the structure of some polymers for the manufacture of certain plastic materials and elastomers.

5. Describe the crystal structure of clays and other ceramics and their bonding arrangements.

Today, we know much about the structure of the physical universe, and through the endeavors of the scientific community, many more facts are constantly being discovered. Engineers in turn design new products based on these new discoveries, often using radically new concepts. For example, look at the phenomenal growth in solid state electronics and high technology. Therefore, technologists must be able to understand some of the new terminology and technology of the scientist and engineer in order to communicate in the manufacturing field. It is useful to understand the basic structure of the atom prior to studying the properties of materials and their behaviors under various conditions. Some of these principles and concepts are included in this chapter.

Although *atoms* are too small to be seen with the aid of ordinary light microscopes, their outline can now be detected with such devices as the electron microscope and the field ion emission microscope. More than 100 elements have been discovered—92 occurring in nature, and the others produced by atomic reactions. The *molecule* is defined as the smallest particle of any substance that can exist free and still exhibit all the chemical properties of that substance. A molecule can consist of one or more atoms of the same kind. For example, oxygen gas is found in nature as a molecule consisting of two oxygen atoms (O_2) or of three oxygen atoms, ozone (O_3). A *compound* is composed of two or more elements combined chemically; for example, hydrogen and oxygen combine to produce water (H_2O). A *mixture* consists of two or more elements or compounds that are physically, not chemically, combined.

Atoms, and in fact all materials, are composed of the same basic components: *protons, neutrons,* and *electrons* (Figure 1.1). The **nucleus** of the atom is composed of positively charged protons and neutrons. The latter have almost the same mass as protons but are without electrical charge. Surrounding the nucleus, orbiting it in what are called shells, somewhat like the orbits of planets in our solar system, are the electrons, which have a negative charge equal and opposite to the charge of the proton in a neutral atom. It is this electrical charge that holds the atom together and that is responsible for much of the chemical and electrical behavior of the elements. Almost all of the atomic mass is contained in the nucleus; electrons have only about $\frac{1}{1839}$ the mass of the proton or neutron.

ATOMIC STRUCTURE

All matter is composed of one or more of the more than 100 chemical elements. Each element is a pure material, such as a metal or nonmetal, unlike any other; yet every

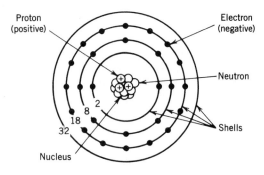

FIGURE 1.1
The basic parts of the atom.

element is composed of protons, neutrons, and electrons; only the number and arrangement differ.

Atomic Bonding

Valence is the capacity of an atom to combine with other atoms to form a molecule. Valences can be positive or negative. When an atom has more or fewer electrons in the outer shell than it has in its uncombined or free state, it is called an ion and it possesses an electrical charge. The charge is negative when extra electrons are present and positive when some are missing. Metal atoms are easily stripped of their valence electrons and thus form positive ions. When the valence or outer shell has a full complement of electrons, it is said to have zero valence. The inert gases have zero valence and do not readily combine with other elements to form compounds.

Three basic types of primary bonding arrangements hold atoms together. They are *ionic* (also called electrovalent), *covalent* (polar or nonpolar shared electrons), and *metallic*. A common secondary bond is *Van der Waals*. The simplest type of linking or bonding of atoms is the ionic bond. Ionic bonding links dissimilar atoms by the attraction of negative and positive ions. Sodium chloride (NaCl) is an example of ionic bonding (Figure 1.2) where a metal (sodium) loses its single valence electron to a nonmetal (chlorine) to complete its valence shell. The chlorine atom that became negatively charged and the sodium atom that became positively charged have combined to become a neutral compound (Figure 1.3). Ionic bonds tend to be somewhat weak because there are discontinuities in the lattice arrangement and there are unequal numbers of combining atoms, so neutrality is not always maintained. These bonds therefore have the characteristics of brittleness, moderate strength, and high hardness, and are good electrical insulators when they are in pure solid form because there is little opportunity for ion or electron movement. (It should be noted, however, that many ionic compounds, such as sodium chloride, are water soluble

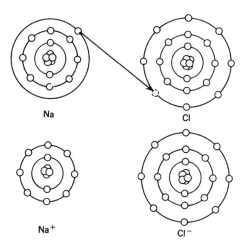

FIGURE 1.2
Ionic bonding of sodium chloride (White, Neely, Kibbe, Meyer, *Machine Tools and Machining Practices,* Vol. II, © 1977 John Wiley & Sons, Inc.).

and that solutions of these compounds will conduct electricity very well.)

Covalent bonding, or the sharing of valence electrons, is a very strong atomic bond whose strength depends on the number of shared electrons. It is interesting to note, however, that a variety of covalent bonding patterns exist. In network covalent substances such as diamond and silicon carbide (carborundum), the sharing of electrons forms a large, strongly bonded unit. On the other hand, molecular covalent substances such as sugars or waxes satisfy their sharing of electrons within the molecule and have very little attraction to other molecules surrounding them. Thus network covalent substances have high melting points and hardnesses, whereas molecular covalent substances have low melting points and hardnesses. Negative

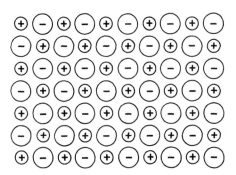

FIGURE 1.3
Lattice structure of sodium chloride (White, Neely, Kibbe, Meyer, *Machine Tools and Machining Practices,* Vol. II, © 1977 John Wiley & Sons, Inc.)

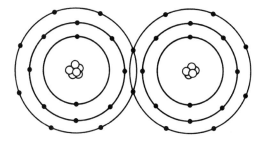

FIGURE 1.4
Chlorine atoms form a covalent bond by sharing atoms.

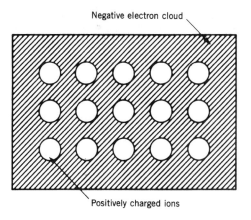

FIGURE 1.6
The metallic bond.

valences greater than four do not occur, but these atoms tend to assume an inert gas electronic structure of eight electrons in the outer shell. (In the case of hydrogen it is one.) Chlorine, having a complement of only seven electrons, therefore tends to join with any element that will fill its outer shell to eight electrons. As we have seen, it will join with sodium to obtain an extra electron and form an ionic substance; it will also join with another chlorine atom to gain an extra electron by sharing it to form a covalent substance (Figure 1.4). The element oxygen has the same tendency, but it shares two valence electrons (Figure 1.5).

The metallic bond is formed among metallic atoms when some electrons in the valence shell separate from their orbital path and exist in a free state as a cloud or gas, surrounding the atoms in a highly mobile state while being shared for bonding-like purposes (Figure 1.6). Because of this, bond strengths can vary, deformation can take place without rupture, and good heat and electrical conductivity can occur because of the easy movement of electrons from one atomic site to the next. This type of bonding makes possible some of the useful characteristics of metals such as plasticity and ductility, which will be discussed later in this chapter.

Van der Waals bonding is found between neutral atoms of elements such as the inert gases and between neutral molecules of covalent molecular compounds. This is only a very weak attractive force, and is important in the study of metals only at very low temperatures.

CLASSIFICATION OF ATOMS

Atoms of the element hydrogen have in their normal state a single electron moving about a nucleus that has a positive charge equal to the negative charge of the electron. Likewise, the atom of the element helium has two electrons moving about its nucleus, which has a positive charge equal to twice one electronic charge. An atom of the element lithium has three electrons and a positive nuclear charge equal to three times that of a hydrogen atom. This is the basis for the numbers on the periodic table (Table 1.1), called *atomic numbers*. These numbers range from 1 for hydrogen to 92 for uranium, and still higher for the transuranic elements. These numbers are important because they determine the chemical behavior of the atoms. The table of elements is called periodic because elements with similar properties appear at regular intervals in the arrangement of the elements. The horizontal rows of elements are known as *periods* and the vertical columns are known as *groups*. The greatest similarities exist in the vertical columns. For example, copper (Cu), silver (Ag), and gold (Au) have similar crystal structures and readily combine as alloys. The vertical group of chromium (Cr), molybdenum (Mo), and tungsten (W) has similar characteristics as alloying elements in steel for promoting strength and hardenability. Also, sulfur (S), selenium (Se), and tellurium (Te) have all been used to promote machinability in steels. Besides the relationship

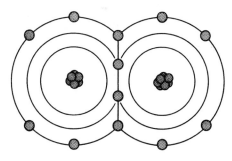

FIGURE 1.5
The covalent bond of the oxygen molecule (White, Neely, Kibbe, Meyer, *Machine Tools and Machining Practices*, Vol. II, © 1977 John Wiley & Sons, Inc.).

TABLE 1.1
The periodic table of the elements

Light Metals							Heavy Metals					Nonmetals					Inert gases
IA																	0
H 1	IIA	IIIB	IVB	VB	VIB	VIIB	VIII		IB	IIB	IIIA	IVA	VA	VIA	VIIA		He 2
Li 3	Be 4	←Brittle metals→					←Ductile metals→				Low melt-ing	B 5	C 6	N 7	O 8	F 9	Ne 10
Na 11	Mg 12									IIB	Al 13	Si 14	P 15	S 16	Cl 17		Ar 18
K 19	Ca 20	Sc 21	Ti 22	V 23	Cr 24	Mn 25	Fe 26 · Co 27 · Ni 28	Cu 29		Zn 30	Ga 31	Ge 32	As 33	Se 34	Br 35		Kr 36
Rb 37	Sr 38	Y 39	Zr 40	Nb 41	Mo 42	Tc 43	Ru 44 · Rh 45 · Pd 46	Ag 47		Cd 48	In 49	Sn 50	Sb 51	Te 52	I 53		Xe 54
Cs 55	Ba 56	* 57–71	Hf 72	Ta 73	W 74	Re 75	Os 76 · Ir 77 · Pt 78	Au 79		Hg 80	Tl 81	Pb 82	Bi 83	Po 84	At 85		Rn 86
Fr 87	Ra 88	† 89															

*	La 57	Ce 58	Pr 59	Nd 60	Pm 61	Sm 62	Eu 63	Gd 64	Tb 65	Dy 66	Ho 67	Er 68	Tm 69	Yb 70	Lu 71
†	Ac 89	Th 90	Pa 91	U 92	Np 93	Pu 94	Am 95	Cm 96	Bk 97	Cf 98	Es 99	Fm 100	Md 101	No 102	Lr 103

In general, it can be stated that the properties of the elements are periodic functions of their atomic numbers. The greatest similarities between elements exist in the vertical columns. There are also some similarities between the A and B groups on either side of the VIII groups. For example, the Scandium group, IIIB, is in some respects similar to the Boron group, IIIA. The most important properties in the study of metallurgy that vary periodically with the atomic numbers are crystal structure, atomic size, electrical and thermal conductivity, and possible oxidation states.

(*Source:*   John E. Neely, *Practical Metallurgy and Materials of Industry,* 2d ed., Wiley, New York, © 1984.)

of the elements in the vertical groups, there are resemblances in several horizontal triads: iron (Fe), cobalt (Co), and nickel (Ni) are all ferromagnetic, for example.

The six metals of the platinum group consist of two horizontal triads: ruthenium (Ru), rhodium (Rh), palladium (Pd); and osmium (Os), iridium (Ir), platinum (Pt). The first subgroup (Ru, Rh, Pd) has atomic weights and specific gravities that are about half those of the second subgroup (Os, Ir, Pt). Furthermore, each metal of the first group has a member in the second group that seems more closely related to it; for example, Ru and Os, Rh and Ir, and Pd and Pt may be considered as pairs. All of these metals are chemically resistant and have high melting points. These six metals along with gold and silver are called precious metals. They are also called the noble metals because of their resistance to atmospheric conditions

and to most acids. In general, crystal structures, melting points, densities, and other properties of elements are functions of the atomic number.

In the foregoing discussion, the electrons were spoken of as moving around the nucleus like planets in orbit around the sun. We know now by the **uncertainty principle** that this cannot be true of the electrons. The exact orbit of an electron about the nucleus has been replaced by a mathematical function called an *orbital.* These are electronic energy levels to which quantum numbers have been assigned in which electrons occupy certain *shells.* Each orbital electron contains discrete amounts of energy. These energy packets are called *photons* (that is, a photon is a quantum of radiant energy). If a group of electrons becomes excited, they will jump to a higher energy orbit. If an electron returns to its original energy orbit or shell, it will release a photon of energy (such as heat) that is characteristic of that jump. The first basic shell can contain no more than two electrons, the second shell contains eight, the third shell contains eighteen, and the fourth shell contains thirty-two. Electrons in the outer shell of metals are called valence electrons and can be easily stripped off in order to form compounds and bonding arrangements. These electrons are responsible for the electrical and chemical behavior of elements.

Metals and Nonmetals

As can be seen in Table 1.1, approximately three-quarters of all elements are considered to be metals. Chemical elements may be roughly classified into three groups: metals, nonmetals, and inert gases. Some nonmetals, such as carbon, germanium, and silicon, resemble metals in one way or another and are sometimes called metalloids. They can have a metallic luster and crystalline microstructure, and they can be semiconductors of electricity, but they cannot be deformed without fracturing because they are brittle, as are all nonmetals. Some of the properties that a metallic must exhibit are:

1. Ability to donate electrons to form a positive ion.
2. Crystalline structure.
3. High thermal and electrical conductivity.
4. Ability to be deformed plastically.
5. Metallic luster or reflectivity.

Metals are the most important of the elements from a metallurgical standpoint, but plastics, rubbers, and other synthetic products are produced for the most part from nonmetals. Solid state electronic components are produced from metalloids such as germanium and silicon.

CRYSTALLINE STRUCTURE

When metals cool from the molten state and solidify, they generally form a crystalline structure called a **space lattice.** If imaginary lines are drawn connecting the centers of the atoms together, it is called a *lattice structure* (Figure 1.7). A recent development has created an exception to the foregoing statement. Metals can now be rapidly quenched from the liquid to the solid state so that there is no opportunity for a crystal structure to develop. These are appropriately called *metallic glasses* or amorphous metals. Only a small amount of this type of metal is being produced at present. Since it has very good electrical and magnetic properties, it is useful for making transformer cores. However, when metals cool at normal rates, they

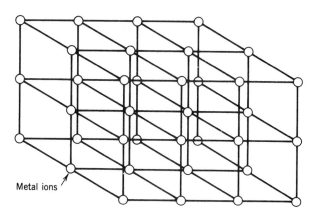

Metal ions

FIGURE 1.7
Lattice structure of atoms in solids.

always form a space lattice that consists of orderly rows of atoms.

Interatomic Distances

Since similar electrical charges tend to repel and unlike charges attract, the electron-filled outer shells of the metallic ions tend to have a repulsive force. If it were not for the free valence electrons vibrating between the atoms to produce forces of attraction, metal atoms would fly apart. However, there is a distinct balance of forces that makes up metallic bonding. These atoms maintain a distance at a location where these two forces are in equilibrium. These interatomic distances can be altered by external forces, such as thermal energy and electrical or mechanical forces. This is why materials expand when heated and contract when cooled. The distance between atoms is different for different materials because it is related to the number of shells and the number of valence electrons. Atoms of different elements have different sizes. For example, an atom of iron is much larger than an atom of carbon. Removal of electrons from the outer shell, as is done in a metal ion, reduces its size.

Unit Cells

A space lattice can be described as having repeating unit building blocks throughout its structure. These are called *unit cells* (Figure 1.8). Metals solidify into 14 known crystal structures, eight of which are shown in Table 1.2, but most of the commercially important metals solidify into one of three types of lattice structures (Figure 1.9)—body-centered cubic (BCC), face-centered cubic (FCC), and hexagonal close-packed (HCP).

If we assume that atoms are like tiny balls that almost touch their nearest neighbors, two common structures

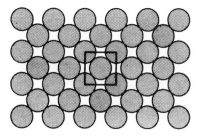

FIGURE 1.8
Body-centered cubic unit cell shown in the lattice (White, Neely, Kibbe, Meyer, *Machine Tools and Machining Practices,* Vol. II, © 1977 John Wiley & Sons, Inc.).

would look like that in Figure 1.10. The metallic bond, to have maximum plasticity and other needed properties, is best when the atoms are close-packed as in body-centered cubic structures that form a cube with one atom in its center. Even better for plastic deformation is the more close-packed structure, face-centered cubic structure. The hexagonal close-packed unit structure does have some planes that are close-packed, but in another direction the atoms are spaced farther apart than in BCC and FCC, causing these metals to have poor formability.

Some materials have two or more of these structures under different conditions such as temperature change. Such a material is called **allotropic.** For example, the three allotropic forms of carbon are amorphous (charcoal, soot, and coal), graphite, and diamond. Iron also exists in three allotropic forms that are temperature dependent (Figure 1.11). (The term *iron* refers to elemental or pure iron unless specified as wrought or cast.) Below 1300° F (704° C) iron is BCC, called ferrite or alpha iron, which is ferromagnetic. It becomes nonmagnetic at the Curie temperature (1440° F or 782° C), sometimes called beta iron. Between 1666° F (908° C) and 2554° F (1401° C) iron is FCC, called austen-

TABLE 1.2
Crystal structures of some common metals

Symbol	Element	Crystal Structure
Al	Aluminum	FCC
Sb	Antimony	Rhombohedral
Be	Beryllium	HCP
Bi	Bismuth	Rhombohedral
Cd	Cadmium	HCP
C	Carbon (graphite)	Hexagonal
Cr	Chromium	BCC (above 26° C, 78.8° F)
Co	Cobalt	HCP
Cu	Copper	FCC
Au	Gold	FCC
Fe	Iron (alpha)	BCC
Pb	Lead	FCC
Mg	Magnesium	HCP
Mn	Manganese	Cubic
Mo	Molybdenum	BCC
Ni	Nickel	FCC
Nb	Niobium (columbium)	BCC
Pt	Platinum	FCC
Si	Silicon	Cubic, diamond
Ag	Silver	FCC
Ta	Tantalum	BCC
Sn	Tin	Tetragonal
Ti	Titanium	HCP
W	Tungsten	BCC
V	Vanadium	BCC
Zn	Zinc	HCP
Zr	Zirconium	HCP

(*Source:* John E. Neely, *Practical Metallurgy and Materials of Industry,* 2d ed., Wiley, New York, © 1984.)

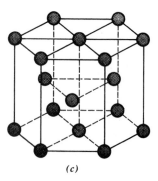

(a)   (b)   (c)

FIGURE 1.9
Three common unit cells of the lattice structure are *(a)* body-centered cubic, *(b)* face-centered cubic, and *(c)* hexagonal close-packed (John E. Neely, *Practical Metallurgy and Materials of Industry,* 2d ed., © 1984 John Wiley & Sons, Inc.).

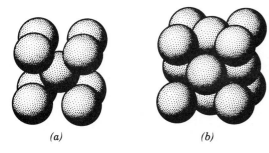

FIGURE 1.10
Model of atoms in unit cells. *(a)* Body-centered cubic (BCC). *(b)* Face-centered cubic (FCC) (White, Neely, Kibbe, Meyer, *Machine Tools and Machining Practices,* Vol. II, © 1977 John Wiley & Sons, Inc.).

ite or gamma iron. Between 2554° F (1401° C) and 2800° F (1538° C) iron is also BCC, called delta iron, which is of little importance in manufacturing. These three states of iron are called **phases;** the ferrite and austenite phases are very important in heat treating and manufacturing.

Note on the cooling curve in Figure 1.11 that the straight horizontal lines represent the time required to

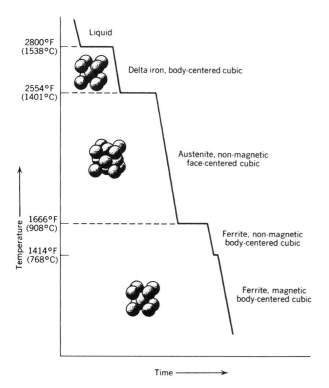

FIGURE 1.11
Cooling curve for pure iron. As iron is cooled slowly from the liquid phase (or heated from solid to liquid), it undergoes these allotropic transformations (White, Neely, Kibbe, Meyer, *Machine Tools and Machining Practices,* Vol. II, © 1977 John Wiley & Sons, Inc.).

effect the transformation from one phase to another. Under cooling conditions heat is released during the transformation and, when being heated, heat energy is absorbed during the transformation.

Dendrite Formation

As solidification of the molten metal begins, the arrangement of the space lattice takes on the characteristic pattern for that metal. Each unit cell builds on another to form crystal growths that resemble pine trees. These structures, called dendrites (Figure 1.12), begin to grow from seed crystals or nuclei particles such as impurities. Aluminum is usually put into molten steel to make it fine grained and to eliminate dissolved oxygen in the molten metal. The aluminum oxide particles are so numerous that many crystals start growing from these nuclei and consequently the metal contains more grains than it normally would. Also, slow cooling promotes large grains and rapid cooling promotes smaller grains.

Grain Boundaries

The dendrites grow outward to form a single crystal or **grain** until it meets another dendrite crystal. The places where these grains meet are called **grain boundaries.** Metals therefore become polycrystalline; that is, the axes of the lattice structures are oriented at random in each grain depending on how the dendrite began to grow

FIGURE 1.12
Dendrite growth in metal as shown in a photomicrograph.

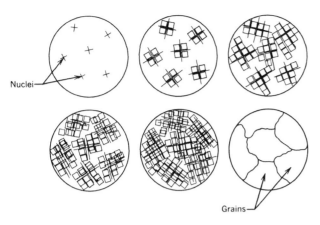

FIGURE 1.13
The formation of grains during solidification of molten metal (White, Neely, Kibbe, Meyer, *Machine Tools and Machining Practices,* Vol. II, © 1977 John Wiley & Sons, Inc.).

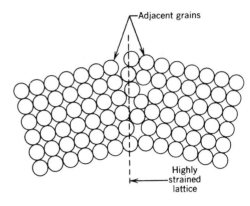

FIGURE 1.15
The grain boundary is in a highly strained condition (White, Neely, Kibbe, Meyer, *Machine Tools and Machining Practices,* Vol. II, © 1977 John Wiley & Sons, Inc.).

(Figure 1.13). All metals when normally cooled are composed of these tiny grain structures, which can be seen with the aid of a microscope when a specimen is cross-sectioned, polished, and etched. Often the grain structure can be seen with the naked eye as small crystals in a broken section of a piece of metal (Figure 1.14). Some metals, when slowly cooled, have extremely large grains. Large grains of zinc can often be seen on galvanized articles.

Since each crystal lattice or dendrite grows in different directions, it can be seen that the grain boundaries do not form a continuation of the lattice, but instead the atoms are jammed together in a misfit pattern (Figure 1.15). Although the nature of the atomic forces in the grain boundaries is not entirely understood, it is generally assumed to be an interlocking border in a highly strained condition. Grain boundaries are only about one or two

atoms wide, but their strained condition causes them to etch darker than the less strained grain, allowing them to be seen with a microscope. Grain size in metals is very important in manufacturing products, as will be explained later in this chapter.

PLASTIC DEFORMATION IN METALS

Two of the most important properties of metals for manufacturing purposes are **elastic deformation** and **plastic deformation.** Stress is the material's resistance to the applied load or force. When metals are placed under tensile, torsion, or compression stress, a slight elongation or compression takes place in the crystal lattice. This movement is called **strain.** The distortion of the crystal lattice is illustrated in Figure 1.16. Elastic deformation is not permanent. As soon as the stress is removed, the structure returns to its former shape. Also, this elongation or compression in one direction will produce an opposite change at 90 degrees to that force. That is, a piece of metal will become thinner if it is stretched lengthwise. This ratio of movement at right angles to the applied force is called **Poisson's ratio.**

Plastic deformation is one of the most useful characteristics of metals. Forging, drawing, forming, extruding, rolling, stamping, and pressing all involve plastic deformation. Plastic deformation in metals can take place only at a stress or load higher than the elastic limit (Figure 1.17). As the applied load is increased, the atoms must slide over each other to produce a permanent change in shape or break the atomic bonds, resulting in total rupture. The behavior of metal crystals under load depends on five things:

FIGURE 1.14
Broken sections of steel showing its crystalline structure. The one on the left is coarse and can be seen clearly. The one on the right is fine grained and requires magnification to see the grains.

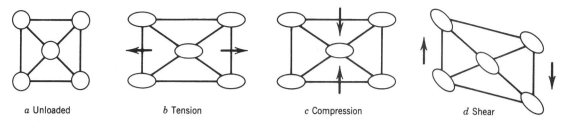

FIGURE 1.16

Distortion of crystal lattice with various kinds of stresses. *(a)* Unloaded. *(b)* Tension.
(c) Compression. *(d)* Shear.

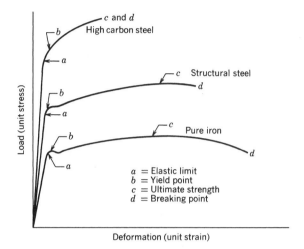

FIGURE 1.17

Load-deformation (stress-strain) diagram for a ductile steel
(White, Neely, Kibbe, Meyer, *Machine Tools and Machining
Practices,* Vol. II, © 1977 John Wiley & Sons., Inc.).

1. The spacing between planes of atoms.

2. Interatomic bond strength.

3. The density of atoms in various planes.

4. The lattice type.

5. Lattice irregularities (vacancies and discontinuities).

Plastic deformation tends to be favored along planes of
highest atomic density (close-packed planes) and the
greatest parallel separation (Figure 1.18). It can be seen in
this figure that the more dense or closely packed the atoms
are, the better chance there is for slip along planes or rows
of atoms. Slip within grains is roughly analogous to a
deck of cards, each card sliding a small amount (Figures
1.19 and 1.20). This movement of atoms tends to elongate
the grain and flatten it (Figure 1.21). Each atom shifts, one
at a time, and the process of slip is progressive, not quite
like cards sliding over each other. It is believed that
imperfections in metals, such as vacancies (spaces where
there are no atoms), dislocations or distortions of the lat-
tices, and impurity atoms are responsible for the phenomenon of slip that allows for permanent deformation of
metals without rupture (Figure 1.22). When a vacancy site
exists, atoms move one at a time, moving the vacancy site
or dislocation along the lattice until there is no more room
to move. When that happens in all of the grains, **work
hardening** results, so that the material is no longer plastic
and it becomes more brittle. If the load increases beyond
the ultimate strength of the material when it is no longer
plastic, the result is failure as a brittle fracture.

Work hardening or strain hardening is a problem to
be dealt with in manufacturing where cold working processes are involved. After a certain amount of cold working, a metal must be subjected to a process called annealing
to restore its grain structure to a softer, more plastic condition. This is called **recrystallization.** Annealing and
recrystallization will be further discussed in Chapter 4,
Heat Treatment of Metals.

FIGURE 1.18

Rows of atoms can glide along slip planes as shown. BCC and FCC have rows of
atoms closely spaced as in *(a)*, making movement easier than in *(b)*. HCP type
metals are less ductile since many slip planes as in *(b)* are spaced closer (Neely,
Mettallurgy, 2d ed., © 1984 John Wiley & Sons, Inc.).

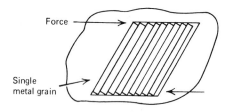

FIGURE 1.19
When stress is applied to a metal grain above its elastic range, slip occurs and the grain flattens and becomes elongated. Slip planes are often rotated when stress is applied (Neely, *Metallurgy,* 2d ed., © 1984 John Wiley & Sons, Inc.).

Grain Size

It can be seen why large grain metals have a greater capacity for slip, to be deformed, and to have more plasticity than fine grain metals. A fine grain steel, for example, is stronger than an equivalent coarse grain steel; that is, it can withstand greater stresses without being permanently deformed. Since each grain has a random orientation (direction of dendrite growth), slip in each grain ends at the grain boundary; small grains have a lesser capacity for slip than large grains do. For this reason, grain size is an important principle to consider when describing the nature of metals. Steel manufacturers provide mechanical properties charts and tables for each carbon, alloy, or tool steel that give, among other specifications, the grain size in terms of standard grain sizes (Figure 1.23). Ordinary machine steel may have a grain size of about 6 to 8. When a specimen of steel is fractured, the broken cross section

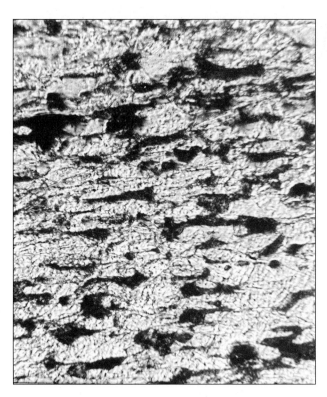

FIGURE 1.21
Micrograph of flattened grains in drawn wire showing partial recrystallization of the ferrite (white) grains.

often reveals the grains without magnification. These can be compared to the Shepherd grain size fracture standards; number 1 is coarsest and number 10 is finest (Figure 1.24).

Hardening Metals

Metals may also be hardened by locking up the slip planes with atoms of other elements or compounds, by alloying, or by trapping very small atoms such as carbon in the

FIGURE 1.20
Slip planes as they appear in a micrograph resulting from the cold working of low carbon steel (*Metals Handbook,* Vol. 7, 8th ed., American Society for Metals, 1972, p. 135. With permission).

FIGURE 1.22
Rows of atoms move, one atom at a time, to fill in a vacancy of space created by dislocated atoms. Interstitial atoms jam up the slip planes of atoms, making the metal that contains them harder and less ductile (Neely, *Metallurgy,* 2d ed., © 1984 John Wiley & Sons, Inc.).

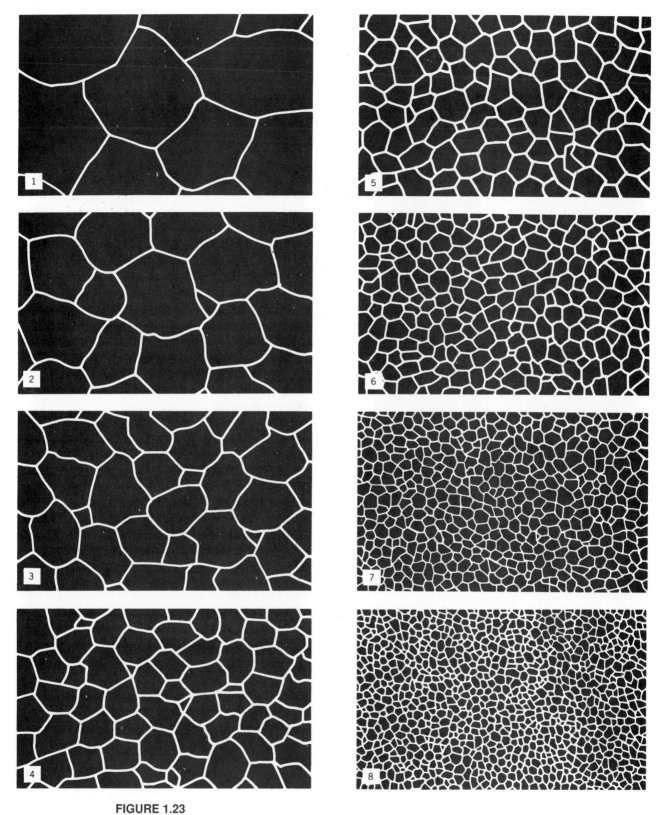

FIGURE 1.23
Standard grain size numbers. The grain size per square inch at 100× (Courtesy of Bethlehem Steel Corporation).

FIGURE 1.24
Shepherd grain size fracture standards. The fracture grain
size test specimen can be compared visually with this
series of 10 standards. Number 1 is the coarsest and
number 10 is the finest fracture grain size (Photo courtesy
of Republic Steel Corporation).

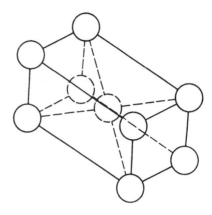

FIGURE 1.25
Body-centered tetragonal structure. This is the distorted
form of the unit cell (Neely, *Metallurgy,* 2d ed., © 1984
John Wiley & Sons, Inc.).

under a tremendous stress that causes the crystal to elongate to form a *body-centered tetragonal* unit crystal formation (Figure 1.25). This structure is called **martensite**
and it is extremely hard and brittle. Hardening and heat
treating metals is also covered in Chapter 4, Heat
Treatment of Metals.

PLASTICS, ELASTOMERS, AND CERAMICS

Of all the nonmetallic materials used in manufacturing,
plastics, elastomers, and ceramics are probably the most
important to understand. No attempt will be made here to
go into great detail about the chemical, molecular, or crystal structure of these materials; we will cover only what is
necessary to gain an understanding of their behavior for
manufacturing purposes.

Crystalline Structure of Plastics

Soluble substances can be divided into two classes, the
crystalloids and the **colloids.** The crystalloids, such as
table salt or sugar dissolved in water, easily pass through
a semipermeable membrane, but colloids are thick, viscous substances, like table jelly or gelatin, which will not
pass through a semipermeable membrane. This colloidal
behavior is seen in the synthetic plastics and rubber and is
caused by extremely large molecules. These macromole-
cules do not form crystals in the same way as in metals,
but many plastics do have crystalline forms. Some polymers, such as polyethylene, exhibit definite crystalline
formations, whereas others, such as polystyrene with its

interstices (spaces or voids) of the iron lattice. When carbon steel is hardened, it is quenched (rapidly cooled) from
a high temperature of about 1500° F (816° C) to room
temperature. A plain carbon tool steel having about 1 percent carbon can contain all of the carbon in the FCC phase
of iron at 1500° F (816° C) in the interstices of the lattice
because there is more open space than in BCC. There is
room for only a very tiny percentage of carbon in the
BCC structure. Therefore, when the steel is suddenly
quenched to room temperature, there is no time for the
carbon atoms to leave the interstices and form a compound of iron carbide (Fe_3C), which is what happens
when steel is slowly cooled. The result of sudden cooling
is that the trapped carbon atoms in the BCC structure are

FIGURE 1.26
An example of structural isomerism.

FIGURE 1.27
A monomer unit can become a branch source unit if it has one or more hydrogen atoms replaced by other elements. The subscript *n* indicates the indefinite length of the polymer.

long chains, are amorphous; that is, under a microscope they might look like a bowl of spaghetti with no apparent order.

In organic chemistry and in the study of structural polymers, the phenomenon of **isomerism** is of some importance. Some compounds having the same number and kind of atoms can join to form different structures having completely different properties. This is roughly analogous to various crystalline phases in metals (allotropism). Many plastics are isomers. A simple example of structural isomerism is the way in which the atoms C, H, O, can be arranged as ethyl alcohol and dimethyl ether (Figure 1.26).

Polymers

Natural organic materials such as leather, wood, cotton, and natural rubber are typically **polymers,** made up of long-chain molecules. Early discoveries in organic chemistry that produced substitutes for organic materials were not widely accepted and were considered to be poor replacements. For example, in 1866 John Wesley Hyatt first produced a synthetic material he called *celluloid* by combining camphor with nitrocellulose (gun cotton) that was made with slightly less nitric acid than usual. This new product was used as a substitute for the ivory that was obtained from the tusks of animals and was quite expensive. Soon products such as combs, handles, and billiard balls were made of this new synthetic material. Celluloid had two bad qualities; it had a tendency to burn rapidly when ignited and it deteriorated (discolored and cracked) with age. Movie film was first made of celluloid (cellulose nitrate), but later a nonflammable cellulose acetate was developed for movie film. Since this early discovery, a wide range of plastics, synthetic rubbers, textile fibers, adhesives, and paints has been developed by chemists. These plastics are considered by some to be among the greatest scientific achievements of modern times.

These substances, often called high polymers, have molecular weights that are often hundreds of times greater than ordinary materials. These giant molecules are made up of small molecules called monomers (Figure 1.27).

These are the building blocks from which the polymer chain is built (Figure 1.28). **Polymerization** is a chemical process in which many small molecules are linked together to form one large molecule. Celluloid was not the first polymerized plastic since the basic component of cellulose nitrate is cellulose, which in cotton or wood is already a high polymer. The first high polymer to be made from small molecules of raw materials was the strong, hard plastic called *bakelite* which was developed by the Belgian chemist Leo Baekeland in 1908. He was able to join the small molecules of phenol and formaldehyde to produce large molecules. Its basic structure is shown in Figure 1.29. There are three basic types of polymerization: *addition, copolymerization,* and *condensation.*

Addition polymerization takes place when similar monomers join to form a chain (Figure 1.30). This is also

FIGURE 1.28
These are the molecule forms (structural formulas) that serve as monomers to produce polyethylene and polyvinyl chloride (PVC) as shown here. Thus, giant chain molecules may be formed from monomers as shown in Figure 1.29.

FIGURE 1.29
The basic chemical structure of bakelite as it is produced from formaldehyde and phenol monomers.

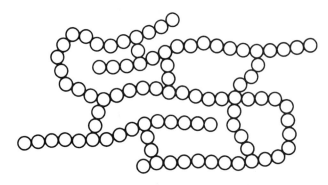

Bakelite polymer

called linear polymerization. Copolymerization takes place when two or more different kinds of monomers are combined. Condensation polymerization is a process of combining long-chain molecules to form more complicated chains. When the reaction takes place there is a residue, usually water. See Figure 1.29. These chains when coiled exhibit plasticity, but some, when they are cross-linked, lose plasticity and become elastic (Figure 1.31). Rubber is an example. Natural rubber latex, when combined with sulfur and then heated, changes the nature of crude rubber. This process of cross-linking rubber latex, a natural polymeric substance (Figure 1.32), is called **vulcanization.**

Natural rubber consists of long chains that must be broken down to short chains by kneading or mastication between hot rolls. Chemically, natural rubber is a polyisoprene, a hydrocarbon that has the empirical formula C_5H_8.

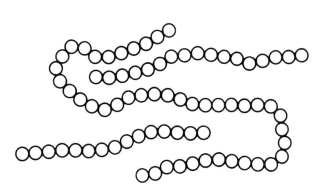

FIGURE 1.30
The spheres in this long-chain molecule represent atoms or monomers. Some chains are rigid, as found in heat-resisting materials, whereas others are elastic as in polyethylene, nylon, and cured rubber.

FIGURE 1.31
Cross-linking is only one of several ways that polymers can be strengthened.

(a)

(b)

FIGURE 1.32
(a) Natural rubber long-chain molecules; wire model, unstretched. *(b)* Vulcanized rubber, cross-linked.

FIGURE 1.33
General formula for vulcanized rubber.

A general formula of vulcanized rubber is shown in Figure 1.33.

Rubber can be made hard, soft, tacky, or resistant to oxidizers by using various additives. Accelerators are used to speed up the relatively slow process of vulcanization.

A great variety of synthetic rubbers have been developed. Many of these have properties different from those of natural rubber and are superior for some specific applications. Most of these synthetic rubbery materials are derived from petrochemicals (petroleum distillates) such as acetylene, butadiene, isoprene, and chloroprene (Figure 1.34). These chemicals and their derivation will be further discussed in Chapter 5.

Ceramics

The term **ceramics** includes a wide diversity of products. Clay bricks, drain pipes, insulators, spark plugs, synthetic diamonds, abrasives, glass, building stone, and concrete are among the many ceramic materials we use today.

The basic ingredients of clay are alumina and silica, that is, hydrated silicate of alumina ($Al_2O_3 \cdot 2SiO_2 \cdot 2H_2O$). Most clays contain impurities such as lime, iron oxide, magnesium, and potash. These impurities impart

FIGURE 1.34
Rubbers and elastomers.

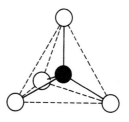

FIGURE 1.35
Unit structure of silicon dioxide (SiO_4). The darkened circle is silicon and the plain circles represent oxygen atoms (Neely, *Metallurgy*, 2d ed., © 1984 John Wiley & Sons, Inc.).

colors to the natural clay. Ferric oxide produces the red color seen in many clays. Shale and slate are two forms of clay that contain many impurities. Kaolin is a white clay in its pure state. It is used for the manufacture of chinaware, porcelain, firebrick, paper, rubber, and paint pigment.

Silicate structures give fired clays their hardness and stability (Figure 1.35). When linked together in a chain, the unit structures become Si_2O_7 (Figure 1.36). When these structures are combined with alumina, they become clay, a very plastic substance when wet, and rigid when dried. Clay is composed of very small crystalline, flat, or platelike grains that tend to be stacked like packs of playing cards, each pack oriented in a different direction. Each crystal plane is separated and surrounded by water that acts as a lubricant and a bond, allowing the grains to slide easily over one another (Figure 1.37). When the water is dried out, as in sun-dried bricks, the lubrication is gone and the material becomes rigid but with low tensile and compressive strength. Clay in this state can easily be wetted and brought back to its plastic state. However, if it is fired, it loses that ability and it becomes somewhat stronger and quite brittle. The plates of hydrated silicate of alumina when fired combine to form new crystalline

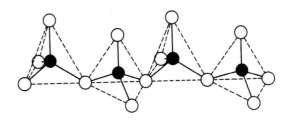

FIGURE 1.36
Silicate units can link through any of the oxygen atoms as shown. This is one way in which a chain of silica tetrahedra can be joined. The dark circles represent silicon atoms and the plain circles are oxygen atoms (Neely, *Metallurgy*, 2d ed., © 1984 John Wiley & Sons, Inc.).

FIGURE 1.37
Platelike grains are like decks
of cards. *(a)* These are wet
and plastic. *(b)* These are
dried, hard, and brittle.

(a)

(b)

phases or glasses that permeate the entire structure, bonding the new crystals together into one fused mass.

Glass

The basic ingredient of glass is quartz (silica) SiO_2. When quartz is cooled, it forms a crystalline structure that is brittle and semitransparent. Glass is slow cooled to keep it from crystallizing and it contains other substances such as lime and sodium. The molecular structure of glass is amorphous, like a liquid; in fact, glass is considered to be

a solid liquid. The raw material is silica sand, which is quite plentiful, making glass a relatively inexpensive material.

Portland Cement

Concrete that is used to make roads, sidewalks, dams, buildings, and other permanent structures is composed of portland cement, sand, and graded gravel. The cement when combined with water hardens to a rocklike consistency, binding the sand and gravel into one homogenous

(a)

(b)

(d)

FIGURE 1.38
The hydration of portland cement. *(a)* Schematic of the formation of silica gel in the presence of water, 1 day. *(b)* Schematic showing fibers beginning to grow, 7 days. *(c)* Electron micrograph showing the growth of fibers at 7 days (3000×). *(d)* Schematic showing the final development of the strong fibers, 28 days. *(e)* Electron micrograph of completely hydrated portland cement at 28 days (5000×). (*c* and *e* are SEM photographs by Hugh Love, Portland Cement Association, Skokie, IL).

mass. This hardening takes place without the presence of air or drying out. In fact, it must be kept from drying out during the hardening process. This process is called *hydration.* Portland cement is basically a calcium silicate mixture predominantly containing tricalcium and dicalcium silicates. These materials are basically derived from shales, clays, and limestone with small amounts of iron oxide and dolomite. The alumina content of clay has an effect on the long-term strength of concrete. Other than impurities, the important ingredients of portland cement could be written as $CaO \cdot SiO_2 \cdot Al_2O_3$. The action of hydration and hardening is that of a silica gel with fibers growing out of cement particles that join with adjacent particles to form a solid interlocked mass. The production and uses of portland cement and concrete will be covered in later chapters.

In Figure 1.38a the schematic shows the tiny cement particles to which mixing water has been added as they would appear magnified many thousands of times. A silica gel begins to form on their outer surfaces. Soon whisker-like fibers begin to grow from the particles, and after 7 days they are interconnected, as can be seen in Figure 1.38b. Also the electron micrograph in Figure 1.38c reveals this condition at 7 days in hydrated portland cement. After 28 days this cement and the concrete made with it is about as hard as it will ever get. In moist conditions, as found in dams, the concrete may get slightly harder and stronger over the years. Figure 1.38d shows how these fibers develop into coarser and stronger bars or plates that interlock to form a solid rocklike mass. The electron micrograph, Figure 1.38e, shows hydrated portland cement at 28 days.

Review Questions

1. Not considering the many subatomic particles, all atoms are made up of three basic components. Name them.

2. Briefly define valence.

3. Atoms with positive or negative valence are called ions. What causes this condition?

4. Name four kinds of bonding arrangements of atoms.

5. What does the type of bond in metals have to do with their elastic, plastic, thermal, and electrical behavior?

6. What is a space lattice in solid materials?

7. What are unit cells in metals?

8. When a metal changes from one unit cell structure to another at certain temperatures, what is it said to be?

9. When a metal begins to solidify, many grains begin to grow. These are called dendrites. At what location do they stop growing?

10. Define stress and strain.

11. What is work hardening?

12. All other conditions being the same, which metal would be more plastic and ductile, a fine grained steel or a coarse grained steel? Which would be stronger (highest yield strength)?

13. When small molecules are linked together to form a large molecule, what is it called?

14. A gummy substance such as natural rubber latex can be made into useful products by the process of cross-linking. In rubber, what is this process called?

15. The first high polymer was produced by joining the small molecules of phenol and formaldehyde. What is this well-known plastic called?

16. How can clay be plastic when wet, brittle when dry, and hard and strong when fired?

17. What is the basic ingredient of glass?

18. Portland cement hardens and bonds together in the presence of water. What is this process called and how does it progress to form a solid mass?

CHAPTER 2

Extraction and Refinement of Common Metals

Objectives

This chapter will enable you to:

1. Describe methods of mining and processing metallic ores.

2. Give an account of the development and methods used in the ironmaking and steelmaking processes.

3. Explain some of the principles and methods used in smelting several nonferrous metals.

4. Show the construction and principles of operation for several types of furnaces and refining vessels.

Metallic ores and other important minerals are found in virtually every continent; however, there is an unequal distribution of many of these minerals. For example, tin (Sn) is a metal that is abundant only in Africa, Bolivia, and Southeast Asia, and is found in a few other locations that are not major producers. Chromium ore is extensively mined in South Africa and in the former Soviet Union; there are only minor sources of this strategic metal in the United States, so most of our chromium is supplied from these two areas.

Our modern civilization is probably more dependent on strategic metals than on oil. These metals include cobalt, chromium, manganese, platinum, columbium, strontium, titanium, and tantalum. Iron and aluminum are relatively plentiful, but even these materials are affected by the increasing cost of energy.

Mining originated in prehistoric times. Some metals were found in small quantities in their "native" or metallic state. Native copper, silver, and gold were found in stream beds and in outcroppings. These were shaped by hammering to make jewelry and crude tools. The need for larger quantities of metals brought about underground and surface mining methods that have been considerably improved in modern times. Processing and smelting ores to extract the metal has been historically a simple and limited process for metals such as iron, copper, silver, tin, lead, and zinc. Only a few pounds of iron could be produced in one day in primitive forges and furnaces. In modern times the daily output of one of our iron smelting furnaces is measured in tons. Other metals were not extracted and refined in commercial quantities until relatively recently. Some metals such as titanium, which are often called space-age metals, have been manufactured in quantity only since World War II. Iron-bearing metals such as pig iron, cast iron, and steel are called **ferrous** and all other metals are called **nonferrous.**

ORES AND MINING

An ore deposit is any geological deposit of nonrenewable material having a sufficiently high percentage of the needed element or metal that can be recovered at a profit. Most elements are thinly distributed over the earth's crust. Aluminum and iron are found in clays (alumino-siliates) in substantial quantities, but it would not be economical to process that material for the metal. Ores that occur in narrow veins are usually mined by underground methods (Figure 2.1). This is done by excavating vertical or sloping shafts and horizontal openings or rooms where the ore is removed, broken up, and hauled to the surface in mine cars or bucket and cable devices. When ores are found in

FIGURE 2.1
Mining machine uses whirling cutting teeth to rip coal from underground seams. Coal is conveyed backward to shuttle cars. Water sprays entrap the coal dust (American Iron and Steel Institute).

large deposits, a surface mining technique, called open-pit mining (Figure 2.2), is used. Ore bodies are almost always covered by soil or other waste material, called overburden, which must be removed prior to mining the ore. In a copper mine near Bingham, Utah (Figure 2.3), an entire mountain has been excavated and removed for its copper ore. In some surface mining operations the topsoil is removed, the ore is removed, and the soil is returned to its former place. This method helps to avoid excessive damage to the ecology of the region.

Iron Ore

Before the large deposits of rich hematite and magnetite ores in the Lake Superior Mesabi Range were depleted, as they are now, the 60 to 70 percent iron in those ores was quite profitable to extract. There are still vast deposits of

iron ore in these regions, but the iron is bound up in a very hard rock formation that is difficult to mine and process and that has a relatively low percentage of iron. This material, called taconite, is now the major source of iron ore in that region. The most common natural iron ores are given in Table 2.1.

Mining Aluminum Ore

Aluminum is the third most abundant element on the earth, after oxygen and silicon; however, only a few deposits of two aluminum-bearing minerals are considered to be profitable to mine. These are gibbsite and boehmite, which are composed of aluminum oxide (alumina Al_2O_3) and water mixed with sand or clay and other minerals that must be removed. This ore is called bauxite, named for a French village, Les Baux, where the first

FIGURE 2.2
In an open pit mine on Minnesota's Mesabi Range, a power shovel dumps a 13-ton dipperful of iron ore into a truck. Although this is a natural ore, it must be beneficiated or improved in some way before it is shipped to blast furnaces (American Iron and Steel Institute).

FIGURE 2.3
An entire mountain has been removed from this gigantic open pit copper mine near Salt Lake City, Utah.

deposit was found in 1821. Bauxite is usually mined by the open-pit method and is found on every continent except Antarctica.

Ore Treatments

The cost of shipping iron ores from the mine to the smelter can be greatly reduced if the unwanted rock and other impurities are removed prior to shipment. This is done with several processes called beneficiation. These processes include crushing, screening, tumbling, flotation, and magnetic separation. The refined ore is enriched to being over 60 percent iron by these methods and is often formed into pellets before shipment. In the case of taconite, about two tons of unwanted minerals are removed for each ton of pellets shipped (Figure 2.4) and the tailings are returned to the mining site. Growing vetches, grasses, and trees on some of these barren landscapes has been a project of biologists and foresters.

Direct Reduction of Iron

In ancient times, the only known method of separating iron from its ore was by the process of direct reduction. This was done in a charcoal forge or furnace. Unlike copper ores, which yielded molten copper in these furnaces, iron would not melt at temperatures below 2799°F (1537°C). The highest temperature that could be reached in those primitive smelters appears to have been about 2192°F (1200°C). Iron ore subjected to that temperature yields not a molten puddle of metal, but a spongy mass mixed with iron oxide and iron silicate with perhaps other impurities that together made up the slag. The iron worker removed this mass of sponge from the furnace, reheated it

in a forge, and then literally squeezed the slag out of it by hammering it on an anvil. The hammering produced a wrought iron with very little carbon interspersed with a few stringers of slag that had not been eliminated. This wrought iron was not much stronger than copper and was a poor substitute for bronze because it could be hardened somewhat only by hammering, as could copper and bronze. However, with the discovery of how to make steel by combining iron with a small amount of carbon and heat treating it, steel became a far superior metal. The old method of smelting iron and steel has long since been abandoned and replaced by using the blast furnace and steelmaking furnaces.

Sponge Iron and Powder

Since the 1920s a new method of direct reduction of ores has been used to produce sponge iron and iron powder, bypassing the blast furnace stage. Many of these "direct-reduced" ores have a lower percentage of impurities than molten iron from a blast furnace. Pellets are formed in this process that contain as much as 95 percent iron and can be economically shipped from remote locations to be remelted in steelmaking furnaces. Iron powder is also produced by direct reduction processes; much of this is used in the powder metallurgy industry. See Chapter 12.

Wrought Iron

The wrought (worked) iron of ancient times had some advantages over steel. It had less tendency to corrode (rust) and it had a fibrous quality (from the stringers of slag) that gave it a certain toughness. Although the ancient process of making wrought iron is obsolete, a modern method, called the Aston process, is used to produce a wrought iron in large quantities. Molten pig iron and steel are poured into an open hearth furnace with a prepared slag that removes most of the carbon. The slag cools the molten metal to a pasty mass that is later squeezed in a hydraulic press to remove most of the slag. Wrought iron is used for iron work such as railings for stairways and for the production of pipe and other products subject to deterioration by rusting.

PIG IRON

Until about 1350 A.D., all of the iron made in the known Western world was made by the direct reduction process. The development of the shaft or blast furnace at that time (although it might have been developed earlier in China) made possible the high production rate of iron that was

TABLE 2.1
Some common ores of industrial metals

Metal	Ore	Color	Chemical Formula
Aluminum	Bauxite	Gray-white	Al_2O_3 (40 to 60%)
Cadmium	Sphalerite (zinc blende)	Colorless to brown to black	ZnS + 0.1 to 0.2% Cd
Chromium	Smithsonite	White	$Zn_4Si_2O_7OH \cdot H_2O$
Cobalt	Chromite	Green	$FeO \cdot Cr_2O_3$
	Cobaltite	Gray to silver-white	$CoAsS$
	Linnaeite (Cobalt pyrite)		Co_3S_4
	Smaltite	Bluish white or gray	$CoAs_2$
	Erythrite	Transparent, crimson, or peach-red	$Co_3(AsO_4)_2 \cdot 8H_2O$
Columbium (Niobium)	Columbite (Niobite)	Black	$(FeMn)(CbTa)_2O_6$
Copper	Chalcopyrite	Brass yellow	$CuFeS_2$
	Chalcocite	Lead gray	Cu_2S
	Cuprite	Red	Cu_2O
	Azurite	Blue	$2CuCo_3 \cdot Cu(OH)_2$
	Bornite	Red-brown	$FeS \cdot 2Cu_2S \cdot As_2S_5$
	Malachite	Green	$CuCo_3 \cdot Cu(OH)_2$
	Native copper	Reddish	Cu
Gold	Native (alluvial or in quartz)	Yellow	Au
Indium	Sphalerite	White to yellow	ZnS + low % of In
Iron	Magnetite	Black	Fe_3O_4
	Hematite	Reddish brown	Fe_2O_3
	Limonite	Brown to yellow	$2Fe_2O_3 \cdot 3H_2O$
	Goethite	Brown	$Fe_2O_3 \cdot H_2O$
	Siderite	Blue	$FeCO_3$
	Taconite	Flintlike	Fe_3O_3
Lead	Galena	Black	PbS
Magnesium	Seawater		$Mg(OH)_2$
Manganese	Pyrolusite	Black	MnO_2
	Manganite	Steel-gray to black	$Mn_2O_3 \cdot H_2O$
	Rhodochrosite	Rose-red	$MnCO_3$
Molybdenum	Molybdenite	Blue	MoS_2
Nickel	Niccolite	Pale copper-red	$NiAs$
	Pentlandite	Bronze-yellow	$(FeNi)_9S_8$
	Garnierite	Bright apple-green	$(MgNi)_3Si_2O_5(OH)_4$
Silver	Argentite	Black	Ag_2S
	Native	White	Ag
Tin	Cassiterite	Brown or black	SnO_2
Titanium	Anatase,	White	
	Brookite, and Rutile	Reddish-brown	TiO_2
Tungsten	Wolframite	Brownish to grayish black	$(FeMn)WO_4$
	Scheelite	Various	$CaWO_4$
	Ferberite		$FeWO_4$
	Huebnerite		$MnWO_4$
Uranium	Uraninite	Black	UO_2
	Pitchblende	Brown to black	U_3O_8
Vanadium	Vanadinite	Brown, yellow, red	$Pb_5(VO_4)_3Cl$
	Carnotite	Yellow	$K_2(UO_2)_2(VO_4)_23H_2O$
Zinc	Franklinite	Black	$ZnFe_2O_4$
	Willemite	Varying	Zn_2SiO_4
	Zincite	Red to orange-yellow	ZnO
Zirconium	Zircon (zirconium silicate sand)	Brown or gray square prisms or transparent	$ZrSiO_4$

FIGURE 2.4
Taconite ore powder, now free of much impurity and rich in iron, is mixed with coal dust and a binder, rolled into small balls in this drum pelletizer, and then baked to hardness (American Iron and Steel Institute).

needed for the Industrial Revolution. The blast furnace (Figure 2.5) made possible a molten, castable iron product that contains about 4 percent carbon and some impurities. The high carbon content caused it to be somewhat brittle and it has little usefulness for tools and cutting instruments or other products. When refined in cupola furnaces, it is an excellent metal for some machinery parts, cooking utensils, and other cast iron products. However, the great-

FIGURE 2.5
One of four blast furnaces at Bethlehem Steel Corporation's Sparrows Point, Maryland, plant can process more than 3000 tons of molten iron per day (Courtesy of Bethlehem Steel Corporation).

est usefulness of the blast furnace is in an intermediate step in the steelmaking process (Figure 2.6).

The operation of a modern blast furnace can be seen in Figure 2.7. The three raw materials needed to make pig iron are iron ore, coke, and limestone. They are put into the furnace at intervals, making the process continuous. About two tons of ore, one ton of coke, and a half ton of limestone are required to produce one ton of iron.

Coke is a residue left after soft coal is heated in the absence of air. The gases are driven off and the remaining hard, brittle, and porous material is the resultant coke, which contains 85 to 90 percent carbon with some impurities. Coke is produced in coke ovens (Figures 2.8 on p. 39 and 2.9 on p. 40) where the volatile elements that are driven off the coal are collected. Many useful products are made from these gases: fuel gas, ammonia, sulfur, oils, and coal tars. From the coal tars come many important products, such as plastics, dyes, synthetic rubbers, perfumes, sulfa drugs, and aspirin.

The operation of the blast furnace is called reduction. This is a process in which oxygen is removed from a compound, in this case, iron ore. The oxygen is combined with carbon to form carbon dioxide (CO_2), and the metallic iron is released from its oxide state, leaving molten iron with about 4 percent carbon. The solid materials coke, limestone, and iron ore are dumped into a hopper at the top of the furnace at regular intervals and heated air is blasted in at the bottom. The air is heated in stoves containing refractory brick that is heated by exhaust gases from the furnace. The stoves alternate between blasting hot air to the furnace to being in a heating cycle. The fuel burns near the bottom of the blast furnace and the rising heat meets the descending charge of coke and ore, producing very high temperatures of about 3000°F (1649°C). As the molten iron is released, it settles to the bottom and a slag forms on top of it. The slag is drawn off at intervals and hauled away in slag cars to a dumping area. This slag is ground up and used as an aggregate for concrete products and asphaltic concrete roads. The molten iron is topped (drawn off) at intervals and collected in an insulated transfer car (Figure 2.10 on p. 40). Then it is taken to a steel-making furnace and charged into it. Some of the iron is used elsewhere and is remelted and refined in foundries for casting purposes. For shipping, it is cast in pigs (Figure 2.11 on p. 41). In earlier times the pigs were made in open sand molds consisting of a groove or trough with many small molds on each side, reminding one of a sow and pigs, hence the name pig iron.

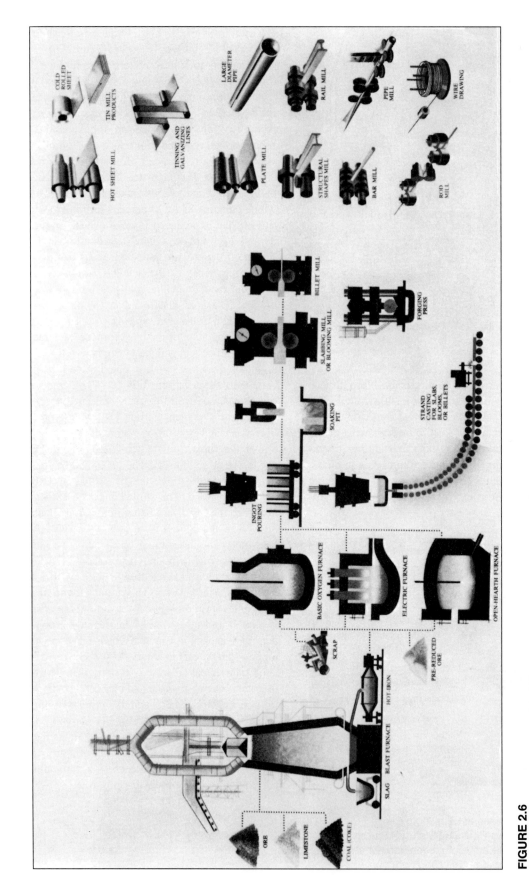

FIGURE 2.6
Flowchart of the iron and steel production process, showing the route from ore to iron to finished steel (Courtesy of Bethlehem Steel Corporation).

FIGURE 2.7
Cross section of a blast furnace. Iron ore is converted into pig iron by means of a series of chemical reactions (Courtesy of Bethlehem Steel Corporation).

STEELMAKING

Pure iron is a soft, easily bent metal having limited usefulness. Steel is iron combined or alloyed with other metals or nonmetals such as carbon. Plain carbon steel actually contains a small amount of the metal manganese as a scavenger to remove unwanted oxygen and to control sulfur. Sulfur is difficult to remove from steel, and iron sulfide (the form it naturally takes in iron) is deleterious to steel, but if manganese is present, it combines more readily than iron with the sulfur, forming manganese sulfide, which is not harmful to the steel. For this reason no steel is made without manganese. The element carbon causes steel to be such a useful material since it can be hardened by heat treatments. Only a small amount of carbon is needed to make steel: .20 percent to .30 percent for low carbon steel, .40 percent to .60 percent for medium carbon steel, and .60 percent to over 1 percent for high carbon steel. Steel can contain up to 2 percent carbon, but over that amount it is considered to be cast iron, in which

FIGURE 2.8
An 850,000 ton per year battery of 80 slot-type, side-by-side coke ovens (Courtesy of Bethlehem Steel Corporation).

The coking process viewed from above (Coke oven battery)

Charging silo

Gas to by-product plant

Operations monitoring station

Larry car

Pusher arm

Coke

Covered quench car

Coke wharf

Charging ports

Coke to blast furnace

Quench tower

FIGURE 2.9
The operation of a coke oven battery at Bethlehem Steel Corporation's Sparrows Point, Maryland, plant is similar to the drawing. The larry car (or charging machine) is loaded with coal from a charging silo, then charges the coal into the individual ovens through ports in the oven roofs. After baking for approximately 18 hours and reaching a final temperature of at least 1800°F (982°C), coke is pushed from the ovens by the pushing machine at left. Hot coke is then moved by way of a quench car to a quench tower where it is flooded with approximately 16,000 gallons of water to cool it before being unloaded at a coke wharf to await further processing (Courtesy of Bethlehem Steel Corporation).

FIGURE 2.10
The blast furnace is tapped after slag removal, and the molten iron flows into ladles or hot-metal cars. Most iron is used in steelmaking (American Iron and Steel Institute).

excess carbon forms graphite. Steel can be given many different and useful properties by alloying the iron with other metals such as chromium, nickel, tungsten, molybdenum, vanadium, and titanium and with nonmetals such as boron and silicon.

Furnaces

The principal purpose of the steelmaking furnace is to remove most of the carbon from pig iron. Then a measured amount of carbon is added to the molten steel to give it the desired properties. Steel scrap is also utilized and other elements are added to improve the properties of the steel. A now obsolete furnace, the Bessemer converter, made possible the high tonnage production of steel in the mid-nineteenth century. Great steel ships, railroads, bridges, and large buildings were built as a result of this new source of steel being available. However, many of the impurities remained in the steel and, since a blast of air was used to burn out the carbon, the nitrogen in the atmosphere also became an impurity that weakened the steel.

The three modern processes and furnaces used to produce steel are *basic oxygen, open hearth,* and *electric.* These furnaces are not used interchangeably. Each system requires different energy sources and different sources of raw materials. The kind of facility is therefore chosen for economic reasons and the availability of raw materials and energy sources.

The basic oxygen furnace (BOF) (Figures 2.12 and 2.13) is designed to produce a high quality steel in a relatively short time, about one 200- to 300-ton heat (a single batch) per hour as compared to the 100 to 375 tons per heat in 5 to 8 hours of the open hearth process. The principal raw material used in the basic oxygen furnace is

FIGURE 2.11
Automatic pig casting machine. Some blast furnace iron is poured to solidify in molds. Iron foundries are major users of cast pigs (American Iron and Steel Institute).

FIGURE 2.12
Basic oxygen furnace being charged with molten pig iron (American Iron and Steel Institute).

FIGURE 2.13
Charging basic oxygen furnace with scrap (American Iron and Steel Institute).

molten pig iron from the blast furnace. Steel scrap is charged into the furnace and lime is used as a fluxing agent that reacts with the impurities and forms a slag on top of the molten metal. The water-cooled lance is lowered to within a few feet of the charge to blow a stream of oxygen at more than 100 pounds per square inch (PSI) on the surface of the molten bath. The oxidation of the carbon and impurities causes a violent rolling agitation that brings all the metal into contact with the oxygen stream. The ladle is first tipped to remove the slag and then rotated to pour the molten steel into a ladle. Carbon, usually in the form of anthracite coal pellets, is added in carefully measured amounts and other elements are also added to produce the desired quality in the steel. A schematic cross section of a basic oxygen furnace is shown in Figure 2.14.

Open-hearth furnaces get their name from a shallow steelmaking area called a *hearth* that is exposed or open to a blast of flames that alternately sweeps across the hearth from one end for a period of time and then from the other end of the furnace (Figure 2.15). The furnace is charged (loaded) from a door in the side facing the charging floor. Theoretically, an open-hearth furnace can operate with blast furnace iron alone or steel scrap alone, but most operations use both in about equal proportions, varying with the price of scrap and the availability of molten iron. Scrap is placed in the furnace by a charging machine, which usually serves a line or series of open-hearth furnaces in a single building (Figure 2.16). Other elements are added to improve the steel. Mill scale forms on the surface of hot steel and falls off as a black scale; it is actually one form of iron oxide (Fe_3O_4), called magnetite as an ore. In steel mills it is used in steelmaking furnaces to reduce carbon content. Aluminum ferrosilicon is added if the steel is to be killed. A **killed steel** is one that has been sufficiently deoxidized to prevent gas evolution in the ingot mold, making a uniform steel. If the steel is not killed, it is called **rimmed steel** because the gas pockets and holes from free oxygen form in the center of the ingot and the rim near the surface of the ingot is free of defects. Most of these flaws, however, are removed by later rolling processes (discussed in Chapter 10). When the contents of the heat or melt are

BASIC OXYGEN STEEL PLANT CROSS - SECTION

FIGURE 2.14
Schematic of a basic oxygen furnace showing the facilities needed to charge scrap and molten iron into the vessel and receive the steel after the oxygen-blowing process is complete (American Iron and Steel Institute).

FIGURE 2.15
Simplified cutaway diagram of a typical open-hearth furnace. Oxygen may be injected through one or more lances. In some cases it is introduced through the burners to improve combustion (Courtesy of Bethlehem Steel Corporation).

acceptable and the temperature is right, the furnace is tapped and the molten metal is poured into a ladle (Figure 2.17). As with the basic oxygen furnace (BOF), carbon is added to the contents of the ladle to adjust the carbon content. Aluminum pellets are added to the ladle to deoxidize the molten metal and produce a finer grain size.

The ladle with its load of molten steel is picked up by a traveling crane and brought over a series of heavy cast iron ingot molds. The steel is teemed (poured) into the molds by means of an opening in the bottom of the ladle (Figures 2.18 and 2.19).

Electric furnaces lined with refractory (heat-resistant ceramic) materials (Figures 2.20 to 2.23 on pp. 44–46) produce a high-quality steel from selected scrap, lime, and mill scale. Most alloy and tool steels are made in electric furnaces, but for these high-quality steels selected scrap must be used. Tin cans and automobile scrap contain contaminants such as tin and copper that spoil the steel for these purposes. Electric furnaces are most competitive where low-cost electricity is available and where very little coal or iron ore is found. The charge is melted by the arcing between carbon electrodes and metal scrap. In some cases, the electrodes are 2 feet in diameter and 24 feet long.

Advanced Melting and Refining

Standard steelmaking furnaces are not capable of removing all impurities and unwanted gases. When higher grade specialty steels are needed, further processing is necessary. There are two types of these refining vessels, those that receive molten metal from conventional steel furnaces and those that remelt to refine it (Figures 2.24*a* to 2.24*h* on p. 47).

FIGURE 2.16
Closeup of charging machine arm thrusting a box of baled scrap through an open-hearth furnace door (American Iron and Steel Institute).

FIGURE 2.17
Tapping an open-hearth furnace. Molten steel fills the ladle; the slag spills over into the slag pot (Courtesy of Bethlehem Steel Corporation).

FIGURE 2.19
Once steel is poured into an ingot mold and allowed to solidify, the mold is stripped off, exposing a glowing hot ingot (Courtesy of Bethlehem Steel Corporation).

FIGURE 2.18
Molten steel from a basic oxygen furnace is teemed into ingot molds (Courtesy of Bethlehem Steel Corporation).

FIGURE 2.20
Scrap is being charged into an electric furnace that produces 185-ton heat of steel every 2½ hours (Courtesy of Bethlehem Steel Corporation).

FIGURE 2.21
In a rare view, the interior of an electric steelmaking furnace resembles an amphitheater. The furnace interior is lined with refractory brick and special water-cooled panels to withstand an interior temperature of 3000°F (1649° C). Steel scrap is dumped into the furnace through the roof and is melted electrically and then refined into high-quality steel (Courtesy of Bethlehem Steel Corporation).

Vacuum Stream Degassing Figure 2.24*a*. Ladles of molten steel from conventional furnaces are taken to vacuum chambers where the molten steel is poured into a pony ladle. There the molten stream of steel is controlled and broken up into droplets when exposed to the vacuum within the chamber. At this point large quantities of undesirable gases escape from the steel and are drawn off by the vacuum pump.

Vacuum Ladle Degassing Figure 2.24*b*. Several kinds of ladle degassing facilities are in current use. One of the most important is shown here. The furnace ladle is placed beneath a heated vacuum vessel that has a descending suction nozzle. The entire vessel is lowered into the ladle of molten steel. The molten steel is forced through the suction nozzle into the heated vacuum chamber. Gases are thus removed in the vacuum chamber, which is then raised so that the molten steel is returned by gravity into the ladle. The process is repeated several times.

Argon-Oxygen Decarburization Figure 2.24*c*. Argon, an inert gas, is used in this refining vessel to dilute the carbon-oxygen atmosphere in the melt. The argon and

oxygen gases have separate tuyeres so they can enter the bath of molten steel simultaneously. In the case of stainless steel, which must have a very low carbon content, this process increases the affinity of carbon for oxygen, producing harmless CO_2. This also minimizes the oxidation of chromium and a less expensive form of chromium can be used. The argon, being inert, also stirs the molten steel, promoting equilibrium between the molten metal and the slag.

Vacuum Oxygen Decarburization Figure 2.24*d*. The vessel, full of molten steel, is placed in an induction stirrer, which may be mounted or permanently installed on a transfer car. Then a separate roof is placed over the ladle. The roof contains an oxygen lance and is equipped for vacuum degassing. The metal is degassed (decarburized) while being stirred. Then this first roof is removed, and a second roof (through which additives and alloys are dropped from a hopper into the melt) is placed over the ladle.

Electron Beam Melting Figure 2.24*e*. This purification process involves a cascading action of molten steel carried

FIGURE 2.22
The entire electric furnace is tilted during a tapping operation in which molten steel flows into a waiting ladle (Courtesy of Bethlehem Steel Corporation).

FIGURE 2.23
A cross section of an electric furnace (Courtesy of Bethlehem Steel Corporation).

on in a vacuum. The drawing shows steel melted in an induction-heated crucible inside a vacuum chamber. This crucible tilts, pouring partially degassed steel into an induction-heated ladle in an adjoining vacuum chamber. The electron beam hits the steel imparting thermal energy to keep it molten as it flows over a water-cooled copper hearth. The steel is continuously cast into ingots.

Vacuum Induction Melting   Figure 2.24*f*. In this process, molten steel, or more commonly steel scrap, is charged into an induction furnace situated in a vacuum chamber. Undesirable gases are removed by a vacuum pump after which the molten refined steel is poured into a trough by tilting the furnace. The steel is then conveyed to a holding ladle in an adjoining vacuum chamber and ingot molds are filled from this ladle.

Consumable Electrode Melting   Figure 2.24*g*. Sometimes called the vacuum arc process or furnace, this vessel remelts steels produced by other methods and allows it to solidify into ingots, usually in a cylindrical shape. A solid steel bar is slowly lowered on a control rod into a vertical water-cooled mold. The electric current forms an arc that melts the bar much like the electrode melts at the arc in electric arc welding. The gaseous impurities are drawn off by the vacuum in the chamber as the molten steel progressively fills the mold. The consumable electrode melting process is also used in refining some nonferrous metals such as titanium and zirconium. (See the flow sheet of zirconium production in this chapter, Figure 2.27.)

Electroslag Melting   Figure 2.24*h*. There is a strong similarity between the electroslag process and the consumable electrode process, but there are some further technological refinements. As in the consumable electrode process, the electrode is suspended in a mold and connected to an electrical power source. Instead of a vacuum, a powdered slag, which quickly melts, is used to refine the steel. The molten droplets from the steel electrode drop through the liquid slag and collect in a molten pool beneath the slag and quickly solidify to form the ingot. Impurities are removed by flotation in the slag and a thin layer of slag also solidifies between the ingot and mold walls, resulting in a smooth surface free of defects.

ALUMINUM EXTRACTION

The bauxite ore from which aluminum is obtained contains many unwanted impurities such as iron oxide, silicon, titania, and water. The ore is first crushed to a powder

FIGURE 2.24
Remelting processes that are used mostly to produce highly sophisticated alloy and specialty steels for the world's new technologies. (a) Vacuum stream degassing. (b) Vacuum ladle degassing. (c) Argon-oxygen decarburization. (d) Vacuum oxygen decarburization. (e) Electron beam melting. (f) Vacuum induction melting. (g) Consumable electrode melting. (h) Electroslag melting. (American Iron and Steel Institute).

47

and mixed in a solution of strong caustic soda (sodium hydroxide, NaOH) and dumped into large tanks or digesters where more caustic soda or soda ash and lime are added to make up for losses.

The alumina (aluminum oxide, Al_2O_3) is dissolved in the caustic soda, forming a sodium aluminate solution and leaving the impurities, which are insoluble. A series of separation processes involving filtration and precipitation separates the alumina from unwanted solids called red mud. This aluminum hydrate is then heated to 1800°F (982°C) in kilns to drive off the chemically attached water. The resulting alumina is a fine white powder that is about half aluminum and half oxygen, bonded so tightly that neither heat nor chemical action alone can separate them.

Smelting Aluminum

An abundant supply of low-cost electricity is required to produce aluminum competitively. For this reason, aluminum smelters are usually located in areas where hydroelectric generating plants are established. One great advantage of using hydroelectric power for smelters is that low-cost electricity can be purchased in off-peak hours and when excess power is available in high water run-off periods.

Smelting is the process by which alumina is reduced—that is, separated into its two constituents, aluminum and oxygen (Figure 2.25). In a modern smelter, the alumina is dissolved in a bath of molten cryolite in electrolytic furnaces called pots. Cryolite (sodium aluminum fluoride, Na_3AlF_6) is a material, mined in Greenland, that allows the passage of an electric current, whereas alumina does not conduct electricity. The smelting process is continuous; low voltage-high amperage current is passed through the cryolite bath from carbon anodes suspended in the melt to a carbon lining in the bottom of the pot, which serves as a cathode. This powerful electric current separates the dissolved alumina into pure aluminum and oxygen.

The molten aluminum collects at the bottom of the pot and is periodically siphoned off while the oxygen combines with the carbon anode and is released as carbon dioxide. A small portion of aluminum fluoride is added from time to time to compensate for small fluoride losses. About two pounds of alumina will yield one pound of aluminum, and about one-half pound of carbon anode is consumed for every pound of aluminum produced.

FIGURE 2.25
The process of making aluminum (Courtesy of Aluminum Company of America).

COPPER SMELTING

The two main classes of copper ores are oxides and sulfides. The oxidized ores can simply be heated with carbon to release the copper metal. However, most available ores are sulfides and perhaps half the world's copper deposits are in the form of chalcopyrite ore ($CuFeS_2$). These sulfide ores require more complex treatment in the melting process.

The ore is first crushed and then ground to a powder in ball mills. The powder is enriched by a flotation process that removes much of the unwanted material. Silica is added and the mix is melted in a furnace (usually a reverberatory type), producing two liquid layers: a lower layer of copper matte (cuprous sulfide plus sulfides and oxides) and an upper layer of silicate slag, which is drawn off, leaving cuprous and ferrous sulfides called matte. Air under pressure is blown through the molten copper matte in a converter. This operation is one of oxidation-reduction similar to the iron-producing blast furnace. The iron is first removed, and in a second blow the sulfur is removed and becomes sulfur dioxide. The molten copper is poured into ladles. At this stage it is about 98 percent to 99 percent pure and is cast into ingots. When further refinement is necessary, the copper anodes are placed in electrolytic cells that produce 99.95 percent pure copper. The impurities collect at the bottom of the tanks in a sludge from which other metals such as gold, silver, and other rare and precious metals are later removed.

Nickel is often associated with copper and iron ores as a nickel sulfide concentrate. The smelting process for nickel ores is similar to that of copper, in which the ore is roasted, or smelted, in an electric furnace to produce a matte. It is then placed in a converter and further refined before casting the nickel-copper anodes. The nickel is then separated from other metals by the electrolytic refining process.

MAGNESIUM EXTRACTION

There are several important ores of magnesium, but the most important source is in seawater, which contains about 0.13 percent dissolved magnesium. Thermal reduction is used to extract magnesium from ores such as roasted dolomite ($MgO \cdot CaO$). The electrolytic method is used to extract magnesium from seawater to which calcium hydroxide ($Ca(OH)_2$) has been added. This causes the magnesium in the water to precipitate as magnesium hydroxide ($Mg(OH)_2$), which is converted to magnesium chloride ($MgCl_2$) by adding hydrochloric acid (HCl). This product is dried and electrolyzed to produce chlorine gas and magnesium metal.

LEAD AND ZINC SMELTING

The most common form of lead ore is the sulfide (PbS), known as galena. It is often associated with zinc and frequently with silver. The ore is ground into small pieces and the impurities are separated by flotation. The lead ore is smelted in reverberatory furnaces or in water-jacketed blast furnaces to produce molten lead, which is drawn off at the bottom of the furnace.

SPACE-AGE METALS

The metals columbium, titanium, hafnium, zirconium, and tantalum alloys are used in the aerospace industry, in jet aircraft, rockets, and missiles, and for nuclear reactors. Columbium (niobium) is a relatively lightweight metal that can withstand high temperatures and can be used for the skin and structural members of aerospace equipment and missiles. Columbium, a superconductor of electricity, presents a possible way to store and transfer large quantities of energy in the future.

Titanium is extensively used for jet engine components (rotors, fins, and compressor parts) and other aerospace parts because it is as strong as steel and 45 percent lighter. Titanium-aluminum-tin alloys possess high strength at high temperatures and, since titanium forms an alloy with many metals, it can be made to have properties that make it useful for various applications.

Hafnium and zirconium are always found together, and their properties are so similar that there has been no commercial reason to separate them except one. Zirconium and hafnium have opposite properties in relation to neutron flow in a nuclear reactor, so they have to be separated for such use. Hafnium has the property of stopping the flow of neutrons by absorbing them, thus stopping the reaction taking place in the reactor core. In contrast, zirconium freely allows neutrons to flow through itself, allowing the fission process to occur. These metals make possible the precise control of nuclear reactors. Zirconium (and its alloys) was originally developed to encapsulate uranium in nuclear reactors, but because of its superior corrosion resistance, it is being used in the chemical industry and for surgical implants. Pure zirconium is a reactive metal that will burn in air with a brilliant white light. It has long been used as the light source in photo flash bulbs.

Tantalum often occurs as the mineral columbite-tantalite. The metal tantalum is very malleable and ductile and has a high melting point 5425°F (2996°C). It is used as a replacement for platinum in chemical, surgical, and dental instruments and equipment. Tantalum is also used in vacuum furnaces and to make electrolytic capacitors.

Tantalum and columbium are probably more closely related in more ways than any other two metals in the periodic table. They occur together in the same ores. Both are highly immune to attack by strong acids and both are more ductile (easily formed) than any other refractory (high-temperature) metal.

EXTRACTION OF SPACE-AGE METALS

Columbite or tantalite ore is usually separated from other minerals by washing and magnetic separation at the mine. Ores processed for use as alloys of steel are normally converted to ferro-columbium-tantalum by reduction in electric furnaces.

The production of pure columbium or tantalum requires a more elaborate treatment in which the ore is ground and fused with caustic soda. The impurities are washed away and the fused product is dissolved in acids to form salts, which are separated and purified by crystallization. Both columbium and tantalum are made by powder metallurgy in which the metal is prepared as a powder by crushing and washing with acids. The powders are compacted by a hydraulic press to produce bars of about one square in. in cross section and about 30 in. long. These are sintered in a vacuum chamber, clamped on each end to a water-cooled terminal, and subjected to the flow of a heavy electric current, raising it to a fusing temperature. After sintering, the bar is cold-worked by rolling or forging.

Titanium ores are converted to titanium tetrachloride and reduced with magnesium or sodium in an inert gas atmosphere. The product of this chemical reaction is a spongy mass of metal that must be melted to form an ingot before it can be further processed. Since titanium is a reactive metal, it will react with oxygen in the air and burn while in the molten state. Therefore, melting must be carried out in a vacuum or in an inert gas atmosphere. The titanium sponge is crushed and then compacted into briquettes in a hydraulic press. These briquettes are welded together in a weld tank to form an electrode, which is melted in a water-cooled copper crucible in a vacuum arc furnace. The resulting ingot is processed in rolling mills in much the same way as steel products are formed.

Zirconium is processed in a manner similar to titanium; however, hafnium and zirconium are separated by precipitation and calcining, which produces zirconium oxide (ZrO_2) and hafnium oxide (HfO_2). The product is chlorinated and then reduced with magnesium, and finally

FIGURE 2.26
Pressed bars of zirconium sponge welded together to form an electrode (Courtesy of Teledyne Wah Chang Albany, Albany, OR).

the magnesium is distilled from the zirconium metal, leaving a porous sponge-like material. The sponge is broken up, crushed, blended, and then pressed into compact bars. These are welded together to form an electrode (Figure 2.26). An ingot is formed in a vacuum arc furnace by the consumable arc melting process (Figure 2.27). The ingot must be machined on the surface to remove scale and impurities. The ingots (Figure 2.28) are then fabricated into products by forging, hot or cold rolling, or extrusion.

Zirconium is also a reactive metal that can burn in the presence of oxygen or in the air. This property and others can be greatly altered in reactive metals by alloying them with other metals. For example, columbium-titanium alloys are produced for superconducting wire and other products. Titanium is alloyed with aluminum and molybdenum to give it a high tensile strength of about 170,000 PSI.

FIGURE 2.27

Flowchart of zirconium production (Courtesy of Teledyne Wah Chang Albany, Albany, OR).

FIGURE 2.28
Zirconium ingots ready to be processed into useful products (Courtesy of Teledyne Wah Chang Albany, Albany, OR).

Review Questions

1. Are titanium and zirconium ferrous or nonferrous metals? Explain.

2. A large proportion of common clay, which is available almost everywhere on earth, is aluminum. Why is bauxite used for the extraction of aluminum instead?

3. Iron ore is usually processed at the mine site to upgrade it and remove unwanted impurities. Why is this done at the site?

4. What is the name of the furnace that first made possible the smelting of iron in a molten state? What product is made in this furnace?

5. What three raw materials are used to make pig iron?

6. Why does all steel contain the element manganese in small quantities?

7. Why is cast iron a poor material to use for making cutting tools such as knives?

8. How much carbon can steel contain?

9. Name the three modern types of steelmaking furnaces.

10. What energy source is a necessity for smelting aluminum?

11. What is the final stage of refinement of 99.95 percent copper called?

12. Space-age metals such as titanium are extracted by complex processes. What is the major energy source for these processes: coal, coke, petroleum, oil, or electricity?

Case Problem

Case 1: Cast Iron versus Steel

You are required to choose a metal for manufacturing carpenter's claw hammers. Nonferrous metals have been ruled out because they are all too costly or have insufficient strength; this leaves cast (pig) iron and steel. Making the hammer heads of cast iron is the least expensive method since the metal is cheaper and the hammer head can be readily cast in sand or reusable permanent molds. Thus, their sale price would be much lower than the price of hammer heads made of steel forgings or castings. What other consideration should be taken in this choice? Would you choose cast iron or steel to make these hammer heads?

Metallurgical Science

Extractive or *process metallurgy* is the technology of extracting metals from their ores and refining them for further processing. This includes mining procedures, furnace operations, and ingot casting, methods covered in Chapter 2. *Physical metallurgy* is the science or study of the properties, structure, behavior, and composition of metals and how to adapt metals to make useful products. In the past, discoveries about the behavior of metals were fortuitous accidents in which the underlying principles and causes were not understood. Perhaps some accomplishments in modern physical metallurgy are still made by trial and error, but for the most part metallurgy is a science in which new metal alloys are planned and developed to meet exacting specifications, such as those for jet engine impeller blades. Scientific equipment such as light and electron microscopes is used to view and analyze the crystal structure of metals, and testing machines are used to pull, push, bend, twist, and impact metal specimens to determine their properties. The behavior of metals is no longer shrouded in mystery and folklore as it was in ancient times; much is known today about metals even at the atomic level. The atomic and crystalline structure of metals was dealt with in Chapter 1. This chapter will be confined to the study of physical metallurgy and the behavior of metals under various conditions.

Objectives

This chapter will enable you to:

1. Understand the behavior of metals in terms of their mechanical properties.
2. Describe the method of testing metals to determine their various properties.
3. Make some material selections on the basis of mechanical and physical properties.
4. Explain how the microstructure of metals can be seen and photographed.
5. Show how mechanical parts, steel structures, and pipelines can be tested for flaws without damaging them for use.

MECHANICAL AND PHYSICAL PROPERTIES OF METALS

In the field of manufacturing in which metals are cast, shaped, or formed, the mechanical properties of the metal selected for processing must be known. Some of these properties are the relationship between stress and strain, elasticity, strength, ductility, malleability, hardness, brittleness, creep, and toughness. Various tests and testing machines are used to determine the mechanical properties. These tests are usually destructive; that is, the specimen or part is destroyed by testing.

Engineers and scientists need to have an understanding and a means of measurement of the physical properties of metals. Some of these properties are the melting point, coefficient of thermal expansion, electrical and thermal conductivity, specific gravity, magnetic susceptibility, and reflectivity. Tests for determining the physical properties of metals are often nondestructive. Nondestructive testing is discussed later in this chapter.

Mechanical Properties

An elastic material is one that can be deformed and then return to its original shape when the load is removed.

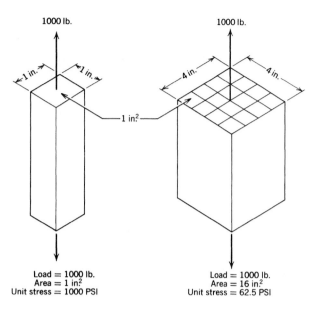

FIGURE 3.1
Unit stress (White, Neely, Kibbe, Meyer, *Machine Tools and Machining Practices,* Vol. II © 1977 John Wiley & Sons, Inc.).

gage). Elasticity is expressed by Hooke's law (δ/ε in the elastic region) as the degree to which an elastic body bends or stretches out of shape (strain), which is in direct proportion to the force (stress) acting upon it (Figure 3.2). Within the elastic range, the metal will return to its original shape without permanent deformation when the applied load is released. Metals are elastic only to a certain stress level, above which they become plastic and permanently deform. Below this **elastic limit,** also called the proportional **limit,** stress and strain are proportionate; that is, for every unit of loading there is always the same amount of strain or movement. This can be seen on the stress-strain diagram as a straight line, called the elastic range (Figure 3.3).

Plasticity is the property that allows metals to be permanently deformed beyond the elastic limit without failure by rupture or splitting. Clay, when wet, is an example of a plastic substance that can be formed easily. However, when metals reach a certain limit of strain, they are no longer plastic but become brittle and suddenly fail. This is called the ultimate strength (see Figure 3.3).

Strength of metals is their ability to resist deforming or changing shape when external forces are applied. There are four types of stresses involved in the strength of materials: **tensile, compressive, torsional,** and **shear** (Figure 3.4). Torsional shear is found in rotating parts such as axles and machinery shafting. A torsional shear measurement, torque, is shown in Figure 3.5. Some metals such as steel have nearly equal compressive and tensile strengths, but cast iron has a relatively low tensile strength compared to its higher compressive strength. Shear strength is lower than tensile strength in virtually all metals (Table 3.1). The tensile strength of a specimen is the total amount of stress loading it can withstand before breaking apart. To determine tensile strength, metal specimens are

Elasticity involves the relationship between stress and strain. Although these terms are often used interchangeably, they have separate and distinct meanings in the science of metallurgy. *Unit stress* is defined as the resistance of a material to external elements such as force, load, or weight measured in pounds per square inch (PSI) (Figure 3.1). It is calculated by dividing applied load to a specimen by the cross sectional area of the specimen. *Unit strain* is the movement or stretching of a metal when a stress is applied. Unit strain is the amount of deformation (tension or compression) in a unit length (original or

FIGURE 3.2
Diagram of the elastic properties of metals as explained in Hooke's law. Four weights, of which one represents zero load, are suspended by identical metal wires. The three loads toward the right of the diagram are heavier in the equal progression. The resultant movements *a, b,* and *c,* called strain, are also equal.

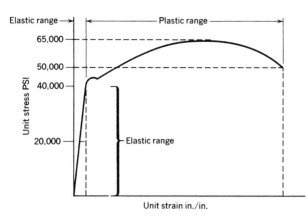

FIGURE 3.3
Stress-strain diagram for a ductile steel (White, Neely, Kibbe, Meyer, *Machine Tools and Machining Practices,* Vol. II, © 1977 John Wiley & Sons, Inc.).

FIGURE 3.5
Digital readout of torque valves on a torque testing machine. (Photograph courtesy of Tinius Olsen Testing Machine Co., Inc., Willow Grove, PA)

"pulled" on a machine called a tensile tester (Figures 3.6 and 3.7) until they break. A continuous record of load and strain movement is often recorded (Figure 3.8) and the information is placed on a graph called a stress-strain diagram. Standard specimens are usually used with a cross-sectional area that can be calculated easily (Figure 3.9 on p. 58). For example, the most common C-1 standard has a diameter of .505 in., making the area .2 square inches. A specific gage length is used to record elongation, usually 2 in. ±.005 in. Tensile strength in pounds per square inch is determined by dividing the maximum load (in pounds) by the original cross section (in square inches) before testing. This is called unit stress. Unit stress equals the load divided by the area.

$$\text{unit stress} = \frac{\text{load}}{\text{area}}$$

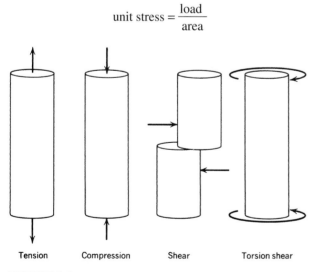

| Tension | Compression | Shear | Torsion shear |

FIGURE 3.4
The four types of stresses (White, Neely, Kibbe, Meyer, *Machine Tools and Machining Practices,* Vol. II, © 1977 John Wiley & Sons, Inc.).

The tensile test not only reveals the yield point, ultimate strength, and strain, it also determines the relative **ductility** of a metal specimen. The amount (percent) of elongation and reduction of area at the point of failure is an indication of ductility (which is the property that allows a metal to deform permanently when in tension). Any metal that can be drawn into wire is ductile. Soft steel, aluminum, gold, silver, and nickel are ductile metals.

Malleability is the ability of metals to be deformed permanently when loaded in compression. Metals that can be rolled, hammered, or pressed into flat pieces or sheets are malleable. Ductile metals are usually also malleable but there are some exceptions. Lead is very malleable, but not very ductile and cannot be drawn into wire very easily, but it can be extruded in wire or bar form. Gold, tin, silver, iron, and copper are malleable metals.

Hardness is generally defined as the resistance to penetration or indentation. The greater the hardness, the smaller the indentation with the same penetrator tip and pressure. However, the property of hardness is related to the elastic and plastic properties of metals and certain hardness tests are based on rebound or elastic hardness. Also, certain scratch tests are used to determine abrasive hardness or resistance to abrasion. Hardness is also related to tensile strength so it can be assumed that a hard steel is also strong and resistant to wear.

The hardness test that uses pressure and a penetrator or indenter is widely used for industrial applications because of its simplicity and ease of operation. The Rockwell (Figures 3.10 and 3.11 on p. 58) and Brinell

TABLE 3.1
Material strength

Material	Modulus of Elasticity (PSI)	Allowable Working Unit Stress			
		Tension	Compression	Shear	Extreme Fiber in Bending
Cast iron	15,000,000	3,000	15,000	3,000	
Wrought iron	25,000,000	12,000	12,000	9,000	12,000
Steel, structural	29,000,000	20,000	20,000	13,000	20,000
Tungsten carbide	50,000,000				

Material	Elastic Limit (PSI)		Ultimate Strength (PSI)		
	Tension	Compression	Tension	Compression	Shear
Cast iron	6,000	20,000	20,000	80,000	20,000
Wrought iron	25,000	25,000	50,000	50,000	40,000
Steel, structural	36,000	36,000	65,000	65,000	50,000
Tungsten carbide	80,000	120,000	100,000	400,000	70,000

(*Source:* John E. Neely, *Practical Metallurgy and Materials of Industry,* 2d Ed., Wiley, New York, © 1984.)

(Figures 3.12 on p. 58 and 3.13 on p. 59) instruments are the most common of this type. Microhardness testers that enlarge the area of testing with a microscope are often used in metallurgical laboratories. Some of these instruments are portable and are designed to be used in the field. One measurement of hardness is not based on penetration but rather on the rebound of a dropped hammer with a diamond point. The higher the rebound, the harder the material. This rebound is based on elastic resistance to penetration. Because of its simplicity of operation, tests can be made very quickly.

A material that will not deform plastically under a load is said to be **brittle** (Figure 3.14 on p. 59). A brittle break or failure in metals can be identified by a crystalline surface, often with a tiny rock candy appearance. Sometimes the individual crystals are so small that the surface has a somewhat smooth appearance to the eyes, but under magnification the crystals can be seen. Of course, metals and some other materials may exhibit other properties such as ductility or malleability along with brittleness. When a metal specimen is pulled apart in a tensile tester, a considerable amount of deformation can be seen in a ductile metal before a brittle failure takes place. However, very hard steel and cast iron show little or no ductility when pulled apart, so we say those metals are brittle, meaning they are completely brittle. A normally ductile metal can exhibit complete brittleness under some circumstances. A sharp notch that concentrates the load in a small area or at one point can reduce plasticity (Figure

3.15 on p. 59). Also low temperatures can have an embrittling effect on some steels. The behavior of metals at low temperatures is discussed later in this chapter.

Notch toughness, also called impact strength, is the ability of a metal to resist rupture from impact loading when there is a notch or stress raiser present. The property of toughness is not necessarily related to ductility, hardness, or tensile strength. However, as a rule, a brittle metal with low tensile strength such as cast iron will fail under low shock loads; hardened, tempered tool steels show a high impact strength; and coarse-grained metals have lower shock resistance than fine-grained metals. A notch or groove will lower the shock resistance of a metal. Test specimens are made with given dimensions and a specific notch shape (Figure 3.16 on p. 60).

Notch toughness is measured on a device called an Izod-Charpy testing machine in which either of the two styles of testing can be used (Figure 3.17 on p. 60). A weighted, swinging arm or pendulum strikes a specimen that is clamped in a vise, breaks the specimen, and continues swinging. It will not swing as high as the starting position because of the energy absorbed in breaking the specimen (Figure 3.18 on p. 60). The difference in starting and ending height is measured in foot-pounds. The pendulum will not swing as high for tough metals as for brittle metals. The difference between the two styles of testing, Izod and Charpy, is in the mounting of test specimens (Figures 3.19 and 3.20 on p. 61). The Izod specimen is mounted horizontally and the Charpy specimen is mounted vertically.

FIGURE 3.6
A specimen with a strain gage attached to it is placed in a
tensile testing machine to be pulled. Data will be plotted
on an *X-Y* graph, shown on the right (Photograph courtesy
of Tinius Olsen Testing Machine Co., Inc., Willow Grove,
PA).

FIGURE 3.7
Tensile testing machines are also equipped to make
compressive tests; here a concrete sample is being tested
(Photograph courtesy of Tinius Olsen Testing Machine
Co., Inc., Willow Grove, PA).

Metals at low temperatures are tested on Izod-Charpy
machines to determine the effect of temperature drop on
various metals. Metals that remain ductile at low temperatures show a slow steady decrease in ductility as the temperature drops. Metals that become brittle at low
temperatures show a temperature range in which ductility
and, more important, toughness drop rapidly. This *transition zone* can be seen in Figure 3.21 on p. 61. When the
notch-bar specimens show half brittle and half ductile failures, the transition temperature has been reached. Metal
for parts that are designed for low-temperature service
must be selected so their operating temperature is well
above their transition zone of embrittlement.

This transition brittle zone of metals has caused
some spectacular failures in the past, before it was
known that some low carbon steels would fail at temperatures lower than 0°F (−18°C). Some of these failures occurred in pipelines, storage tanks, bridges, and
ships that broke in half at sea. These failures are almost

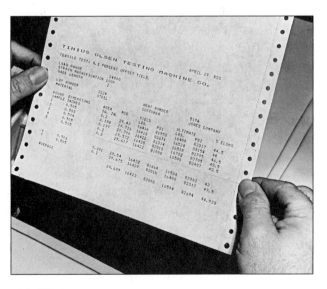

FIGURE 3.8
Tensile testing data system. A printout of several tests
made on a tensile tester (Photograph courtesy of Tinius
Olsen Testing Machine Co., Inc., Willow Grove, PA).

FIGURE 3.9
This standard test specimen has a cross-sectional area of
.002 in. and a gage length of 2 in. Round numbers make
calculations for percent reduction of area and percent
elongation easier (Neely, *Metallurgy,* 2d ed., © 1984 John
Wiley & Sons, Inc.).

FIGURE 3.11
Testing a part on a Rockwell-type hardness tester (© 1985
King Tester Corporation).

FIGURE 3.10
Rockwell-type hardness testing machine (© 1985 King
Tester Corporation).

FIGURE 3.12
A large Brinell hardness tester with a separate readout
panel (Photograph courtesy of Tinius Olsen Testing
Machine Co., Inc., Willow Grove, PA).

FIGURE 3.14
Brittle fracture of a shaft.

FIGURE 3.13
The Brinell microscope is used for measuring indentation diameters for simple Brinell hardness testers. Some Brinell testers, such as the one in Figure 3.12, are fully automatic and do not require a microscope (© 1985 King Tester Corporation).

instantaneous, with the split moving along pipelines at 600 feet per minute (FPM). Plain carbon steels having coarse grain have higher transition temperatures than fine-grained and alloy steels. Certain alloying elements in steel, particularly nickel, tend to lower the transition temperature. Stainless steel containing nickel is used to contain liquefied gases at temperatures below −150°F (101°C). Aluminum and monel are also used at cryogenic temperatures without embrittlement. High strength, low alloy steels (HSLA) that contain such elements as columbium, cerium, molybdenum, and silicon were developed for Arctic pipelines.

The **modulus of elasticity** is actually an expression of the stiffness of a metal. It is also called Young's modulus. If, within the elastic range, the unit stress is divided by the

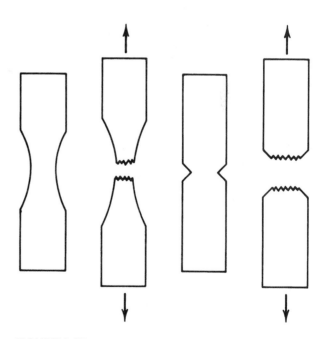

FIGURE 3.15
Notching and its effect on plasticity. When pulled, the specimen (left) becomes deformed before breaking, showing ductility. The notched specimen of the same material (right) exhibits complete brittleness.

Izod test specimens

Charpy test specimen

Tension impact specimen

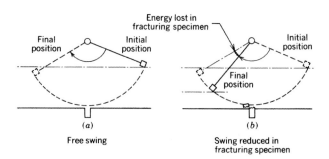

FIGURE 3.16
Test specimen specifications for Charpy and Izod test (Photograph courtesy of Tinius Olsen Testing Machine Co., Inc., Willow Grove, PA).

corresponding unit strain at any given point, the result will be the modulus of elasticity for that material; thus,

$$\text{modulus of elasticity} = \frac{\text{unit stress}}{\text{unit strain}}$$

The modulus of elasticity for some common metals is given in Table 3.1 on p. 56. Carbon, alloy, and hardened or soft steels all have about the same modulus, which varies between 29 million and 30 million PSI. This means a steel piece of a given size and shape will twist or bend about the same amount with a given load within the elastic range. Therefore, by replacing a mild steel part with an alloy part, its tensile strength may be increased but not its stiffness and it will still deflect the same amount if it is the same size. However, if it is exchanged for a metal having a higher modulus, such as tungsten carbide, which is 50 million PSI, then it would not bend or deflect so easily. Surprisingly, cast iron will deflect to some degree within its elastic range, showing a stiffness that is about half that of steel.

FIGURE 3.17
Izod-Charpy testing machine (Photograph courtesy of Tinius Olsen Testing Machine Co., Inc., Willow Grove, PA).

Creep is a continuing plastic flow at a stress below the yield strength of a metal. For the most part, creep is a high-temperature phenomenon that increases as the temperature increases. Table 3.2 gives some stress loads at various temperatures required to cause elongation by creep. Some metals have been designed for high-temperature service such as for aircraft jet turbine engine

FIGURE 3.18
The method by which (Izod) impact values are determined (White, Neely, Kibbe, Meyer, *Machine Tools and Machining Practices,* Vol. II, © 1977 John Wiley & Sons, Inc.).

FIGURE 3.19
Mounting of the Izod specimen (Photograph courtesy of
Tinius Olsen Testing Machine Co., Inc., Willow Grove, PA).

FIGURE 3.20
Mounting of the Charpy specimen (Photograph courtesy of
Tinius Olsen Testing Machine Co., Inc., Willow Grove, PA).

FIGURE 3.21
Appearance of Charpy V-notch fractures that were
obtained in a series of tempered martensite of hardness
30Rc (Illustration courtesy of LTV Steel Company).

blades that show no signs of creep at temperatures that
would melt many metals. Creep tests are made at elevated
temperatures under a certain stress in creep-testing
machines (Figure 3.22) over a period of many hours.

Fatigue in metals can occur when metal parts are
subjected to repeated loading and unloading, particularly
in cyclic reversals of stress, such as is seen in a rotating
shaft having transverse or one-side loading. Fatigue failures can occur at stresses far below the yield strength of a
material with no sign of plastic deformation. When
machine parts that are subject to cyclic loading are
designed, fatigue strength may be more important than
yield or ultimate strength. Fatigue is not time dependent

and can be initiated by any number of factors such as
machining tool marks or welding undercuts and localized
stress caused by a welding bead. Fatigue can be accelerated by a corrosive atmosphere and by higher frequency
stress reversals. A typical fatigue break pattern of a rotating shaft can be seen in Figure 3.23. Fatigue break patterns on a shaft usually can be seen in three stages: first, a
slow progression of striations that has a smooth surface;
second, coarse striations (like on a clam shell) Figure
3.24) that progress more rapidly; and third, the brittle,
crystalline section that suddenly fails when the shaft
strength finally becomes less than the applied stress.
Metals are tested for their fatigue strength (endurance test)
on a fatigue testing machine that rotates a specimen under
various loads and conditions. Fatigue life can be considerably increased by surface hardening or by carburizing or
shot peening.

Physical Properties of Metals

The **thermal conductivity** and electrical conductivity of
metals are related. If a metal such as silver or copper is

TABLE 3.2
Creep strengths for several alloys

Alloy	70°F Tensile Strength (PSI)	Creep Strength (PSI)		
		800°F— Stress for 1 percent Elongation per 10,000 hr	1200°F— Stress for 1 percent Elongation per 100,000 hr	1500°F— Stress to Failure
.20 percent carbon steel	62,000	35,100	200	1,500
.50 percent molybdenum .08 percent to .20 percent carbon steel	64,000	39,000	500	2,600
1.00 percent chromium .60 percent molybdenum .20 percent C steel	75,000	40,000	1,500	3,500
304 stainless steel 19 percent chromium 9 percent nickel	85,000	28,000	7,000	15,000

(*Source:* John E. Neely, *Practical Metallurgy and Materials of Industry,* 2d ed., Wiley, New York, © 1984.)

FIGURE 3.22
Creep testing machine in Oregon Bureau of Mines testing laboratory.

FIGURE 3.23
Fatigue on a shaft.

FIGURE 3.24
Fatigue of Ti-6A1-4V, 28,000 cycles at 8000 lb. (10 Hz).
Transmission electron microscope replica (5400 ×)
(Oregon Bureau of Mines).

a good conductor of electricity, it is also a good conductor of heat. Pure metals are better conductors than their alloys. This is the reason that copper or aluminum used for electrical wiring must be uncontaminated. Electrical copper grades are 99.97 percent pure. Resistance to the flow of electricity can be altered by several factors.

1. Resistance increases as the temperature increases.

2. It increases as a result of cold-working metal.

3. It increases with the amount of impurities and alloying elements present.

In the case of electrical elements, resistance is an advantage in which the electric current produces and liberates heat.

Metals may look alike but have very different physical properties. For example, one can scarcely tell the difference between zinc, nickel, stainless steel, and silver by appearance, but if you held a bar of equal length of each in your hand and heated the other ends, you would quickly drop the silver bar but could hold the others for several minutes before feeling any heat. Figure 3.25 compares the thermal conductivity of some common metals.

Most metals expand when heated and contract when cooled. This change in dimension is called the coefficient of **thermal expansion.** Each metal expands and contracts at a different rate when heated or cooled. This physical property of metals can cause difficulties in construction and manufacturing. Engineers plan use of expansion joints in paved highways and bridges to avoid buckling. When a metal device is to be subjected to repeated heating and cooling, dissimilar metals that expand at different rates can cause abnormal stresses and subsequent damage when rigidly fastened together. If a machinist measures a finished dimension while the part is still hot from machining, the part may be under the tolerance limit when it is cool and it will have to be scrapped. For example, brass expands at the rate of 0.0001 in. per degree Fahrenheit per unit length (or diameter). A machined brass sleeve, 3 inches in diameter, has a tolerance of plus .001 in. or minus .000 in. The temperature rises 100°F (38°C) above room temperature. When it cools back to room temperature, it has shrunk undersize (.00001 in. × 100 × 3 = .003 in.) and must be scrapped. The coefficient of thermal expansion for some common metals can be seen in Figure 3.26.

Some metals have a high degree of *reflectivity* and are used as light reflectors and in insulation as heat deflectors. Aluminum, nickel, chromium, tin, and bronze are among the metals having high reflectivity, whereas lead has virtually none; yet reflectivity is one of the char-

FIGURE 3.25
Comparison of thermal conductivity (White, Neely, Kibbe, Meyer, *Machine Tools and Machining Practices,* Vol. II, © 1977 John Wiley & Sons, Inc.).

Aluminum
Aluminum Alloy 2024
Copper
Brass 85% Cu   15% zn
Phosphor Bronze 95 Cu 5sn
Iron
Stainless Steel 302
Magnesium
Nickel
Silver
Zinc

Thermal conductivity CGS/units at 70° C

FIGURE 3.26
Coefficient of thermal
expansion per degree
Fahrenheit per unit length
(White, Neely, Kibbe, Meyer,
*Machine Tools and Machining
Practices,* Vol. II, © 1977
John Wiley & Sons, Inc.).

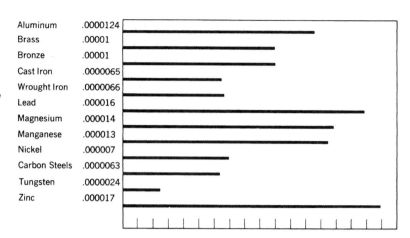

Aluminum	.0000124
Brass	.00001
Bronze	.00001
Cast Iron	.0000065
Wrought Iron	.0000066
Lead	.000016
Magnesium	.000014
Manganese	.000013
Nickel	.000007
Carbon Steels	.0000063
Tungsten	.0000024
Zinc	.000017

acteristics that distinguishes metals from other substances.

Ferromagnetism is characterized by an attraction between certain metals that have the ability to retain a residual magnetic force. The ferromagnetic metals are iron, cobalt, and nickel. Of these, iron is the most commercially important metal used extensively in electrical machinery. Soft iron does not retain much residual magnetism when removed from a magnetic field. This quality makes it useful for electromagnets and electric motors in which the field is turned on and off or reversed repeatedly. Powerful permanent magnets are made by alloying metals such as cobalt, nickel, iron, and aluminum.

METALLURGICAL MICROSCOPY

Metal surfaces are examined by means of metallurgical microscopes and associated techniques of surface preparation. Structures in metals not readily visible to the naked eye may be seen at high-power magnification from 10 times the diameter (10×) to 3000× for light microscopes and many more thousands of diameters with the electron microscope. Techniques of photomicrography in which a camera is attached to the microscope make possible photographic illustrations of specimens (Figure 3.27). Metallurgical microscopes, unlike those used to see through a specimen, use reflected light from an internal source (Figure 3.28). Inverted stage microscopes are widely used to study metals (Figure 3.29 on p. 66).

Specimens to be analyzed are cut from a metal part with a metallurgical saw (Figure 3.30 on p. 66). The specimen is mounted in plastic in a press (Figures 3.31 and 3.32 on p. 66) and rough ground (Figure 3.33 on p. 67). Next, the encapsulated specimen is ground on successively finer grit abrasive surfaces (Figure 3.34 on p. 67), after which it is polished on a rotating cloth-covered disk (Figure 3.35 on p. 67) or by

other techniques such as electropolishing or vibratory polishing. When the specimen has a mirror finish, it is etched in a reagent that is most commonly a dilute acid, varying with the type of metal being prepared. Etching causes surface characteristics of the metal to be visible; some of these are grain boundaries, grains, impurities, and microstructures in

FIGURE 3.27
The Versamet metallographic microscope has the advantage of having a display screen, so more than one person may observe the magnified specimen. Also photographs may be taken of the enlarged specimen. (*Versamet* is a trademark and a copyright of Unitron Inc.)

To
eyepiece

← Light

To
objective

FIGURE 3.28
Illumination in a metallurgical microscope (Neely, *Metallurgy,* 2d ed., © 1984 John Wiley & Sons, Inc.).

the grains such as pearlite, bainite, ferrite, iron carbide (cementite), and martensite (Figure 3.36 on p. 68).

Ferrite is more or less pure iron. Pearlite is a relatively soft form of alternate layers of cementite and ferrite. Bainite is harder than pearlite and is actually a tough form of martensite. Acicular martensite is an extremely hard form of steel. The microscopic study of metals can help us to readily determine the condition of the specimen. Of course, much more information about metals is gained by metallurgists by studying microstructures in metals with powerful microscopes.

NONDESTRUCTIVE TESTING

As the name implies, nondestructive testing in no way impairs the part for further use. It does not measure mechanical or physical properties, but instead identifies defects such as voids, inclusions, or cracks that will later lead to failure of the part. Also, improper handling of materials that causes changes in the microstructure can lead to failure. For example, a nick or dent in a highly stressed part such as a spring can cause early failure. At present, even these aberrations can be detected on a production basis. Nondestructive methods can also verify hardness or heat treatment in many cases. Since testing for defects in the early stages of the manufacturing process can substantially reduce rejects, nondestructive testing is closely related to quality control. (See Chapter 7.) In the study of fracture mechanics, the damage-tolerant philosophy has assumed that all metals have microscopic flaws such as discontinuities and microcracks on the surface that will lead to ultimate failure. With inspection and testing methods, a safe service life or stress-life curve can be determined. Along with a safety factor, the design is known as a safe-life design. With the increasing use of composites and HSLA steels in automobiles and other

FIGURE 3.29
BMEC inverted stage microscope. The specimen is placed upside down on the stage on the top of the microscope. A 35mm camera can also be attached to this microscope (BMEC is a trademark and a copyright of Unitron Inc.).

FIGURE 3.31
Metallurgical press. After the specimen is placed in the press, granulated thermoplastic is poured in. It is then heated under pressure, forming a solid cylindrical shape with the specimen embedded on one side (Beuhler, Ltd.).

FIGURE 3.30
Metallurgical saw. The specimen is cut off with a thin abrasive blade to which coolant is applied. This keeps the specimen from becoming overheated, causing damage to the metallurgical structure (Beuhler, Ltd.).

FIGURE 3.32
Encapsulated specimen.

FIGURE 3.33
Rough grinding the specimen (Beuhler, Ltd.).

FIGURE 3.35
Polishing table (Beuhler, Ltd.).

FIGURE 3.34
Finish grinding the specimen on HandiMet® (Beuhler, Ltd.).

equipment, nondestructive testing will be needed even more in manufacturing and maintenance operations.

The most common types of nondestructive testing are:

1. Magnetic-particle inspection
2. Fluorescent penetrant inspection
3. Dye penetrants
4. Ultrasonic inspection
5. Radiography (X ray and gamma ray)
6. Eddy current inspection

Magnetic-particle inspection methods are used to detect surface flaws in ferromagnetic materials such as iron and steel (Figures 3.37 to 3.40 on p. 69). One method starts with sprinkling dry iron powder on a magnetized part. A more widely used method starts with pouring a liquid in which the iron powder is suspended over the magnetized workpiece; this is then viewed with a fluorescent light. Surface cracks only a few millionths of an inch wide can be seen. This method of testing is also used in automated manufacturing systems with advanced laser scanning techniques to sort out questionable parts from good ones.

Fluorescent penetrants are available in sealed pressure spray cans to apply for visual examination of any material such as metals, ceramics, and plastics. A fluorescent light is also used with this process to detect cracks. Testing systems range from huge automated systems to portable test kits for field inspection. Zyglo is the Magnaflux copyrighted name for this method (Figure 3.41 on p. 70). Dye penetrants also depend on visual inspection but do not use fluorescent light. Essentially, a dye is caused to seep into cracks and then is wiped off, and a developer is applied. A bright color then indicates the location of cracks (Figures 3.42 and 3.43 on p. 70). This method, called Spotcheck®, is also useful for field testing.

Ultrasonic inspection makes use of high-frequency sound waves that are sent through a test piece by means of a transducer, which is usually a piezoelectric crystal that converts electrical energy to mechanical energy (Figures 3.44 on p. 70 and 3.45 on p. 71). A transceiver that has the opposite function of converting mechanical energy (vibrations) to electrical pulses picks up the ultrasound waves

FIGURE 3.41
With Zyglo-Pentrex, this vanadium alloy stainless forged hip-ball joint used in a medical prosthesis is revealed to have a crack, detected in manufacture. This could result in severe injury to a patient (Courtesy of Magnaflux Corporation, Chicago).

FIGURE 3.43
Defect is now visible on the gear (Courtesy of Magnaflux Corporation, Chicago).

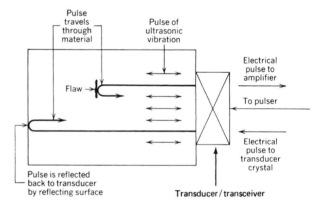

FIGURE 3.44
How ultrasonic sound waves are used to locate flaws in material (White, Neely, Kibbe, Meyer, *Machine Tools and Machining Practices,* Vol. I, © 1977 John Wiley & Sons, Inc.)

FIGURE 3.42
Spraying Spotcheck developer on a gear (Courtesy of Magnaflux Corporation, Chicago).

and displays them on an oscilloscope (Figure 3.46) as signal pulses called pips. In the pulse-echo system, part of the sound wave is reflected back from the other side of the material to the transducer/transceiver and is seen as a second pip on the oscilloscope. If there is a flaw, a smaller pip will be seen between the two, indicating the distance from the surface to the flaw. The through-transmission method uses a transducer on one side and a transceiver on

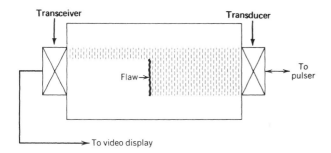

FIGURE 3.45
Through-ultrasonic transmission (White, Neely, Kibbe, Meyer, *Machine Tools and Machining Practices,* Vol. I, © 1977 John Wiley & Sons, Inc.)

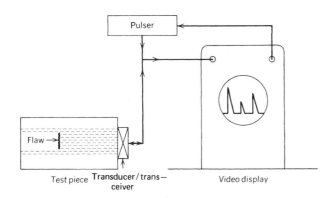

FIGURE 3.46
Ultrasonic pulse-echo system (White, Neely, Kibbe, Meyer, *Machine Tools and Machining Practices,* Vol. I, © 1977 John Wiley & Sons, Inc.)

FIGURE 3.47
Portable ultrasonic machine instruments can be used for routine manual inspection under field conditions. The instruments are very useful for structural weld testing and corrosion surveys (Courtesy of Magnaflux Corporation, Chicago).

the other. Ultrasonics is one of the best ways to check for cracks in pipelines and for structural weld testing (Figure 3.47). This method of inspection is very adaptable to automated manufacturing systems and is used widely to inspect for internal defects, cracks, laminations, variation in thickness, and weld bonds.

Radiography, X-ray, and gamma ray inspection have been used widely on pipelines and manufactured parts to detect flaws, especially internal defects such as porosity (Figures 3.48 and 3.49). There are several disadvantages of this system. Since X-ray and gamma radiation are both hazardous, technicians using this process are required to be trained. Radiography is also expensive and time consuming and there is a considerable time delay until the film is developed. The test is also two dimensional, possibly missing flaws viewed at another angle. However, a new process has been developed in which a video image and tape is made in real time (instantly). In the past, attempts at real-time imaging of X rays had poor resolution and were not practical.

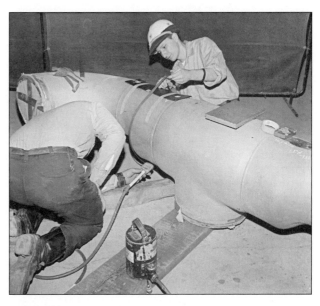

FIGURE 3.48
An experienced technician is preparing to make a gamma ray inspection of a section of pipe (Courtesy of Magnaflux Corporation, Chicago).

netic field that is read on a meter or oscilloscope (Figure 3.50). Variations such as seams, thicknesses, and other physical differences can be detected quickly with this method. Eddy-current testing is quite adaptable to automated systems. In fact, computerized data analysis systems for eddy-current testing have been developed in which analog data are converted to digital printout or magnetic tape storage.

FIGURE 3.49
X-ray method of testing. A specialist makes a test of pilot-run parts to evaluate manufacturing techniques and procedures. (Courtesy of Magnaflux Corporation, Chicago).

FIGURE 3.50
Eddy current testing instrument for nondestructive comparison of magnetic materials. It is used to separate materials according to their metallurgical properties, such as alloy variations, heat treat condition, hardness differences, internal stress, and tensile strength (Courtesy of Magnaflux Corporation, Chicago).

Eddy-current inspection systems can test only electrically conducting materials. When a test piece is placed in or near a coil in which alternating current produces a magnetic field, an eddy current is induced in the test piece. This current causes a change in impedance in the mag-

Review Questions

1. What is the difference between elasticity and plasticity? Define each.

2. How can tensile strength of a specific metal be determined?

3. What property does a metal possess when it can be easily drawn into a wire, but cannot be rolled into a sheet without splitting?

4. Define hardness as it is measured on a Rockwell or Brinell testing instrument.

5. When a material breaks suddenly with no sign of deformation, what is it said to be?

6. Some very ductile metals, such as certain low carbon steels, can change their properties at low temperatures at their particular transition zone. How can this information help an engineer design machinery that will be operating in subzero temperatures?

7. Steel has a modulus of elasticity of 29 million PSI and tungsten carbide of 50 million PSI. If a steel and a tungsten carbide bar of the same dimensions were each loaded on one end and supported on the other as cantilever beams, which would bend the most?

8. Name the phenomenon that occurs when a metal is loaded well within the elastic range at a temperature of 800°F (427°C) and it slowly deforms permanently.

9. Welds, machining tool marks, and stress reversals, such as those in rotating parts, can cause a progressive failure even on a lightly loaded part such as a shaft. There is no sign of ductility, only a smooth surface having striations like a clamshell. How would you identify this kind of failure? What is it called?

10. Which is the better conductor of heat or electricity, a pure metal or an alloy?

11. By what means is it possible to see the internal structure of metals at high magnification? How is this done?

12. How can machinery parts and pipeline welds be tested for flaws without destroying them? Name several processes.

Case Problems

Case 1: Case of the Twisted Shaft

The shafting shown in Figure 3.51 was taken from a mill transfer chain that was frequently overloaded, causing the shaft (made of low carbon steel) to fail by torsion shear. It has obviously been loaded in torsion beyond its elastic limit into the plastic range and has taken a permanent set. Since the sprocket and bearing bores are not easily changed in order to use a larger diameter shaft, what kind of replacement shaft would solve the problem? What

FIGURE 3.51
See Case 1.

different mechanical properties must it possess so it will not again fail by torsion shear?

Case 2: Broken Shaft

A hammer mill that was located outside a building was driven by an 800 hp electric motor whose main shaft was 8 in. in diameter. It had operated for years; the only maintenance needed was to lubricate bearings and replace hammers and grates. One winter morning the temperature dropped to −10°F. When the machine operator turned the switch to start the machine, which was unloaded and free-running, the main shaft broke in two instantly in a brittle-type fracture. What could have been the problem and how could it be avoided in the future?

Case 3: Piston Clearance

An aluminum piston 4 in. in diameter that moves in the cylinder of a cast iron engine block must have clearance diameter of .003 in. when hot. The temperature rise is from 70 to 160°F. What should the clearance between the bore diameter and the piston be if the linear coefficient of expansion per unit length (inch) per degree F is .00001244 for aluminum and .00000655 for cast iron when the temperature is at 70°F?

CHAPTER 4

Heat Treatment of Metals

From ancient times, the immense value of metals to the progress of civilization could be attributed to their ability to be shaped easily at ordinary temperatures and even more easily at high temperatures. Second only to their elastic and plastic behavior is the ability of carbon-bearing iron, called steel, to become very hard when heated and suddenly cooled as in a quenching medium like water. Other heat treatments such as temper drawing, annealing, and stress relieving also have a profound effect upon metals. The high quality of swords made in Damascus has been legendary since Alexander the Great was said to have cut the Gordian knot with one and the Saracens wielded them against the Crusaders. The process of making those swords was a well-kept secret that was never discovered and has since been lost. It is now believed that the unique forging process and, most important, the heat-treating process used produced this unusually hard and tough grade of carbon steel.

Today, there are many kinds of alloy and tool steels, each of which often requires special heat treatments in the manufacturing process to give them the qualities necessary to fit the required specifications. In addition, many nonferrous metals such as aluminum and titanium are heat treated to increase their strength and durability. The manufacture of machinery, automobiles, modern aircraft and space vehicles, and consumer products would not be possible without the technology of heat treatment of metals.

Objectives

This chapter will enable you to:

1. Use an iron-carbon phase diagram to explain the basic principles involved in heat treating steel.

2. Use an I-T diagram and cooling curves to explain the various microstructures found in carbon steel as a result of cooling rates.

3. Describe how steel is hardened and tempered.

4. Explain several methods or techniques to surface harden steel and the significance of selective surface hardening in manufacturing.

5. State the principles involved in precipitation hardening.

6. List the various types of annealing and stress relief and explain their purposes and uses for industrial applications.

7. Explain the factors involved in recrystallizing low carbon work-hardened steel.

THE IRON-CARBON PHASE DIAGRAM

In order to understand the basic principles involved in heat treating iron and steel, a person should have an understanding of the iron-carbon phase diagram (Figure 4.1). In Figure 1.11, the cooling curve diagram shows the phase changes in iron from liquid to solid delta iron at 2800°F (1538°C) and then to gamma iron or austenite as it cools to about 1600°F (869°C). The iron then undergoes transformation to alpha iron or ferrite as it cools further. Since that diagram represents pure iron, the same changes in solid iron can be seen at the left-hand vertical line on the iron-carbon diagram. The percentage of carbon in the iron increases toward the right side of the diagram, ending at 6.67 percent carbon. The diagram ends there because 6.67 percent C by weight is the amount of carbon in the compound, iron carbide (Fe_3C), and no other useful information would be given by including more carbon. Most of the diagram is devoted to cast irons because 2 percent carbon and above produces cast iron; 2 percent carbon and under produces steel.

The area within the dotted lines (Figure 4.2) is the steel portion of the iron-carbon diagram and is the area

FIGURE 4.1
Simplified iron-carbon diagram showing temperatures in equalibrium or very slow cooling (White, Neely, Kibbe, Meyer, *Machine Tools and Machining Practices,* Vol. II, © 1977 John Wiley & Sons, Inc.).

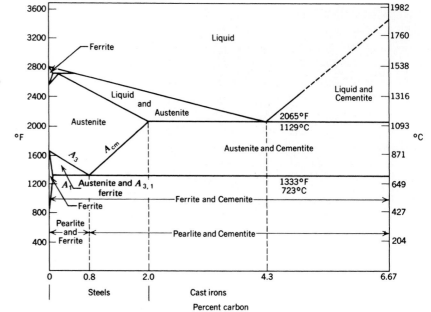

most important to heat treaters. **Eutectic** is a word meaning the lowest melting point. On either side of the eutectic point of the cast iron portion of the diagram, the metal solidifies slowly over a temperature range in which the metal is somewhat "mushy," but at the eutectic point, it solidifies instantly like a pure metal. The word **eutectoid** in the steel portion is similar but not the same as eutectic. Since the metal is now in a solid state, the word does not refer to solidification but rather to the lowest transformation temperature from austenite to ferrite. Again, on either

side of the eutectoid the transformation is gradual over a temperature range, but at the eutectoid, it is almost instantaneous. As noted in Chapter 1, up to 2 percent carbon can be contained in austenite as a solution. We can see this in Figure 4.1 in the area above the A_3 and A_{cm} lines.

When austenite containing .8 percent carbon (eutectoid) is slowly cooled, there is just enough carbon present to produce a 100 percent pearlitic microstructure (Figure 4.3). Pearlite consists of alternating plates of an iron-carbon compound called cementite (Fe_3C) and relatively

FIGURE 4.2
The iron-carbon diagram showing the steel portion (White, Neely, Kibbe, Meyer, *Machine Tools and Machining Practices,* Vol. II, © 1977 John Wiley & Sons, Inc.).

FIGURE 4.3
SAE 1090 steel slowly cooled, showing very coarse pearlite (500×).

FIGURE 4.5
Low-carbon steel with full ferrite grains (500×).

pure iron, called ferrite (Fe). Less carbon than eutectoid, when slowly cooled, produces whole grains of pearlite and grains of ferrite in proportions depending on the amount of carbon present. When these hypoeutectoid steels cool from a point on the A_3 line to the A_1 line, they undergo a transformation over a range of temperature drop in which the carbon in the austenite forms the pearlite grains of eutectoid steel, drawing the carbon from the prior austenite grains and leaving ferrite at those grain boundaries (Figure 4.4). Full regular grains of ferrite will form, depending on the cooling rate and carbon content (Figure 4.5). When hypereutectoid steel (steel containing more carbon than eutectoid) is slowly cooled from a point on the A_{cm} line, the excess carbon over .80 percent begins to form cementite at the prior austenite grain boundaries since the A_3 line is the point of solubility of cementite. Within the grains the eutectoid composition is formed and at the $A_{3,1}$ line, the grains become 100 percent pearlite with a network of cementite at the grain boundaries (Figure 4.6).

Now microstructures other than those shown in the study of slow-cooled plain carbon steel in the iron-carbon

FIGURE 4.4
Ferrite at the grain boundaries in SAE 1040 hypoeutectoid steel (500×).

FIGURE 4.6
SAE 1095 hypereutectoid steel showing cementite at the grain boundaries. Massive cementite and ferrite both etch white (500×).

phase diagram will form if the cooling rates are more rapid. These other microstructures are best explained in an isothermal-transformation (I-T) diagram (Figure 4.7). Sometimes these diagrams are referred to as a TTT or 3T (time-temperature-transformation) diagram. It can be seen on the I-T diagram that when austenite is cooled to a point below its stable temperature (A_1 on the iron-carbon diagram), it will begin to transform into any of a number of transformation products, depending on the cooling rate. If it is held at a constant temperature (isothermal) for a given length of time, it will transform into one particular kind of microstructure. If, on the other hand, it is cooled so suddenly that the cooling curve No. 1, as shown in Figure 4.8, does not cut into the "nose" of the I-T diagram, no transformation can take place that would form microstructures that are relatively soft, such as pearlite or bainite;

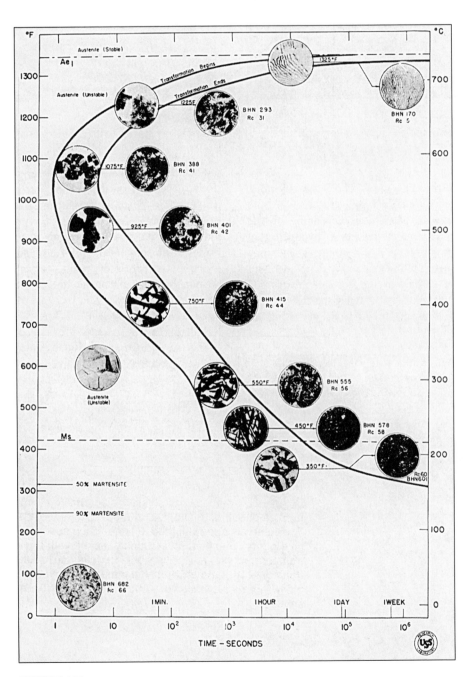

FIGURE 4.7
I-T diagram of .89 percent carbon steel (© United States Steel).

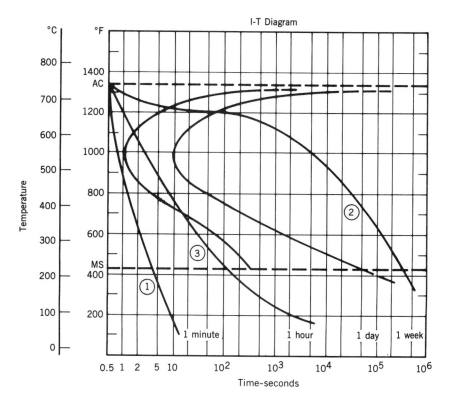

FIGURE 4.8

Three cooling curves on an I-T diagram of .90 percent carbon steel.

instead, the austenite would be cooled to an area below the M_s line where martensite is formed. Therefore, the formation of martensite, if cooled rapidly enough, is temperature and not time dependent as are the other microstructures. Martensite is very hard and brittle and must be tempered to be useful (see the section on tempering carbon steel later in this chapter). The .90 percent C steel I-T diagram (Figure 4.8) shows cooling curve No. 2 with a very slow cooling rate as when steel is annealed to soften it. This produces a soft coarse pearlitic microstructure in this carbon steel that is easily machined. Cooling curve No. 3 represents a steel that is air-cooled or normalized in which mixed microstructures are formed.

Hardening Plain Carbon Steel

Plain carbon steel contains no alloying elements other than carbon and small percentages of elements such as manganese that are necessary in steel manufacture. It is used for knives, files, and fine cutting tools such as wood chisels since it will hold a keen edge. Carbon steel is much easier to heat treat than the alloy and tool steels.

The process of hardening steel is carried out in two operations. The first step is to heat the steel to a temperature that is slightly above the A_3 and $A_{3,1}$ lines on the iron-carbon phase diagram (Figure 4.9). This operation is called austenizing by metallurgists. The austenized steel (face-centered cubic crystal structure) contains all the car-

bon in the interstices (spaces or voids in the lattice structure). The second step is to cool the red-hot metal so quickly that it has no chance to transform into softer microstructures, but still holds the carbon in solution in the austenite. This operation is called quenching (Figure 4.10). **Quenching media,** such as brine, tap water, fused salts, oil, and air, all have different cooling rates. Slower cooling is necessary for tool steels and rapid rates are needed for plain carbon steel. Rapid quenching can produce cracking in thicker sections and therefore is normally

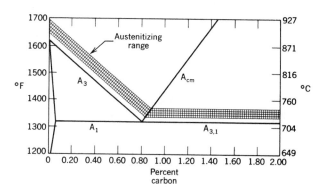

FIGURE 4.9

Diagram showing hardening temperature of steels (austenitizing range) (White, Neely, Kibbe, Meyer, *Machine Tools and Machining Practices,* Vol. II © 1977 John Wiley & Sons, Inc.).

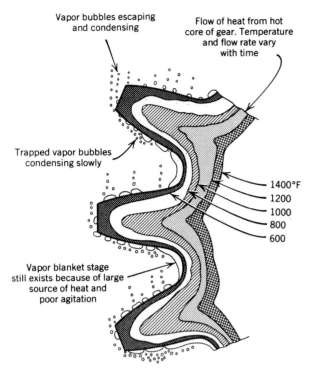

FIGURE 4.10
Temperature gradients and other major factors affecting the quenching of a gear. The gear was quenched edgewise in a quiescent (not agitated) volatile liquid (From *Metals Handbook,* Vol. 2, 8th ed., American Society for Metals, 1964, p. 16. With permission).

used on small or thin sections with low mass and for plain carbon steels.

The austenized steel, when quickly cooled to a point below the M_s temperature, attempts to transform to its natural crystal structure of body-centered cubic, which can contain almost no carbon in its interstices. The result is a distorted, elongated cubic structure that is very hard and is called martensite (Figure 4.11).

TEMPERING (DRAWING) PLAIN CARBON STEEL

Martensite is very brittle until it is tempered or drawn, as some heat treaters call it. A tool that is hardened and not tempered would break in pieces when first used. Tempering is simply a process of reheating the hardened steel to a much lower temperature than that used for hardening. In this process the steel loses some of its hardness, depending on the temperature, and gains toughness, losing some brittleness. Hardening and tempering temperatures

FIGURE 4.11
The microstructure of martensite (500×).

are given in Table 4.1. In Table 4.2 it can be seen that the higher the tempering temperature is, the softer the metal is. Oxide colors that form on the clean surfaces of steel in a given temperature range show heat treaters the approximate temperature of the metal. This color method of tempering was used by blacksmiths to determine the temperature before plunging the part into a water tank to stop the heating action. It is still used to some extent in small shops, but more exact methods are used in the manufacturing of heat-treated steel parts.

HEAT TREATING EQUIPMENT

Almost any method can be used that will heat steel enough to get it red hot for the purpose of hardening it. Blacksmiths used forges and charcoal or coke. A heating torch is sometimes used, but the most reliable method is to use a temperature-controlled furnace or a molten salt or metal bath (Figures 4.12 to 4.14). These high-temperature salt baths used for heating metals to the hardening temperature are sometimes called pot furnaces. Metal baskets are used to lower small parts into the furnace. In manufacturing, the heat treating process is often done in a continuous furnace that has a heat-resistant belt slowly moving parts through the furnace. The parts are correctly heated for quenching as they emerge from the furnace (Figure 4.15 on p. 83). All heat treating furnaces are built of a refractory (heat-resistant material), usually firebrick, and are controlled by an electronic relay system that turns the furnace off and on. The temperature is sensed by a bimetal thermocouple that is placed directly in the furnace (Figure 4.16 on p. 84).

TABLE 4.1

Temperatures and colors for heating and tempering steel

Colors		Degrees Fahrenheit	Process
Heat Colors	White	2500	
		—	
		2400	High Speed Steel Hardening (2250–2400°F)
		—	
	Yellow to white	2300	
		—	
		2200	
		—	
		2100	
		—	
	Yellow	2000	
		—	
		1900	
		—	
	Orange to red	1800	Alloy Steel Hardening (1450–1950°F)
		—	
		1700	
		—	
		1600	
	Light cherry red	—	
		1500	
		—	Carbon Steel Hardening (1350–1550°F)
	Cherry red	1400	
		—	
		1300	
	Dark red	—	
		1200	
		—	
		1100	
		—	
	Very dark red	1000	
		—	
		900	
		—	
	Black red in dull light or darkness	800	High Speed Steel Tempering (350–1100°F)
		—	
		700	Carbon Steel Tempering (300–1050°F)
		—	
Temper Colors	Pale blue (590°F)	600	
	Violet (545°F)	—	
	Purple (525°F)	500	
	Yellowish brown (490°F)	—	
	Straw (465°F)	400	
	Light straw (425°F)	—	
		300	
		—	
		200	
		—	
		100	
		—	
		0	

(*Source:* Pacific Quality Steels, "Stock List and Reference Book," No. 85, Pacific Machinery and Tool Steel Company, 1981.)

TABLE 4.2
Temper color chart

Degrees			
C°	F°	Oxide Color	Suggested Uses for Carbon Tool Steels
220	425	Light straw	Steel-cutting tools, files, and paper cutters
240	462	Dark straw	Punches, dies
258	490	Gold	Shear blades, hammer faces, center punches, and cold chisels
260	500	Purple	Axes, wood-cutting tools, and striking faces of tools
282	540	Violet	Springs, screwdrivers
304	580	Pale blue	Springs
327	620	Steel gray	Cannot be used for cutting tools

(*Source:* White, Neely, Kibbe, Meyer, *Machine Tools and Machining Practices,* Vol. I, Wiley, New York, © 1977.)

FIGURE 4.13
Large heat treating furnace (Bureau of Mines).

SURFACE HARDENING PROCESSES

Flame Hardening and Induction Hardening

The processes of flame hardening and induction hardening in manufacturing are also usually continuous (Figures 4.17 on p. 84 and 4.18 on p. 85). Surface hardening of metal parts for greater wear resistance has been common practice in the manufacture of automotive components, industrial machinery, and hardware for many years. Surface hardening creates a hard case around the relatively soft core of the steel part. Steels that contain sufficient carbon and also some cast iron can be quickly heated on the surface. When the steel contains very little carbon, the surface chemistry may be altered by diffusing carbonaceous or nitrogenaceous substances into the surface followed by heating and quenching in most cases. Some of these processes are car-

FIGURE 4.12
Small electric furnace.

FIGURE 4.14
Drawing of one type of salt bath furnace. Electrodes immersed in the salt produce the heating effect.

burizing, nitriding, carbonitriding, and cyaniding. Low carbon steel can be carburized and surface hardened to a depth of about .003 in. by heating it with a torch to about 1700°F (927°C) and rolling it in a carbon compound such as Kasenit® followed by reheating and water quenching. In order to harden to $\frac{1}{16}$ in. deep, the part must be packed in the carburizing compound and held at that temperature for about 8 hours. Nitriding produces a harder case with a lower temperature and less distortion. Other methods produce a more uniform, harder case than carburizing in a shorter time. Some of the disadvantages to these methods of surface hardening are that the entire part must often be heated and quenched, altering its entire chemical structure. Rising energy costs and the need for increased production efficiency have brought about the development of new methods.

Selective Surface Hardening

Although induction hardening has been used for many years to harden small parts or ways on machine tools, newer processes make use of this hardening process in a selective manner in which wear surfaces are hardened only in stripes, moving progressively along the surface (Figures 4.19 and 4.20). This is done on flat surfaces, inside cylinders, and for bearing races on shafts. In these quick-heating processes the heated area is self-quenched by the adjacent cold metal, resulting in a shallow hardened area in the form of a line (stripe) or spiral. Electron beam equipment is also capable of producing selective hardened areas, but it is usually done in a vacuum. Laser systems can operate in ambient conditions for selective heat treating (Figure 4.21). Their primary advantage is accessibility to selective areas on a metal part where other systems cannot be easily applied. A longer part life is claimed for laser-hardened parts. One disadvantage to laser hardening is that a nonreflective coating on the part to be hardened is sometimes needed. High-frequency resistance heating is an alternative method that can produce a hardened line or stripe in one "shot" without the traversing operation. The cycle time is also shorter and is quite adaptable to conventional automation. Tempering is not normally used in surface-hardening processes.

Precipitation Hardening

Some metal alloys, both ferrous and nonferrous, can also be hardened by solution heat treatment and aging processes called precipitation hardening. Some stainless steels that cannot be hardened by ordinary heat treatment have an alloying element or compound added that begins to precipitate as small particles. After the metal has been

FIGURE 4.15
Continuous heat treating furnace.

FIGURE 4.16
Thermocouple sensor. The twisted wires of a thermocouple produce a feeble electric current when heated. Furnace temperatures are controlled from this sensor.

heated to put the elements in solution and then quenched, these small particles lock up the slip planes. The period of time these particles take to precipitate (often two or three days) is called aging. The 2000 series of aluminum can be solution heat treated and aged to produce a light, tough and strong metal that is used extensively in aircraft manufacture.

Aluminum containing about 4 percent copper is heated to about 1000°F (538°C) for a period of time to bring the copper into solution (Figure 4.22 on p. 86). When the aluminum is quenched in water to room temperature, the solubility of the copper is drastically decreased and a compound of copper aluminide forms that slowly disperses along grain boundaries and slip planes to harden the alloy (Table 4.3). Artificial aging is often practiced to speed up the process but it is not recommended in some cases, since it can decrease corrosion resistance. Some copper, nickel, magnesium, and titanium alloys, as well as alloys of some other metals, can be hardened by the process of precipitation heat treatments.

ANNEALING, RECRYSTALLIZATION, AND STRESS RELIEF

The term **anneal** is often used when referring to any one of several processes: full anneal, normalizing, spheroidize anneal, process anneal, and stress relief. When tool steel that has been hardened by heat treatment must be softened for the purpose of machining or subsequent heat treatment, the full anneal is used. This process consists of heating the tool steel to a temperature about 50°F (28°C) above the A_3 or $A_{3,1}$ lines (Figure 4.23 on p. 87). The part is then allowed to remain in the furnace while it cools slowly to room temperature. Alternately, the part could be placed in a noncombustible insulating material such as ashes. The cooling rate depends on the type of tool steel being annealed. Several hours of cooling is usually required for plain carbon tool steel; other kinds often take much longer. Full anneal is necessary when hardened structures such as iron carbides, other metallic carbides, and martensite are present. These are found in the higher carbon steels (about .40 to 2 percent) and some alloy and tool steels.

Lower carbon steels, between .30 and .60 percent carbon, that have been hardened, can be normalized to soften them sufficiently for machining. The normalizing process does not completely soften these steels but leaves them just hard enough for efficient machining. A metal that is too soft will not machine well because it is too "gummy," leaving a rough torn surface finish, whereas a somewhat harder metal will have a much smoother surface. Normalizing is also used to refine castings and forgings

FIGURE 4.17
Induction hardening. Typical work coils for high-frequency units (*Metals Handbook,* Vol. 2, 8th ed., American Society for Metals, 1964, p. 172. With permission).

FIGURE 4.20
Internal surfaces are selectively hardened by high-frequency heating.

FIGURE 4.18
Hardness pattern developed in sprocket teeth when standard flame tips were used for heating. When space permits this method to be used, hardening one tooth at a time results in low distortion (*Metals Handbook,* Vol. 2, 8th ed., American Society for Metals, 1964, p. 200. With permission).

since it recrystallizes the coarse, irregular grains to produce a stronger, fine-grained metal. It is used to prepare tool steel parts for hardening heat treatments by removing machining and other stresses. Normalizing is usually carried out in a furnace by slowly raising the temperature to about 100°F (56°C) above the A_3 or just above the A_{cm} lines. Then the part is removed from the furnace and allowed to cool slowly in still air.

Both the full anneal and the normalizing process require the temperature to be raised above the transformation lines on the diagram in Figure 4.1 in order to dissolve the carbides. Stress-relief anneal and spheroidizing anneal in steel may be carried out at a temperature lower than the A_1 line represents. When medium to high carbon steels must be annealed to their softest possible condition, a spheroidized anneal is often used. Full anneal for some of these high carbon steels will not soften them sufficiently for some cold working and machining operations. For example, commercially available drill rod contains almost 1 percent carbon and is usually spheroidized. In that state it is softer and more ductile than low carbon cold-rolled steel. Spheroidizing is carried out at a temperature just

FIGURE 4.19
Selective high-frequency resistance heating on a flat surface. A water-cooled "proximity conductor" is placed close to the surface to be heated, through which a high frequency current is applied.

FIGURE 4.21
Laser hardening a worm gear. This is an example of selective heat treating. Instead of the entire part being hardened, which could cause distortion, just the contact faces of the gear are hardened (Coherent, Inc., Palo Alto, CA).

FIGURE 4.22
Process of solution heat
treating and artificial aging of
2014-T6 aluminum alloy, ¼
in. thick (Neely, *Metallurgy,* 2d
ed., © 1984 John Wiley &
Sons, Inc.).

TABLE 4.3
Solution heat treatment with aging times and temperatures for some commercial hardenable aluminum alloys

		Soaking Time[a] *for Various Thicknesses (min)*			
Designation	*Soaking Temperature (° F)*	*Up to .032 in.*	*Over .032 to .125 in.*	*Over .125 to .025 in.*	*Over .250 in.*
2014-T6	925 to 950	20	20	30	60
2017	925 to 950	20	20	30	60
2117	890 to 950	20	20	30	60
2024	910 to 930	30	30	40	60
6061-T6	960 to 1010	20	30	40	60
7075	860 to 960	25	30	40	60

Designation	*Aging Temperature (° F)*	*Aging Time (hr)*
2014-T6	345 to 355	2 to 4
	355 to 375	½ to 1
2017	Room temperature	96
2117	Room temperature	96
2024	Room temperature	96
6061-T6	315 to 325	50 to 100
	345 to 355	8 to 10
7075	245 to 255	24
	315 to 325	1 to 2

(*Source:* John E. Neely, *Practical Metallurgy and Materials of Industry,* 2d ed., Wiley, New York, © 1984.)
[a]Soaking time begins after the part has reached the soaking temperature.

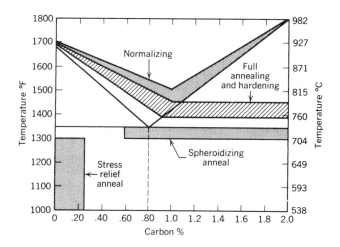

FIGURE 4.23

Temperature ranges used for heat treating carbon steel (White, Neely, Kibbe, Meyer, *Machine Tools and Machining Practices,* Vol. I, © 1977 John Wiley & Sons, Inc.).

under 1300°F (704°C) in a furnace for about 4 hours. Carbides slowly form into globules or tiny spheres in the microstructure instead of the harder lamellar pearlite (Figure 4.24).

Stress relief anneal is closely related to prior cold working of metals. In this process it is not carbides that are softened, but the ferrite grains that have been hardened by cold working. Therefore, this process is confined to low carbon steels and many nonferrous metals. In industrial processes such as wire drawing, cold-worked metal that must be softened so further cold working can be done is subjected to process anneal. This is substantially the same thing as stress relief anneal except it is often carried out in closed containers with an inert gas to prevent oxidation of the surface. It is often called bright annealing. Cold-worked grains become flattened, signifying a loss of ductility. When the metal has been raised to its recrystallization temperature, the flattened grains reform into a soft whole grain structure (Figures 4.25 and 4.26).

Recrystallization takes place when cold worked metals are heated to their specific recrystallization temperatures (Table 4.4). The stored energy from cold working is then released, allowing the distorted grains to reform by migration of grain boundaries and joining of adjacent grains. This process is called grain growth. The following factors affect recrystallization:

1. A minimum amount of deformation is necessary for recrystallization to occur.
2. The larger the original grain size, the greater the amount of cold deformation that is required to give an equal amount of recrystallization with the same time and temperature.

FIGURE 4.24

Micrograph of spheroidized carbon steel (500×).

FIGURE 4.25

Microstructure of .10 percent carbon steel, cold rolled. Both ferrite and pearlite grains have been flattened by plastic deformation (*Metals Handbook,* Vol. 7, 8th ed., American Society for Metals, 1972, p. 10. With permission).

FIGURE 4.26
The same .10 percent carbon steel as in Figure 4.25 but annealed at 1025°F (552°C), the recrystallization temperature of ferrite. Ferrite grains are mostly reformed to their original state, but the pearlite grains are still distorted because the transformation temperature was not reached (1000×) (*Metals Handbook,* Vol. 7, 8th ed., American Society for Metals, 1972, p. 10, With permission).

TABLE 4.4
Recrystallization temperatures of some metals

Metal	Recrystallization Temperature °F
99.999% aluminum	175
Aluminum bronze	660
Beryllium copper	900
Cartridge brass	660
99.999% copper	250
Lead	25
99.999% magnesium	150
Magnesium alloys	350
Monel	100
99.999% nickel	700
Low carbon steel	1000
Tin	25
Zinc	50

(*Source:* John E. Neely, *Practical Metallurgy and Materials of Industry,* 2d ed., Wiley, New York, © 1984.)

3. Increasing the time of anneal decreases the temperature necessary for recrystallization.

4. The recrystallized grain size depends mostly on the degree of deformation and, to some extent, on the annealing temperature.

5. Continued heating, after recrystallization (reforming of grains) is complete, increases the grain size.

6. The higher the cold working temperature, the greater the amount of cold work required to give equivalent deformation.

Stress relief is often needed for castings and weldments. Large welded structures such as tanks are sometimes stress relieved by covering them on the outside with thermal insulation blankets and heating them on the inside with propane burners to the recrystallization temperature.

Thermal stress relief is preferred for most manufacturing processes. However, vibratory stress relief (VSR) is often used for cast or welded structures that are too large to fit into a heat-treat furnace. Dimensional instability can result from contraction caused by welds necessitated by nonuniform cooling in castings. Another cause of dimensional instability can be removing or inducing stresses through machining. In order to be able to use VSR effectively, three principles must be understood:

1. Residual stresses can cause metal components to be more rigid than stress-free ones.

2. Effective loading of a structure by means of resonance requires close control of a vibrator's frequency.

3. Proper instrumentation is required to display the pertinent VSR data.

In order to relieve stresses, the entire part must be made to resonate and induced vibration must be very near to its natural resonant frequency, which is generally in the 50 Hz to 100 Hz range. The dynamic loading causes plastic flow where any concentration of high-level residual stress is present, such as in welds. Flexure at these places combined with the residual stress is sufficient to momentarily exceed the yield point. Electrically powered portable vibrators are attached to the part to be stress relieved with clamps and, in addition, elastic isolation cushions support the part at strategic locations. It seems that there is no upper limit on size for this stress-relieving technique; however, there have been some serious difficulties in its use in the past. Welds are sometimes shot-peened to relieve residual weld stresses. The difference from other methods is that shot-peening induces surface compressive stresses to counteract the tensile stresses in the weld.

Review Questions

1. To what temperature should hypoeutectoid (less than eutectoid) plain carbon steel be heated for hardening purposes as seen on the iron-carbon diagram?

2. If a .80 percent plain carbon steel is austenized (heated to a red heat) and quenched in water to room temperature so that the cooling curve does not cut into the "nose" of the I-T diagram, what will the resultant microstructure be?

3. On the right side of the iron-carbon diagram, what kind of metal is found when the carbon content is over 2 percent?

4. State the difference between the eutectic and eutectoid points.

5. What happens to a quench-hardened tool or machine part that was not tempered when it is subjected to shock loads or bending stress?

6. What does the process of temper drawing do for hardened carbon steel?

7. Name two kinds of equipment that are used for heating steel to a precise hardening temperature.

8. How can a steel that contains insufficient carbon to be hardened be surface hardened?

9. By what means can a small area or stripe be selectively hardened on a large steel part?

10. How can some nonferrous metal alloys such as aluminum or monel be hardened?

11. What annealing method should be used for softening a hardened high carbon tool steel?

12. A low carbon steel part has been partly formed toward its final shape and now is in a work-hardened condition. Any further cold working will cause it to tear. At this point, heating to a red heat would distort it too much to recover its shape with further work. How can this part be restored to a soft condition so it can be cold worked to its final shape?

FIGURE 4.27
See Case 1.

Case Problems

Case 1: The Case of the Badly Worn Pin

The soft steel pin in Figure 4.27 was removed by a millwright in a manufacturing plant from a hydraulic cylinder clevis at the end of a cylinder rod. It was originally a straight cylindrical pin, but it wore almost completely through before a mill shutdown allowed it to be removed. The holes in the clevis, which is made of soft cast steel, were also elongated accordingly. The clevis was periodically lubricated but the extreme forces still caused excessive wear. Considering what you have learned in this chapter, what could be a possible solution to this problem of excessive wear?

Case 2: Strength versus Weight Problem

A structure for an aircraft part (made of 1000 series aluminum) has a certain weight limitation. After all weight reduction by engineering design had been done, the part was still too heavy. No metal other than aluminum is permissible in this case. How can the part weight be further reduced and still have sufficient strength?

CHAPTER 5

Extraction and Refinement of Common Nonmetallic Materials

Objectives

This chapter will enable you to:

1. Identify the raw materials derived from petrochemicals that are used to produce many synthetic materials.

2. Explain the properties of some thermoplastics and thermosets that are used to manufacture products.

3. State the uses for certain additives and finishes for plastic materials.

4. Relate the means by which elastomers are derived.

5. Name the uses and explain the significance of asphalts and lubricants in the manufacturing industry.

6. Demonstrate how ceramic materials are produced.

7. Assess the significance of wood materials in manufacturing.

S tone, clay, and wood are among the oldest engineering materials used by humans to construct buildings and aqueducts, to make containers to store foodstuffs and contain liquids, and to make tools and weapons. The Romans discovered a way to make fluid stone that would solidify into a hard mass. It was called pozzolan cement, because it was first made from the volcanic ash at Pozzuoli, Italy. After this ash was combined with slaked lime it would begin to harden. Modern portland cement, from which concrete foundations, sidewalks, and dams are made, is similar in some ways to that ancient Roman cement.

The making of glass by fusing silica sand with another substance dates back to at least 3000 B.C. Glass was used for making jewelry, containers, small windowpanes, and other artifacts; its use for plate glass windows, however, has appeared only relatively recently because of the difficulties of producing large, flat sheets with optical quality.

Although wood has been displaced by other materials to some extent in recent times in the manufacture of furniture and in building construction, it continues to be important as a structural material. Wood has the advantage of being a renewable resource and, if forestry programs of replanting harvested trees are maintained, there will always be a nondepletable source.

Chemicals from which the commercially important plastic resins are derived are themselves derived from natural products that are now quite plentiful. Those raw materials include coal, water, limestone, sulfur, salt, and petroleum. An Arizona desert wildflower called the popweed (known scientifically as *lesquerella*) may someday partially replace petroleum as a source for plastics. Combining popweed oil with a compound derived from castor oil (from the castor bean) produces a polyester that is made entirely from natural sources. Popweed oil can also be polymerized by itself into a soft rubbery material. Jojoba *(Simmondsia californica),* a desert shrub native to the Southwest, produces a waxy oil that is being tested for use as a substitute for the sperm whale oil presently used in some industrial lubricants. Meadowfoam, grown commercially in the Willamette Valley in Oregon, produces a waxlike oil from its seed. It is also a substitute for sperm whale oil, and is presently used as an oil in cosmetics. Such renewable resources will be needed in the future as petroleum oil becomes more scarce.

Petrochemicals, that is, petroleum products such as plastics and elastomers, have only recently appeared on the scene and have rapidly displaced the use of wood, metal, and glass in some areas. The great advantage of plastics is that a special material can be created for a specific purpose that will fit the requirements better than any natural material. Thus, surgical implants, quiet long-lasting gears, and nondeteriorating rubber are some items that are produced by chemical synthesis of raw materials. All plastics are classified as either **thermosetting** or **thermoplastic,** each group having distinctly different qualities under conditions of repeated heating and cooling.

All of these materials will be discussed in detail in this chapter. However parts manufactured from these materials will be discussed in Chapter 17, Plastics and Composites Processing.

PETROCHEMICALS

Chemicals made from crude oil, natural gas, and liquefied natural gas are the basic raw materials for many of the synthetic fibers, sealants, paints, synthetic rubbers, plastics, and other products. These products are typically polymers that are made from small molecules called **monomers.** These are the building blocks of these synthetic materials and they derive their names from the Greek *mono* (one), *poly* (many), and *meros* (parts). Polymer chains are composed of monomer cells (raw materials) that are combined to produce these useful materials (Table 5.1).

By the process called catalytic cracking, heat is used to induce a chemical reaction in the crude oil with the help of **catalysts,** usually composed of refractory oxides of aluminum, silicon, and magnesium. High-temperature boiling components of petroleum are broken down into gasoline and oils. The remaining gaseous material is made up of compounds having from one to four atoms. Among these are ethylene, propylene, and the butylenes. Natural gas, from which the liquefied gases butane and propane are derived, is also a source of propylene, ethylene, hydrogen, and methane obtained by means of the process of thermal cracking. Acetylene gas, made from the hydration of calcium carbide, is also a source of plastic and rubbery materials. Most acetylene gas is now produced from methane rather than from calcium carbide. Chloroprene and polychloroprene (neoprene) are derived from acetylene.

Nitrogen fertilizer made from natural gas and pesticides made from other petrochemicals are both vital to food production. About 60 percent of the fibers used for clothing are synthetic and are mostly based on petrochemicals; exceptions are rayon, acetate, and acetic anhydride. Many of today's medicines are derivatives of petrochemicals; examples are antihistamines, antibiotics, sulfa, penicillin, and aspirin. Automobiles use plastic gears and body parts; our homes have synthetic materials in such things as carpets, siding, heating and cooling systems, furnishings, and housewares. The petrochemical materials and derivatives covered in this chapter will be the plastics, rubbers, adhesives, lubricants, and asphalts.

TABLE 5.1
Products from petrochemicals

Products	Petrochemical Monomer (Raw Materials)
Antifreeze fluid, Mylar	Ethylene glycol
Butyl rubber	Isobutene
Epoxy resins, polycarbonates (Lexan®) acetone, phenol, bisphenol-A	Isopropylbenzine
Ethylene glycol antifreeze, Mylar (Dacron®) ethyl alcohol	Ethylene
Mylar, poly-para-xylene (Parylene®)	Xylenes
Neoprene rubber	Chloroprene
Nylon synthesis, neoprene, and polybutadiene rubber (source)	Butadiene
Orlon: chloroprene and neoprene rubber	Acetylene
Phenol, phenolic resins, styrene, and polystyrene	Benzene
Phenolic plastics	Toluene
Phenolic resins, epoxy resins	Phenol
Polypropylene, isopropylbenzine, isopropyl alcohol	Propylene
Polystyrene	Styrene
Polystyrene (nonsoluble)	Divinylbenzene
Production of acetylene	Methane
Synthetic rubber	Isoprene

(*Source:* John E. Neely, *Practical Metallurgy and Materials of Industry,* 2d ed., Wiley, New York, 1984.)

PLASTICS (THERMOSETTING)

All plastics are formable when heated, hence, the prefix *thermo.* The difference between the two types of plastic is that the thermoplastics, after once being hardened in the heating process, can be reheated and formed into new shapes a number of times without much change in properties. Thus, thermoplastic scrap can be reheated and reused. However thermosetting types take on a permanent set after they are heated (and sometimes placed under pressure), after which they cannot be remelted or reformed. Additional heating will only destroy them by breaking the permanent cross-links in the polymer chain. Some types of thermosets do not require heat to harden, but are cross-linked by combining liquids in definite proportions. The mixture then hardens in a preset time. Some examples are epoxy, silicone, polyurethane, and polyester. Thermoset scrap cannot be reprocessed; however, the ever increasing amount of automobile thermoset plastic scrap is being considered as a recyclable material and a low-cost natural resource. According to research done at

Lehigh University's research center, this plastic scrap can be ground and used as a filler or aggregate in acrylic polymer concrete to produce medium-strength products. These could compete with portland cement and could perhaps be used as construction materials for some applications where light weight and low density are needed.

As a general rule, the thermosets must be combined with fillers and reinforcing agents to give them the mechanical properties needed for molded parts. For example, most of the epoxies are quite brittle, but when combined with glass fibers they become strong and tough materials.

Plastics have several general characteristics in common. They are among the lightest in weight of the engineering materials. They are all relatively good heat insulating materials since they have a relatively low thermal conductivity, and nearly all plastics are good electrical insulators. However, a recently developed polyacetylene plastic is a good conductor of electricity and shows great promise in production of plastic batteries, possibly solar panels, and plastic wire for motors and generators. This family of plastics can be considered as "synthetic metals." For the most part, plastics can be obtained in a wide range of colors, either transparent or opaque; a few exceptions, such as Bakelite®, have a natural color that is often acceptable for a product. Plastic items can usually be produced in one operation such as casting, molding, or extrusion. Although many plastics are not nearly as strong as metals, some have been developed that are almost as strong as some metals, and sometimes they exceed metals in other properties such as toughness and resistance to abrasion.

Alkyds

The alkyds are polyester derivatives, produced by the reaction of an alcohol and an acid. The name alkyd comes from the words *ALCohol* and *acID*. Their most important use is in paints and lacquers. Also molded automobile ignition parts and light switches are made from this thermoset because they do not break down at higher temperatures and are not affected by moisture, conditions found around automobile engines.

Allyl Plastics

The allyl plastics, produced from an alcohol, are very hard, scratch-resistant, clear plastics used for optical glass—prisms and lenses in eyeglasses (Figure 5.1), radar domes, and window glass. Although a very costly thermoset, it is far superior to the clear acrylic plastic that is also used for transparent shapes. It resists most solvents

FIGURE 5.1
Safety glasses with plastic lenses.

such as gasoline, benzene, and acetone, and has a high tensile and compressive strength.

Aminos

Urea and melamine belong to the family of amino plastics that have an unlimited range of colorability. The urea-formaldehyde plastics are produced by reacting urea with formaldehyde in the presence of a catalyst and mixing in a filler. These are used for knobs, handles, and household fixtures. Urea-formaldehyde plastics are used as adhesives for bonding plywood, foundry sand cores, and other materials. Melamines are similar but more complex plastics of the amino family and are harder, more shock-, heat-, and water-resistant than the urea types, and have good resistance to acid and alkali attack. Because of their water resistance, they are used to bond exterior plywood. They are used for electrical parts such as circuit breakers and terminal blocks, and because of their ability to resist high temperatures and their chemical stability, the melamines are used for dinnerware and abuse-resistant furniture surfaces.

Epoxies

These plastic materials are quite brittle (although a few types are tough and elastic), and have good electrical properties and chemical resistance. They cure at normal temperatures and pressures. Though they have good bonding properties as adhesives, their shock load strength is poor. Epoxies can be cast in molds and are used in paints and for surface coatings for metals.

Phenolics

The phenolic-formaldehyde plastics are low-cost molding compounds with high strength, high temperature resistance, good dimensional stability, and good electrical properties. The most common type, Bakelite, is made from phenol and formaldehyde with a filler material such as wood flour. Bakelite is used for electrical switches and automobile distributor caps. Canvas, linen, or paper is sometimes used as a lamination to reinforce the plastic. Examples of these phenolic laminates are Micarta® and Formica®. The major drawback of the phenolics is the lack of colorability. They are commonly brown or black.

Polyesters

Polyesters belong to a class of plastics related chemically to the alkyd type and contain two different types: "saturated" and "unsaturated." The saturated types are best known as film and fibers such as Dacron® and Mylar®. The unsaturated types are used in molding and casting. There are also many subtypes in each group. Polyester can be cast in molds without heat and pressure, but a disadvantage is that it has considerable shrinkage when hardening. This plastic is widely used with glass fiber and other reinforcements to provide high strength for such applications as boats, tanks, and automobile bodies. Polyesters can be either thermosetting or thermoplastic depending on formulation.

Polyurethanes

Polyurethane plastic materials may be either thermosetting or thermoplastic. Since these are rubbery plastics, they are often made in a foamed structure for mattresses and insulation and are a substitute for foamed rubber. The thermosetting types can be cast as tough, wear-resistant elastomeric materials. They can be rigid or flexible depending on formulation.

Both urethane elastomers and rigid products are produced by the casting, injection molding, and coating process. Since they have good chemical and weather resistance along with high resilience and elasticity, they are used in a number of products such as oil seals (Figure 5.2) that can last up to eight times longer than rubber. Some of these seals are graphite impregnated to make them self-lubricating. Urethanes also have load-dampening ability and very good abrasion resistance, making them ideal materials for industrial solid truck tires (Figure 5.3). They are also used for caster wheels

FIGURE 5.2
New urethane compounds for hydraulic seals will last eight times longer than rubber and provide leak-free sealing up to 10,000 PSI (Disogrin Industries).

(Figure 5.4), gears, drive belts, floor coverings, and roofing materials. Metal objects such as furniture are sometimes covered with a tough, resilient skin of polyurethane that imparts both aesthetic form and a colorful, useful surface on the part, while the steel core provides the needed high strength.

A synthetic leather produced by DuPont, called Corfam®, is a polyester-reinforced urethane. It can be made porous or sealed with silicone. It was developed for shoe-upper parts, but is used in many other applications, such as to make industrial gaskets and shaft seals.

FIGURE 5.3
Urethane industrial solid truck tires (Maine Rubber International).

FIGURE 5.4
These solid polyurethane caster wheels are available in many different sizes, shapes, and colors (The Hamilton Caster & Mfg. Co.).

Silicones

Silicone plastics are a type of synthetic rubber that is one of the highest temperature-resistant plastics; they may be used at temperatures up to 500°F (260°C) (Figure 5.5). Because of their excellent electrical properties, they are widely used in the electrical industry for encapsulation and

components. Silicones are also used as sealants and caulking since they do not lose their rubbery qualities or crack with age and they have good bonding characteristics.

Molds can be made with room-temperature vulcanizing silicone rubbers that require a separate catalyst as a hardener that is mixed in at the time of use. Silicone products are also used as lubricants. Since they do not tend to break down in hostile environments, the silicones have been very successful for use as prosthetic devices, such as artificial heart valves, and for reconstructive surgery near the surface of the body.

PLASTICS (THERMOPLASTIC)

Thermoplastic materials have the cost-saving advantage over thermosets in that they are injection molded (Figures 5.6 and 5.7) and therefore have faster cycling times than compression- or transfer-molded thermosets (Figures 5.8 to 5.10 on pp. 97 and 98). When thermosets can be injection molded, deflashing and other secondary operations are often unnecessary. Thermoplastics are generally less brittle than most thermosets and can also be bonded by the process of ultrasonic welding. Also, since they retain some flexibility, thermoplastics can be designed with snap-fit parts, eliminating screws and other fasteners.

Inertness COHRlastic silicone rubber is odorless,

FIGURE 5.5
Comparative heat aging graph of silicones (CHR Industries, Inc.).

FIGURE 5.6
Modern injection molding machine with microprocessor that automatically controls cycling times (Hergert's Industries, Inc.).

FIGURE 5.7
The hopper on the injection molding machine must be filled frequently while the machine is in operation (Hergert's Industries, Inc.).

Many thermoplastics can be used as a "living hinge" because they can flex thousands of times without breaking. Although thermosets have been the workhorse of the electrical industry applications because of their good electrical properties and dimensional stability, thermoplastics are rapidly taking over in this application. Electrical and electronic standards are changing, highlighted by the trend toward miniaturized parts that must endure ever higher operating temperatures. Tiny, more complex parts are more easily produced by injection molding techniques, so new thermoplastics have been and are being developed to fulfill the requirements for new products.

Fillers are often added to thermoplastics to gain properties such as dimensional stability, high temperature resistance, heat or electrical conductivity, or magnetic properties. Plasticizers are added to improve their flexibility and stabilizers are added to prevent deterioration by light, heat, and oxidation. Coloring and fluorescent additives give these thermoplastics more customer appeal. Surface finishes, such as paint or electroplated metals, are also applied to plastics to enhance their appearance.

Acrylonitrile-Butadiene-Styrene (ABS) Copolymer

ABS copolymers are very tough, very strong thermoplastics with very high tensile strengths and have an unusual combination of high rigidity and high impact strength even at very low temperatures. These copolymers can be obtained in different combinations of heat resistance, flexibility, and toughness. Their major disadvantages are flammability and solubility in many solvents.

The first of the styrene plastics was introduced in 1938 and was called polystyrene. It has a very low strength, but is one of the least expensive plastics, and for this reason is used for disposable products such as knives, forks, spoons, and water glasses. Styreneacrylonitrile (SAN) is a styrene copolymer that is stronger and has a better chemical resistance than polystyrene. However, ABS is, by far, the most outstanding plastic, making a wide range of specific materials available. Since it has high impact resistance, it is used in luggage, safety helmets, and such automobile parts as grilles. It can be plated in large volumes and can be blended with polyvinyl chloride (PVC) to become flame resistant.

Acetals

Acetal thermoplastics, also called engineering resins because of their high strength, have excellent engineering properties, very high tensile strength, stiffness, good

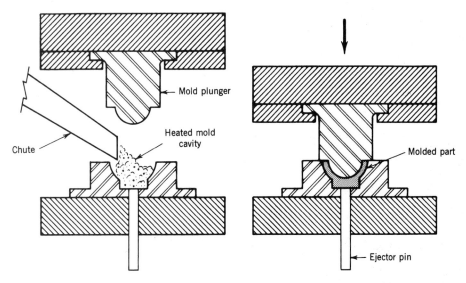

FIGURE 5.8
Compression molding. Granulated plastic is placed in a heated mold (left) and
compressed (right) until object is formed. It is then removed with an ejector pin.

dimensional stability, low moisture absorption, and resistance to creep under load. There are two types of acetals, homopolymer and copolymer. They are very similar in most respects, but the homopolymer is stiffer, stronger, and more impact resistant. It also has a self-lubricating quality that makes it useful for bearings and gears. The DuPont acetal Delrin® is a homopolymer that is used for such products as gears, butane cigarette lighters (because

butane lighter fluid will not affect it), and car door handles (Figure 5.11).

Mineral-filled acetal copolymer resins have potential applications in such products as stereo cassette cases, spools for video cassettes, toys, and zippers. In the plumbing industry, faucet cartridges, shower mixing valves, and shower heads molded of acetal homopolymer are replacing those made of brass and zinc.

FIGURE 5.9
Transfer molding. Plastic is heated in an upper chamber and forced into the mold (left) by a second plunger. The object is formed and ejected (right).

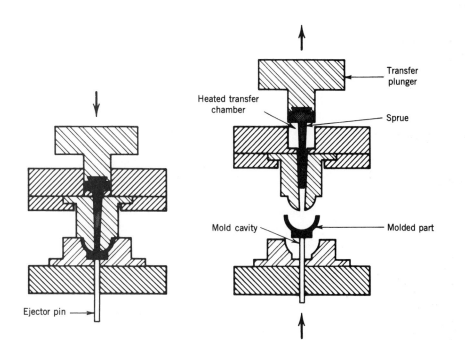

FIGURE 5.10
Principle of injection molding. Granulated thermoplastic is moved forward by the ram and passes through a series of heaters that melt the plastic. It is then injected with pressure into a closed mold where the molten plastic quickly solidifies. The mold is then opened and the part with the sprue is ejected. This cycle is then automatically repeated.

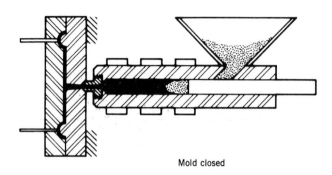

Acrylics

Widely used for their light transmission properties and optical clarity, acrylic plastics have excellent resistance to outdoor weathering and dimensional stability; that is, they tend to keep their shape over a wide range of temperatures. However, they will soften at about 200°F (93°C). The acrylics are not very scratch resistant and have a low shock resistance; therefore, they are not as

FIGURE 5.11
Delrin® in interior automotive door handles is not only lighter in weight than die-cast metals, it also eliminates the need for any finishing operations (Engineering Polymers Division, DuPont Company).

satisfactory for making eyeglasses as are the allyl thermosets. Acrylics can be made in brilliant transparent colors and are used for the manufacture of automobile taillights, jewelry, and novelty items. Lucite® and Plexiglas® are acrylics that are made in the form of tubes, sheets, and rods. Acrylic parts are easily machined from these standard shapes.

Cellulosics

In their natural form, cellulosics do not melt, but with proper chemical treatment they can be molded in the same standard equipment as other thermoplastics. Cellulosic plastics are among the toughest of all plastics. There are five basic groups: nitrate, acetate, propionate, butyrate, and ethyl cellulose. Cellulose nitrate is the toughest of these but has the drawback of being very flammable. Celluloid is one product made from this material. Cellulose acetate is nonflammable, but like the nitrate, embrittles with age. The propionate and butyrate types do not have that problem and have better chemical resistance. The ethyl type possesses a very high shock resistance, so it is used for plastic products that may be handled and accidentally dropped, such as flashlight cases and toys. Cellulose plastics are also made into thin sheets and are sold under the name cellophane. Cellulosics tend to creep or cold flow when under load.

Fluoroplastics

The fluoroplastics family includes polytetrafluoroethylene (TFE), chlorotrifluoroethylene (CTFE), fluorinated ethylene propylene (FEP), and polyvinylidene fluoride (PVF_2). TFE has the greatest chemical resistance of all plastics. However, TFE is difficult to mold. FEP is more easily molded but still possesses the properties of TFE, yet it is more expensive than the other types to produce. CTFE is similar to the other fluoroplastics but has a much lower permeability to water and gases and is the least permeable of all plastics. PVF_2 has a greater stiffness than the other types and is used for rigid pipe and in load-bearing applications. Also it is the easiest type to extrude.

Fluoroplastics have the lowest coefficient of friction of any plastic and they have a very high chemical resistance; none of the common chemicals will react with them at all. In fact, adhesives for the most part will not adhere to them, although some adhesives were developed to bond fluoroplastics to metals. Fluoroplastics are useful for nonstick cooking surfaces, such as baking pans and frying pans that need not be greased. Paints made of these materials have exceptional resistance to weather and sunlight. Fluoroplastics are used for labware, valve seats, O rings,

and bearings. TFE fibers are sometimes used as fillers in acetal resins to improve thermal and frictional qualities.

Ionomers

Ionomers are plastics that have both ionic and covalent bonds. These clear plastics have high optical clarity, excellent toughness, and high chemical resistance, and are easily formed. Films are hard to tear, but once a tear begins, propagation is easy.

Nylon

Nylon is the name given by DuPont to a family of polyamides. It was first used for the production of fibers for such products as webbing, parachutes, and stockings. The most widely used nylons are Type 6, Type 6/6, Type 6/10, Type 11, and Type 12. All nylons have the properties of toughness, fatigue resistance, low friction, and inertness to aromatic hydrocarbons. Nylons 6 and 6/6 possess some incredible qualities when combined with other materials such as glass fibers. They can replace metals in many applications, such as automobile speedometer gears (Figure 5.12), bicycle wheels, and carburetors (Figure 5.13). Nylon 6/10 is more flexible than nylons 6 or 6/6 and has less tendency to absorb moisture (which can alter

FIGURE 5.12
Heat-stabilized Zytel® (nylon) has been used for speedometer take-off, driven, and governor gears since 1954. In this difficult environment they have performed as reliably as the metals they have replaced, and they weigh much less and are much less expensive (Engineering Polymers Division, DuPont Company).

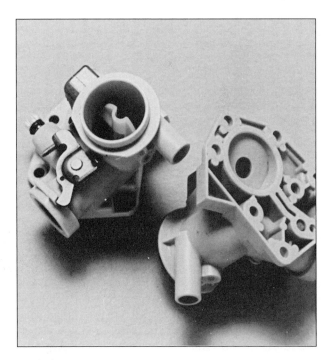

FIGURE 5.13
This lawnmower carburetor molded of Minlon® 10B (a reinforced nylon) is a first among four-cycle engines (Engineering Polymers Division, DuPont Company).

the physical dimensions of a plastic part) and thus can be used in water-mixing valves. Nylons 11 and 12 are commonly used in film and tubing.

Polycarbonates

Polycarbonate resins can be used for many of the same products that can be made of nylon since they have similar properties. They are extremely strong and are very difficult to break. For this reason they are used for unbreakable bottles and as a window glass replacement where crime and vandalism are problems. The General Electric polycarbonate Lexan® is an excellent glazing product, which is made with a special surface coating that resists ultraviolet light which causes yellowing and impact degradation in polycarbonates. This nonyellowing coated product is used for architectural glazing and for solar energy systems. When polycarbonate sheets are laminated and fused with proprietary layers to a thickness of .75 in., they outperform bullet-resistant glass against medium-power weapons. Polycarbonate foams are other important products that are used for thermal insulation and sound dampening suitable for automobile and business machine use.

Polyester

These thermoplastic polyester molding compounds are listed chemically as polybutylene-terephthalate (PBT) and polytetra-methylene terephthalate (PTM). Compared with the thermosets, thermoplastic polyesters possess better molding characteristics and more rapid cycle times because of their rapid crystallization rate. Glass fiber is added to improve heat resistance and strength. Applications include automotive electrical parts, gears, housings, and pump impellers.

Polyimide

This engineering resin can exist both as a thermoset and as a thermoplastic, but unlike other thermoplastics it does not melt. Polyimides are difficult to injection mold, requiring high molding temperatures 660°F (349°C), a long cycle time, and high injection pressures. These resins can withstand 500°F (260°C) continuously and 900°F (482°C) for a short time without breaking down. Polyimide resists most chemicals, except for strong alkaline solutions and strong inorganic acids, and it has poor weathering resistance. Polyimides have almost no creep even at high temperatures and, when certain fillers such as glass or graphite fibers are used, the thermal expansion is close to that of metals. They are used for bearing sleeves, seals, and thrust washers.

Another engineering resin, polyamide-imide, can also resist high temperatures of 500°F (260°C) for long periods. Because of its toughness, high strength (30,000 PSI at room temperature), and heat resistance, it has many applications in the automotive industry and in making cookware and electrical equipment.

Polyolefins

The polyolefins include polyallomer, polypropylene, and polyethylene. Polyethylene is a very strong, flexible resin that has an unusually high flex life, which makes it an excellent material for "living hinges." Polyethylene is used for squeeze bottles, ice cube trays, film for boil-in packaging, and vending machine food containers. High-density polyethylene is used to make pressure pipe (Figure 5.14). Since it can be made either flexible or stiff, it is used for a great variety of housewares and toys. Polyethylene terephthalate (PET) is a more recently developed plastic resin that is being increasingly used for beverage bottles. Since beverage bottles usually contain no reinforcing fibers, the thermoplastic resin is highly recyclable and millions of pounds per year of PET bottles and scrap are recovered and reprocessed. A polycarbonate polyethylene terephthalate blend (PC/PET) was developed and is valued for its high impact strength at ambient and

FIGURE 5.14
Construction workers on street in New York compare section of 48-in. diameter high-density polyethylene pressure pipe, the largest size previously available, with a section of newly available record 63-in. diameter "MLSO-ppd/305" high-density polyethylene pressure pipe (M. L. Sheldon Plastics Corp., New York, N.Y.).

low temperatures as well as its resistance to chemical attack.

Polypropylene is used for electrical insulation because of its outstanding electrical properties. Poly-allomer possesses many of the properties of polyethylene and polypropylene, but it can be processed more easily.

Polysulfone

The polysulfone engineering resin has a very high constant use temperature of 400 to 500°F (204 to 260°C) and good impact strength. It is resistant to salt solutions, acids, alkalies, and most solvents and can be processed by extrusion, injection molding, and compression molding. Such resins can also be ultrasonically welded and machined. Because of their good electrical insulating properties, they are appropriate for use in printed circuit boards, switches, lamp housings, and electrical insulation.

Polyvinyl Chloride

There are two basic types of polyvinyl chloride (PVC), rigid and flexible. Rigid PVC is used in pipes and pipe fittings and other construction materials. Flexible PVC has replaced rubber in many applications, such as in making gaskets, vacuum tubing, garden hoses, shower curtains, floor mats, and toys. PVC can also be foamed and coated with other solutions, can be fabricated by most processes, and may be joined by heat sealing and solvent welding. Plastic products manufacture will be covered in Chapter 17.

ADDITIVES AND FINISHES

Various additives, pigments, and fillers are combined with polymer resins by mixing and kneading machinery. It is now possible to compound these materials continuously as a manufacturing process (Figure 5.15).

Additives

Since the first plastic materials were synthesized, the list of uses and applications for products has grown significantly. In many cases plastic products are replacing metal products and competing with other products made from natural materials such as wood and leather. This widespread and diversified use of plastics is partly possible because of additives that alter and strengthen the plastic resins to make them more easily processed and more useful. Some of the additives used are as follows.

FIGURE 5.15
Flow chart showing continuous compounding of engineering plastics (Buss-Condux, Inc.).

Antioxidants   These materials impart ultraviolet stability for such resins as ABS, polyethylene, polypropylene, and polystyrene, and they impart melt-flow retention, making them easier to mold.

Blowing agents   When plastics must be processed at low temperatures, as required in sponge plastic and rubber applications, these additives allow production speeds to be increased and create voids to give porosity or low density.

Colorants   Pigments and dyes for plastics give them their brilliant colors and are important for giving them sales appeal. Some pigments are designed to provide a solid, opaque color, whereas others tint the clear resin to impart a translucent coloration. Luminous colors are provided for some products, such as bicycle reflectors, to make them easily seen. A new luminous-green pigment was developed that is said to be nontoxic and safe for toys and glows in the dark for about 8 hours after being exposed to a light source.

Fillers   Filler materials can reduce the cost of plastic products by reducing resin usage. They also lower the part weight and still retain, or even increase, physical properties such as tensile and compressive strength. Common fillers are wood flour, quartz, limestone, clay, and metal powders.

Plasticizers   Some plastics such as vinyls are normally hard, brittle materials, but by adding a plasticizer they can be made soft and flexible.

Reinforcements   Resins tend to have rather low physical strengths, but by adding reinforcements their strengths can be markedly increased. Some reinforcing materials for plastics are glass fiber, mica, asbestos, jute, sisal, graphite, carbon, and ceramic. Containers such as large tanks are formed by winding filaments of various materials such as glass, boron, and graphite with a plastic binder such as epoxy over a form that rotates while the filament moves in circumferential, longitudinal, and helical patterns, or a combination of these to produce a very high-strength product.

Stabilizers   Some plastic resins such as styrene and vinyl are subject to degradation or breakdown in the presence of oxygen or heat. PVC compounds are stabilized by barium-cadmium-zinc compounds and phenols can be added to styrene to stop degradation.

Plastic Finishes

Almost every type of molded plastic part can be decorated to bring about wear, scratch, and chemical resistance, as well as for aesthetic appeal. The molding process can impart a decorative surface with raised or depressed designs in geometric patterns such as basket weave, pebble, or leather texture. Texturing allows the manufacturer to mask or hide flaws, flow-lines, sink marks, pinholes, and swirls, thus lowering the rate of part rejection. Other finishes are made with lacquers, enamels, and decorative overlays. Plastics can be metallized on the surface by the process of vacuum metalizing. This is done by vaporizing metal filaments onto the plastic surface and condensing them in a vacuum chamber. Gold, silver, nickel, chromium, and pure aluminum may be deposited by this method. Another widely used method of depositing metals on the surface of plastic pans is called sputtering (Figure 5.16). This method ionizes the metal in a vacuum chamber by using an electrical discharge between two electrodes. This forms a metallic cloud inside the chamber that fogs the plastic parts with a thin layer of metal. Sputtering techniques are widely used to plate copper and other metals on printed circuit boards (PCBs) for the electronics industry. Glass-epoxy laminates are usually used for PCBs, and the copper-plated circuits printed on them are easily soldered to the electronic components.

Engineering Properties of Plastics

When a plastic material must be selected from the many available types for the manufacture of a new product, the choice must be based on the various engineering proper-

FIGURE 5.16
Bright metallic surface on this plastic toy appears to have been applied by the process of sputtering or by condensation.

ties of the plastic. For example, if a gear were made of a brittle material such as acrylic with no reinforcing material, it would shatter when a load was applied, but using a high-performance plastic such as one of the acetals and nylons would make excellent gears for many purposes. However, if a less expensive, disposable plastic product were needed, it would be noncompetitive and wasteful to select an expensive engineering plastic, and a low-cost material such as polystyrene should be selected. Some products must be made highly resistant to impact; some of these are automobile windshields, doors, bumpers, telephones, and toys. The engineer should probably choose the plastic for these items from among the high-impact polymers such as the polycarbonates, polyesters, and the newer polyphenylene oxide-based polymers and polyetherimides. Plastic resin manufacturers make the properties of their plastics available to engineers and designers. See Table 5.2 for an example of a properties chart.

Standard tests are used to assure the engineer that the selected plastic material will meet the requirements of the product. These tests were set up by the American Society for Testing and Materials (**ASTM**). The many tests for plastics can be divided into several general areas. The most common are mechanical properties, thermal properties, electrical properties, physical properties, and special characteristics. Mechanical properties include tensile strength, hardness, compressive strength, impact strength, and flexural strength (Figures 5.17 and 5.18 on p. 104). Thermal properties include Underwriters Laboratories (UL) ratings,

TABLE 5.2
Properties chart of polyamide-imide[a]

	Dixon TL Series Polyamide-Imide			
	TL-01[b]	TL-02[c]	TL-03[d]	ASTM Test Method
I. Mechanical Properties				
Tensile strength (psi)	19,600	26,900	27,300	D-638
Elongation %	6	12	4.1	D-638
Flexural strength (psi)	26,400	30,700	40,500	D-790
Compressive strength (psi)	21,000	32,000	—	D-695
Impact strength (ft-lb/in.)	1.1	2.5	1.6	D-256
II. Thermal Properties				
Heat distortion temperature (° F at 264 psi)	525	525	550	D-648
UL temperature rating (° F)	410	410	410	
Coefficient of thermal expansion (linear—in./in./°F)	2.4×10^{-5}	2.0×10^{-5}	1.1×10^{-5}	D-696
Thermal conductivity (BTU/hr/ft²/° F/in.)	2.5	1.7	—	UL94
Flammability rating	V-O	V-O	V-O	
Limiting oxygen index %	42	42	42	
III. Electrical Properties				
Dielectric constant, 1 MHz	—	3.5 to 4.0	—	
Dissipation factor, 1 MHz	—	.001 to .009	—	D-150
Dielectric strength (V/mil)	—	600	—	
IV. Physical Properties				
Specific gravity	1.45	1.4	1.6	
Density (lb/in.³)	.05	.05	.06	D-792
Hardness	M109	M119	—	D-785
Water absorption %	.22	.28	.18	D-570
Radiation resistance	Resistant	Resistant	Resistant	
V. Friction and Wear Ratings				
Coefficient of friction	Good	Fair	—	
Wear life	Excellent	Fair	—	
VI. Chemical Resistance	Excellent	Excellent	Excellent	

[a]All statements, information, and data given herein are believed to be accurate and reliable but are presented without guaranty, warranty, or responsibility of any kind, expressed or implied.
[b]Dixon TL-01 is a self-lubricating grade with excellent continuous use temperature.
[c]Dixon TL-02 is an unfilled material for structural applications.
[d]Dixon TL-03 is a fiberglass-filled structural material with high stiffness and low thermal expansion.
(*Source: Dixon Data: Flow Control Products Unit,* Dixon Industries Corporation, Bristol, Rhode Island, © 1984.)

heat distortion, coefficient of thermal expansion, thermal conductivity, and flammability rating. Electrical properties are dielectric constant, dielectric strength, and dissipation factor. Physical properties are specific gravity, density, hardness, water absorption, and radiation resistance. Special characteristics include weatherability, optical clarity, chemical resistance, and colorability.

Plastics designers selecting plastics for food packaging must consider the government regulations set forth by the National Institute for Occupational Safety and Health (NIOSH). The institute also makes recommendations concerning toxicity in plastics manufacture to the Occupational Safety and Health Administration (OSHA) regulatory agency.

FIGURE 5.17
Izod-Charpy impact tester used for determining impact strength of plastics, ceramics, and light metals. This device uses a sensing system with a rotary transducer and a digital display to replace the analog dial. Its range is between 0 and 1 and 0 and 45 ft lb (Testing Machines Inc.).

FIGURE 5.18
Slip and friction tester. This device can make several tests, such as tensile tests, measurements of the coefficient of plastic surfaces, and peel tests for plastic laminates (Testing Machines Inc.).

ELASTOMERS AND ADHESIVES

Elastomers are materials, synthetic or natural, possessing rubbery qualities such as high resilience, extensibility, and elastic properties. The term rubber was originally used only for the natural product that is obtained from a thick, milky fluid (latex) that oozes from certain plants when they are cut. Most latex comes from the Pará rubber tree *(Hevea brasiliensis)* that grows in South America, Southeast Asia, and Sri Lanka. Natural rubber in the form of cured latex is a sticky, gummy substance that has a limited usefulness for rubber products.

In 1839 Charles Goodyear discovered that this latex could be greatly strengthened, and its elastic properties were improved when he added a small amount of sulfur to it and heated the mixture. The process is called vulcanizing. By varying the sulfur content and adding certain pigments such as carbon black, fillers, and softeners, rubber can be made to any hardness from very soft to very hard, such as ebonite. Latex is compounded with additives and vulcanizers in a mixing machine to form a homogeneous mass. The mix is then placed in a mill with chilled rolls (to prevent premature vulcanization). The rubber compounds are made into sheet form on calender rolls (Figure 5.19). The manufacture and vulcanizing of rubber products is covered in more detail in Chapter 17. When rubber products such as in automobile tires need to be strengthened, textile cords or fabrics are used as reinforcing materials.

Artificial Elastomers

Synthetic rubbers were first created in the early 1930s and were subsequently developed because of the uncertainty of the supply of natural rubber. This first rubbery synthetic was derived from acetylene gas and was a long chain molecule called polychloroprene, better known as neoprene. Natural rubber is somewhat more resilient than neoprene, but it tends to deteriorate in sunlight (ultraviolet) and swells when in contact with petroleum oil; neoprene has better resistance to these factors. However, chemical additives such as antioxidants, plasticizers, and stabilizers are now added to natural rubber and other elastomers to overcome these difficulties. One synthetic rubber, polyisoprene, seems to have almost the same properties as natural rubber, and some others such as the rubbery plastics have certain specific advantages over natural rubber.

Adhesives and Sealants

Adhesives bond two surfaces together as a kind of fastener. Sealants are materials that adhere to a crack or joint for the purpose of making a fluid or airtight seal. Sealants

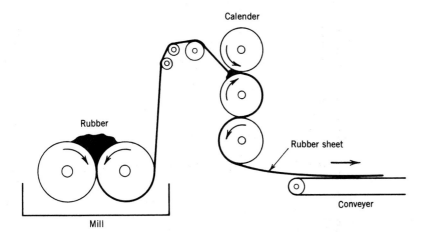

FIGURE 5.19
Rubber calender rolls.

are very similar to adhesives in that they must adhere to the sealing surface even though they usually do not have the high bonding strength of adhesives. Some types of sealants can remain soft and flexible for years, maintaining a good seal.

Prior to the development of synthetic adhesives, bonding agents were glues made from the hides and hooves of animals or parts of fish. Pastes were made of tree gums, starch, or casein. All of these products were easily loosened by the action of moisture. Since there is little molecular attraction between these materials and the parts to be bonded, they are not very strong adhesives. Now, almost any material can be bonded to another if the right adhesive is used. Most adhesives used today are plastics, either thermoplastic or thermoset.

Because of their flexibility, structures bonded with adhesives resist damage from vibration better than riveted or bolted assemblies. A riveted joint must resist high stress at each rivet hole and cracking can result, but adhesive bonding is made over the entire surface, reducing stress concentration. Also connectors, such as rivets or bolts, can be a cause of corrosion, allowing galvanic currents to flow, which allows corrosion to take place. In contrast, adhesives prevent galvanic current flow because they normally have a high electrical resistance.

Nitrile rubber epoxy resins are used for bonding honeycomb structures for aircraft (Figure 5.20). Two-part epoxy kits are available for general use in which a resin and a catalyst are measured out and mixed together. When mixed, this cement hardens in minutes or hours, depending on the formation. Epoxy resins with fiberglass cloth are used on the hulls of ships and boats to give them a strong leakproof surface. Automobile body repairs are often made with fiberglass and resin.

The cellulosic plastics are used in model cement and some household cement. Polyvinyl acetate is a white glue used in woodworking. Polyethylene is often used as a hot-melt adhesive. Polysulfides, polyurethanes, and silicones are used as sealants for caulking of buildings and boat hulls. Phenolic resins have been used for years as a bonding material for abrasive grinding wheels. Most adhesives will harden in the presence of air (aerobic conditions), but some types such as the cyanoacrylates will harden only when they are not exposed to oxygen in the atmosphere (anaerobic conditions) and only when spread out as a thin film between two surfaces. Cyanoacrylates, popularly known as super glue, will instantly bond smooth flat surfaces with a high bonding strength but will not bond uneven or porous surfaces. Table 5.3 lists some of the kinds of adhesives and their uses.

As with other plastics and rubbers, additives are used in adhesives to impart needed properties such as resilience and toughness. When bonding agents are required to conduct electricity, conducting additives such as carbon black, graphite, or metal flakes or powders are used. Certain destructive tests are used to determine joint strength and efficiency for various adhesives and bonding materials. Among these are shear, tensile, compression, impact, and peel tests. Cleavage (a variation of the peel

FIGURE 5.20
Bonded honeycomb aluminum structures for aircraft wing parts.

TABLE 5.3
Adhesives and their uses

Name	Advantages	Disadvantages	Uses
Anaerobic cyanoacrylates (popularly called super glue)	High strengths (2000 psi) in a few minutes	High cost, thicknesses greater than .002 in. cure slowly or not at all, does not cure in presence of air	Production speed in bonding small parts, especially metal to metal and metal to plastics
Acrylic	Clear with good optical properties, ultraviolet stability	Moisture sensitive, high cost	Bonding plastics
Cellulose esters and ethers	Very soluble	Moisture sensitive	Bonding of organic substances such as paper, wood, and some fabrics
Epoxies: modified, nitrile, nylon, phenolic, polyamide, vinyl	Low-temperature curing, adheres by contact pressure, good optical qualities for transmitting light, high strength, long shelf life	High cost, low peel strength in some types	Bonding of microelectron parts, aircraft structural parts, cutting tools, and abrasives
Phenolics (Bakelite)	Heat resistance, low cost	Hard, brittle	Abrasive wheels, electrical parts
Polyimides	Resistance to high temperature (600° F or 315.5° C)	Cost, high cure temperature (500 to 700° F or 260 to 371° C)	Aircraft honeycomb sandwich assemblies
Polyester (anaerobic)	Hardens without the presence of air	Limited adhesion, cost	Locking threaded joints
Polysulfide	Bonds well to various surfaces found in construction	Low strength	Caulking sealant

test), tack (adhesion), creep, and fatigue tests are also very common (Figure 5.21). There are several specifications applicable to adhesive bonding: two of them are the Military, Federal Adhesive Specifications and those of the American Society for Testing and Materials (ASTM).

PETROLEUM PRODUCTS (LUBRICANTS AND ASPHALTS)

In 1859 Colonel E. L. Drake struck oil in the little Pennsylvania town of Titusville. Its early impact on the economy was to replace whale oil as fuel for lamps. This,

no doubt, helped to save the whales for posterity but, as everyone today knows, petroleum now has many more uses; it fuels our industries and our transportation systems and provides the raw materials for many of the products we use. Without the lubricants, oils, and greases, all of our industrial machinery would stop running and transportation would come to a halt.

As it comes from the oil wells, petroleum crude is composed of a large variety of hydrocarbon molecules in many chemical combinations. By the process of distillation and condensation, these constituents are separated into several classes. These are gases, gasoline, distillate fuel, and lubricating oil, with the residual heavy end products being used in the production of heavy lubricating-oil

Name	Advantages	Disadvantages	Uses
Polyurethane (rubbery adhesive)	Strong at very low temperatures, good abrasion resistance	Low resistance to high temperatures, sensitive to moisture	Flexible and rigid foam protective coatings, vibration damping mounts, coatings for metal rollers and wheels
Rubber adhesives: butadienes, butyl, natural (reclaim), neoprene, nitrile	High-strength bonds with neoprene (4000 psi), oil resistance (nitrile)	Low tensile and shear strength, poor solvent resistance	Adhesives for brake linings, structural metal applications, pressure sensitive tapes, household adhesives
Silicones	High resistance to moisture, can resist some temperature extremes	Low strength	Castable rubbers, vulcanized rubbers, release agents, water repellents, adhesion promoters, varnishes and solvent type adhesives, contact cements
Urea-formaldehyde resins	Low cost, cures at room temperatures	Sensitive to moisture	Wood glues for furniture and plywood
Vinyl: polyvinyl-butyralphenolic vinyl plastisols, polyvinyl acetate and acetals, polyvinyl chloride	High room temperature peel, ability to bond to oily steel surfaces (vinyl plastisols)	Sensitive to moisture	Bonds to steel (brake linings), safety glass laminate, household glue, water and drain pipe (PVC), household wrap

(*Source:* John E. Neely, *Practical Metallurgy and Materials of Industry,* 2d ed., Wiley, New York, © 1984.)

stocks, bunker fuels, and asphalt. Chemically, the carbon and hydrogen molecules in petroleum exist in many combinations, giving various properties to oils from different sources of crude. In addition, all petroleum oils contain small amounts of sulfur and nitrogen, which affect the stability and corrosiveness of the oil. The hydrocarbon molecules found in petroleum can be grouped into three classes: paraffins, naphthalenes, and aromatics. Of these, the paraffins and naphthalenes have the greatest stability and are the principal constituents of lubricating oils that are used in automobile crankcases and in other machinery. The aromatic oils have good solvency characteristics and are used for certain chemical formulations and for ink oils.

Greases

Greases are actually lubricating oils held by absorption and capillary action in the fibrous or granular structure of a thickening agent (clay or lamp black). The oil used for greases that are intended to withstand high pressures at relatively slow speeds is usually one of higher viscosity than one intended for normal or higher speeds. The most important quality of greases is their adhesiveness to metallic surfaces and their "stay-put" properly. Greases are extensively used to pack wheel bearings and steering linkages in automobiles and for lubricating manufacturing machinery.

FIGURE 5.21
The glueability tester tests the strength of a bond formed between adhesive and material being tested under specific, carefully controlled circumstances (Testing Machines Inc.).

Asphalts

Asphalt is a black, sticky substance that is also separated from crude petroleum. However, some asphalts occur naturally in pits or lakes that are residues of crude oil in which the lighter fractions (more volatile parts) have evaporated over a period of many thousands of years. Asphalts can generally be divided into two categories, paving asphalts (asphalt cements) and liquid asphalt materials. Thermoplastic asphalt cements are solid but can be heated to a liquid condition for use as roofing tar and for paving. Liquid asphalts are not heated but are diluted with solvents to provide the proper consistency. The liquid asphalts harden over a period of time as the lighter fractions evaporate, but the molten (heated) asphalt hardens when it cools.

Asphalts are used in the manufacture of waterproof papers and textiles and for waterproof linings for irriga-tion canals and reservoirs. However, plastics are begin-ning to replace asphalt products in many of these applications. Asphaltic concrete is used extensively for road and highway construction and for parking lots and driveways.

CERAMIC MATERIALS

Ceramic materials include clay, silica, glass, natural stone, and portland cement (with which concrete is made). Ceramics are held together by ionic and covalent bonds, which are very strong but rigid and much less ductile than the metallic bond, making them subject to brittle failure. Because they are good thermal and electrical insulators and can withstand high temperatures, ceramic materials are extensively used in the electrical industry and for furnace linings.

Clays

Clays vary considerably according to where they are mined. Ordinary clays containing impurities such as iron oxide (which gives them their reddish color) are used to manufacture brick for building construction and firebrick for furnaces. Kaolin, a purer white clay composed mostly of alumina and silica, is used in the manufacture of earthenware, fine china, porcelain, paper products, and firebrick. Fire clay has less than 10 percent impurities and is used for the manufacture of clay products and for joining firebrick in furnace construction.

Porcelain enamels are composed of quartz, feldspar, borax, soot ash, and other substances. These are used to coat iron, steel, or aluminum for wear and decorative purposes for products such as stoves, refrigerators, pots, and pans.

Some recently developed kinds of ceramic materials are known as fine ceramics or high technology ceramics. These include insulation materials for electronic substrates, transition-metal oxides (semiconductors), heat-resistant materials for engines, cutting tools, and corrosion-resistant materials. Fine ceramics, along with plastics technology, automation, and robotics appear to be among the rapidly growing areas of high technology.

Glass

Glass is a supercooled rigid liquid. It is in a metastable state; that is, it can go to a stable lower energy state of crystallization only by passing through an intermediate higher energy state. In effect, glass has become too cold to freeze from the liquid. Common types of glasses are based on silicon dioxide (SiO_2) which occurs abundantly in nature as quartz and cristobalite and as a part of many of the silicate minerals. Natural volcanic glass, called obsidian, was used by primitive peoples for arrowheads and spear points. Quartz and silica sand are common raw materials for the production of commercial glasses and silica. Additives can alter the glass, giving it special properties. For example, the common soda-lime glass has a lowered softening temperature and is used for windows, containers, and lamp bulbs. Borosilicate glass, with a low thermal expansion, does not tend to crack when unevenly heated; consequently, it is used for laboratory glassware, large telescope mirrors, and household cookware such as Pyrex®. Lead-alkali-silicate glass has an even lower softening point and is known as flint glass for optical purposes and crystal glasses for tableware. For a high refractive index, the lead oxide (PbO) content may be as much as 65 percent.

Glass has excellent colorability and is a versatile material that has excellent chemical resistance, with the exception of hydrofluoric acid, which is used to etch glass objects. A considerable amount of glass plate is used in modern building construction and glass fiber is used extensively as a reinforcement material in plastic resins and in the communications industry.

Portland Cement

The invention of portland cement is generally credited to Joseph Aspdin, an English mason who obtained a patent for it in 1824 and named his product portland cement because the concrete made of it resembled a natural limestone on the Isle of Portland. Materials used in the manufacture of portland cement are lime, silica, alumina, and iron. Each of these can come from any of a number of source materials. For example, alumina comes from clay, shale, slag, aluminum ore refuse, and a number of other source materials.

Selected raw materials are pulverized and proportioned to produce the desired chemical composition. The prepared mix is fed into the upper end of a sloping rotating kiln. A powdered coal, fuel oil, or gas is burned at the lower end of the kiln, producing temperatures of 2600 to 3000°F (1427 to 1649°C). The resulting portland cement clinker is cooled and then pulverized and a small amount of gypsum is added to regulate the setting time of the cement. This finely ground product, which will pass through a sieve with 40,000 openings in a square inch, is the finished portland cement, ready to be shipped. Typical steps in the manufacture of portland cement are illustrated in the flowchart (Figure 5.22). Each manufacturer of portland cement uses a trade or brand name under which the product is sold.

Concrete

Concrete is commonly seen in sidewalks, streets, dams, and building foundations. It is a mixture of portland cement, sand, and gravel plus water and, sometimes, additives. These ingredients (sand and gravel are called aggregates) are mixed together in specific proportions to produce a fluid mass called concrete. It is then poured or pumped into molds where it is left to harden. The forms or molds may be removed in a few days. Ordinary concrete takes about 28 days to completely cure, providing it is kept wet, at which time it is nearly as hard and strong as it will ever be.

Specialty Concretes

Among the many admixtures used to improve portland cement concrete is flyash, a finely divided residue resulting from the combustion of coal. It has a pozzolanic effect in that it acts like a cement in the presence of lime. Its main value lies in its ability to reduce the percentage of portland cement needed in concrete, thus reducing the cost. It also imparts a higher chemical resistance than ordinary concrete. Like flyash, blast furnace slag can also reduce the percentage of cement needed in concrete; however, when granulated and activated by sodium hydroxide, it can completely replace portland cement in some cases. Silica fume, a powdery by-product of electric arc furnace production of ferrosilicon alloys and metallic silicon, also behaves like natural pozzolan and it gains strength earlier in the curing cycle than flyash and slag concretes do.

Polymer Concretes   A growing technology involves the introduction of polymers into concrete. Latex modified portland cement concretes have high strengths, reduced porosity, and water resistance. Polymer-cement concrete has better abrasion, higher strength, and impact resistance than conventional concrete, making it useful for industrial floor applications. Polymer-portland-cement concrete is mixed with either a monomer or a polymer in a liquid, powdery, or dispersed phase that is allowed to cure along with the cement or is polymerized in place.

Polymer concrete is a composite material in which an aggregate mixture is combined with a monomer that is polymerized in place; no portland cement is used. Monomers used in polymer concrete include polyester-styrene resins, furan, vinyl esters, and epoxy. Also sulfur-concrete and sulfur-impregnated concrete are included in a broad definition of polymer concretes. Polymer concrete materials are excellent for patching because of their rapid curing time (15 to 60 min). A major disadvantage is that they cannot withstand higher temperatures.

FIGURE 5.22
Flow chart showing the steps in the manufacture of portland cement by the dry process using a preheater. (1) Quarry and blending of raw materials. (2) Proportioning and fine grinding of raw materials. (3) Kiln system. Preheating, burning, cooling, and clinker storage. (4) Finish grinding and shipping (Oregon Portland Cement Company).

Fiber Reinforced Concrete   Steel reinforcing bars have long been in use in portland cement concrete. Concrete mixtures using fibers act as secondary reinforcement in concrete and can greatly increase its tensile strength. Small fibers can prevent crack growth in concrete structures. Typical fibers include asbestos, cotton, rayon, glass, steel, acrylic, nylon, polyester, polyethylene, and polypropylene.

WOOD PRODUCTS

Forest industries, in which wood products such as lumber, plywood, masonite, and particleboard are manufactured, supply a large portion of the building materials for homes and offices and for furniture manufacture. Paper products produced from wood pulp are extensively used in packag-

ing and in printing. Wood is composed of cellulose, lignin, and small amounts of inorganic materials that make up the ash when wood is burned.

The engineering properties of wood vary a great deal in different species of trees and even within one species because of imperfections such as knots and checks. Moisture markedly decreases the mechanical properties of wood. Because of this and the shrinkage of "green" wood when it dries, construction-grade lumber and other wood products are either air or kiln dried before they are finished (planed or sanded) to size. Wood may be classified into two broad categories, hardwood and softwood.

Hardwoods

Deciduous trees shed their leaves each year and are classed as hardwoods. The definition generally refers to the difficulty of sawing rather than its hardness as the term is used in engineering. In fact, balsa, although it is very lightweight, soft, and porous, is classed as a hardwood because it comes from a deciduous tree.

Oak trees are found in many varieties in the United States, but they generally can be divided into two classes, white oak and red oak. White oak is more resistant to decay and finer grained than red oak, and it is more difficult to work. Red oak is used extensively in furniture manufacture, flooring, and plywood. Oaks, sugar and black maples, and birch grow in the northern forests from Minnesota east and south to Virginia. Sweet gum, ash, and tupelo are found in Texas and some other southern states. Hickory, oak, black walnut, and basswood grow in the prairie states. Oregon maples are not as hard or strong as sugar and black maples. Walnut is found in most eastern and northern states and in some states in the far West.

Softwoods

Coniferous trees have cones and needles instead of flat leaves and are sometimes called evergreen trees because they do not lose their needles in the fall but keep them throughout the year. The coniferous trees are classed as softwoods even though some varieties have a very tough, hard wood. For example, the Sitka spruce was once used for structural frames in aircraft because it was considered to have the highest strength-to-weight ratio of any wood. Because of its resonant qualities, it is also used for musical instruments.

The pines are the most widely used softwoods. Western white pine, sugar, ponderosa, and lodgepole pine are found in the western and northwestern parts of the United States. Southern yellow pine is grown in the southeastern part of the United States.

Douglas firs and white fir are grown in the far West and are widely used in building construction. Firs are somewhat harder and stronger than pines. Spruce, hemlock, redwood, and cedar are generally found in the western United States and are used in building construction, boats, mine timbers, and various wood products.

Review Questions

1. From what raw materials are most plastics derived?

2. What is the source of natural rubber?

3. What polymer derived from acetylene gas is called polychloroprene (neoprene)?

4. From what basic source of raw material do we get most of our pesticides, fertilizers, clothing fibers, and many medicines such as antibiotics?

5. What is the major difference between thermosetting and thermoplastic materials?

6. Which type of material, thermosetting or thermoplastic, can be easily injected by pressure into a mold?

7. If you need a plastic material for a product that must resist a variety of chemicals and the effects of sunlight, have a nonstick surface, and withstand high temperatures such as would be found in a baking oven, what would you choose: (a) silicone plastic, (b) polyurethane plastic, (c) Bakelite, or (d) fluorocarbon plastic?

8. Name two items in which glass-reinforced nylon 6/6 can replace metal where considerable loading is involved.

9. Some plastics are colored or tinted with pigments whereas others have external surface finishes made with lacquers, enamels, or overlays. How can a metal film be applied to a plastic product?

10. In what process is natural latex combined with sulfur and heated to produce the flexible substance we call rubber?

11. Modern aircraft are made with fewer rivets and more adhesive-bonded structures than in the past. Name two advantages in using bonded structures over riveted ones.

12. Of the three classes of hydrocarbon oils, the aromatics are used for solvents and ink oils. What is the principal use of the other two types, paraffins and naphthalenes?

13. The tarry substance used for roofing and highway paving is a petroleum derivative. What is it called?

14. The use of clay for bricks, porcelains, and electrical insulation is well known. New uses for special ceramics, called fine ceramics or high-technology ceramics, are being developed Name two applications for these special ceramic materials.

15. Traditional construction materials include brick, stone, metal, and glass. Also plastic materials are being used for many products in the construction industry. Besides these, what are the two major material types used in construction?

Case Problems

Case 1: Plastic Color Problem

A small plastic product made of Bakelite is black and has had a good sales record for years. However, sales have begun to drop off significantly. The competition is using bright colors in their product and they are taking over the market. In what way can the first company improve its product, add color, and increase productivity to lower the price?

Case 2: Air Cylinder Stick-Slip Problem

A small air cylinder was manufactured to move a machine part. The piston seal was made of neoprene rubber, which had a tendency to take a set or stick to the cylinder wall when it was not operated for a given time. This caused the mechanism to stick-slip, that is, to hesitate and then jump forward too fast. Rubber O rings, cups, and other types of rubber seals did not solve the stick-slip problem, which was obviously a seal problem. From what synthetic material would you choose to make this piston seal in order to eliminate this sticking problem?

CHAPTER **6**

Selection and Application of Materials

Objectives

This chapter will enable you to:

1. Identify many metals and alloys by their numerical classification systems.
2. Describe some of the characteristics of many commonly used metals and alloys.
3. Name some methods used to identify and analyze various metal alloys and materials.
4. State some of the methods and factors involved in selecting materials for manufacturing purposes.

When only carbon steel, cast iron, wrought iron, and a few nonferrous metals were commonly used, identifying them was a relatively simple process. Spark testing on a grinder was enough to separate the three ferrous metals and, for the most part, the nonferrous metals could be identified by their color. Of course, nonferrous metals such as gold, silver, and copper began to be alloyed (combined) as early as the time of Archimedes. As the story goes, after he made the discovery of the natural law of buoyancy, he immediately applied it to solving the riddle of whether a golden crown contained pure gold or was cheapened with an alloying element having lesser value.

In our modern world there are literally thousands of alloys, both ferrous and nonferrous. Selecting and identifying metals is therefore a far more complex process today, requiring classification systems and sophisticated testing equipment. Even more important than metals identification is the need for methods and equipment to test and analyze other materials of industry, especially those that present radiation hazards and contain toxic materials.

With so many new engineering materials being developed in the areas of plastics, composites, metals, and ceramics, it can be difficult to choose manufacturing materials. Although a new product might have more sales appeal when made of plastic, it might not have enough strength or heat resistance. Other products might be too costly to fabricate from metal but could be made from a selected plastic at a high production rate by the injection molding process and still have satisfactory engineering properties. This chapter covers some of these factors in material selection.

CLASSIFICATION SYSTEMS FOR METALS

The most common numerical system used to classify carbon and low alloy steels in the United States was developed by the Society of Automotive Engineers (**SAE**) and the American Iron and Steel Institute (**AISI**) for steels used in manufacturing. The American Society for Testing Materials (**ASTM**) developed specifications for carbon and alloy structural steels. Tool steels are classified under their own system, which covers seven major types. Stainless steels are classified under the chromium and chromium-nickel types and cast irons are identified by ASTM numbers. Nonferrous metals are also classified under specific systems for each metal or alloy. Color coding of steels has long been used as a means of identification, but there is no universal code; each manufacturer, distributor, or local manufacturing plant has its own code. It is a very useful system on a local basis as long as the color coding is not cut off, leaving the remainder of the stock unbranded.

The SAE classification system for alloy steels uses a four- or five-digit number (Table 6.1). The first number on the left designates the type of steel. For example, plain carbon steel is denoted by the number 1, 2 is a nickel steel, 3 is a nickel-chromium steel, and so forth. The second digit indicates the approximate percentage of the predominant alloying element. The third and fourth digits, represented by X, always represent the amount of carbon in hundredths of a percent. For plain carbon steel it is normally between .08 and 1.70 percent. SAE1040 is plain carbon steel containing .40 percent carbon. SAE4140 is a chromium-molybdenum alloy steel con-

taining .40 percent carbon and about 1 percent of the major alloy, molybdenum. The AISI system is basically the same as the SAE system with the addition of certain capital letter prefixes. The two systems are usually combined. The AISI prefixes are:

B —Acid Bessemer

C —Basic open-hearth carbon steel

CB—Either acid Bessemer or basic open-hearth carbon steel at the option of the manufacturer

D —Acid open-hearth carbon steel

E —Electric furnace alloy steel

Tool Steels

Tool steels are mostly high carbon or high alloy in carbon steel. However, a few types do have low carbon or alloy to give them special properties. As the name implies, these special steels were designed for tools, dies, molds, machinery parts, and for certain specific applications. They all require heat treatments such as hardening and temper drawing to prepare them for use. Symbols of one or more than one letter have been assigned to each of the seven major groups of tool steels. Each group is designated by its special purpose or by the quenching medium used (Table 6.2). The classification is as follows:

1. Water-hardening tool steels

   W—high carbon steels

2. Shock-resisting tool steels

   S—medium carbon, low alloy

TABLE 6.1
SAE-AISI numerical designation of alloy steels*

Carbon steels	
Plain carbon	10xx
Free-cutting, resulfurized	11xx
Manganese steels	13xx
Nickel steels	
.50% nickel	20xx
1.50% nickel	21xx
3.50% nickel	23xx
5.00% nickel	25xx
Nickel–chromium steels	
1.25% nickel, .65% chromium	31xx
1.75% nickel, 1.00% chromium	32xx
3.50% nickel, 1.57% chromium	33xx
3.00% nickel, .80% chromium	34xx
Corrosion and heat-resisting steels	303xx
Molybdenum steels	
Chromium	41xx
Chromium-nickel	43xx
Nickel	46xx and 48xx
Chromium steels	
Low-chromium	50xx
Medium-chromium	51xx
High-chromium	52xx
Chromium-vanadium steels	6xxx
Tungsten steels	7xxx
Triple alloy steels	8xxx
Silicon–manganese steels	9xxx
Leaded steels	11Lxx (example)

(*Source:* John E. Neely, *Practical Metallurgy and Materials of Industry,* 2d ed., Wiley, New York, © 1984.)
* x represents percent of carbon in hundredths.

TABLE 6.2
Types of commonly used tool steels

Type of Steel	Examples
Water hardening: Straight carbon tool steel	W1, W2, W4
Manganese, chromium, tungsten: Oil-hardening tool steel	O1, O2, O6
Chromium (5.0%): Air-hardening die steel	A2, A5, A10
Silicon, manganese, molybdenum: Punch steel	S1, S5
High-speed tool steel	M2, M3, M30 T1, T5, T15

(*Source:* John E. Neely, *Practical Metallurgy and Materials of Industry,* 2d ed., Wiley, New York, © 1984.)

3. Cold-work tool steels

    O—oil-hardening types

    A—medium alloy air-hardening types

    D—high carbon, high chromium types

STRUCTURAL STEELS

Standard ASTM specifications are used to designate carbon and alloy structural steels. These steels are produced in steel mills as standard shapes (Figure 6.1) and are used in the construction of buildings, bridges, and pressure vessels, and for other structural purposes. Engineers who design steel structures and welders are more likely to use the ASTM specifications for steels than designers of machinery, heat treaters, and machinists who would probably need the SAE-AISI system to refer to the steels they would be using. These products are usually in the form of bar stock, strip, plate, and round bars and are used extensively in manufacturing steel products. Some bar stock is alloy steel such as AISI 4140, which is used to produce tough heat-treated parts such as gears and machine parts. AISI 8620 is a carburizing grade used where an extremely tough material with a high surface hardness is required. However, much of the steel used in manufacturing is of mild steel (low carbon) AISI 1020. Low carbon steel comes from the mill with either a cold finish or hot rolled surface. Cold finish (CF) is simply a broad term for cold rolled (CR) or cold drawn (CD). These cold worked steels are tougher and stronger than are hot rolled (HR) steels that have the same carbon content.

Cold-drawn steels have a relatively smooth finish without any scale. Hot-rolled steels are covered with a black mill scale that is formed when the steel is at rolling or forging heat by the action of oxygen. This scale, which is chemically Fe_3O_4, must be removed before the steel can be subjected to cold-finishing operations. Cold-finished steels are more expensive than hot-rolled steel, so they are not normally used when less accurate dimensions are acceptable and where heating or welding is to be done. Steel shafting is available as low carbon or resulfurized cold-drawn steel and as alloy steel ground and polished (G and P) which has a brighter surface finish than cold-drawn steel and is held to closer dimensional tolerances. It is also a much higher strength steel and is much more expensive than cold-drawn, low carbon steel. Cold-finished steel strip is produced in rolling mills for use in punch press work for blanking and drawing operations. It is also provided in a variety of surface finishes and colors that are not damaged by the manufacturing process. Tool steels and alloy steels are usually referred to by a trade name, but they usually are also identified by number.

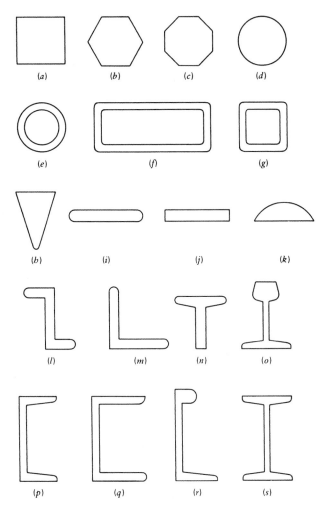

FIGURE 6.1

Steel shapes used in manufacturing. *(a)* Square HR or CR. *(b)* Hexagonal. *(c)* Octagon. *(d)* Round. *(e)* Tubing and pipe (round). *(F)* HREW (hot rolled electric welded) rectangular steel tubing. *(g)* HREW square steel tubing. *(h)* Wedge. *(i)* HR flat bar (round edge spring steel flats). *(j)* Flat bar (CR and HR). *(k)* Half round. *(l)* Zee. *(m)* Angle. *(n)* Tee. *(o)* Rail. *(p)* Channel. *(q)* Car and ship channel. *(r)* Bulb angle. *(s)* Beams—I, H, and wide flange *(Neely, Metallurgy, 2d ed., © 1984 John Wiley & Sons, Inc.).*

HSLA Steels

Low carbon steels such as the ASTM-A36 have been in use since the 1920s, but a relatively new series of construction steels called high strength/low alloy (HSLA) have greatly increased in importance. They were originally developed primarily for automotive weight reduction because a thinner sheet, plate, or structural shape or a smaller diameter bar will almost always do the work of a thicker, heavier, plain carbon steel. These steels are appropriate for many other applications where high strength, formability, and weldability are needed. Some of these HSLA steels in plate, bar, and structural shapes include A709 and A737 for bridges; A441, A572, and A633 for low-temperature use; and A242 and A588 for weather-resistant construction steels. Full ASTM specifications may be found in materials and welding data handbooks.

STAINLESS STEELS

Stainless steels, which are alloys, are resistant to the corrosive action of the atmosphere and most reagents because of the relatively large proportion of chromium present in them. Chromium in quantities of less than 11 to 12 percent in steel will inhibit but not eliminate corrosion, but in those containing more than 12 percent corrosion is completely eliminated under normal circumstances. However, many grades of stainless steel have a limited resistance to chlorine and sulfur ions. Therefore, they should not be used in the presence of seawater and hydrochloric or sulfuric acids. Some stainless steels were developed to resist these corrosives as well as most others.

There are two major series of stainless steel, the chromium (400 series) and the chromium-nickel (300 series). In the 400 series there are two types, the hardenable (by quenching and tempering) and the nonhardenable

(Table 6.3). The 300 series includes the austenitic types that contain 8 percent or more nickel, which gives it higher resistance to corrosion and high temperatures, but this series cannot be hardened by quenching. Since it has such high ductility, it can be work hardened until it is almost as hard as the martensitic type. Ferritic stainless steels in the 400 series contain almost no carbon and are actually a chromium-iron alloy. It is the least expensive of the stainless steels and is used for pots and pans and building trim. The martensitic types can be hardened by heating and quenching in oil or air and are then tempered. Included in the chromium-nickel types are the precipitation hardening stainless steels that harden over a period of time after solution heat treatment. Stainless steels are more difficult to machine than mild steels and they tend to work harden quickly when cut with dull tools. With proper tooling and cutting fluids, however, they lend themselves to good production rates, especially the free machining types (Table 6.4).

CAST IRONS

Cast iron is essentially an alloy of iron, carbon, and silicon. Theoretically, it can contain from 2 to 6.67 percent carbon, but commercial cast iron rarely has more than 4 percent carbon. Since austenite can contain only 2 percent carbon in solution, any more than that amount will pre-

(cont. p. 122)

TABLE 6.3
Classification of stainless steel

Alloy Content		Numerical Designation[a]
Chromium types	Martensitic, hardenable	Types 403, 405, 410, 414, 416[b], 420, 420F, 431, 440A, 440B, 440C
	Ferritic, nonhardenable	Types 125R[b], 185R[b], 400[b], 409[b], 430, 430F, 430TL, 442, 446
Chromium-nickel types	Austenitic, nonhardenable except coldwork	Types 301, 302, 302B, 303[b], 303Se, 304, 304L, 308, 309, 310, 311 DQ[b], 316, 316L, 317, 317L, 321, 347, 348, 18-9 LW[b]
	Semiaustenitic, precipitation hardening	17-7 PH[b], PH 15-7 Mo, PH 14-8 Mo[b]
	Martensitic, precipitation hardening	17-4 PH[b], 15-5 PH VAC CE[b], PH 13-8 Mo
Chromium-nickel manganese, nitrogen strengthened		Nitronic 20[b], Nitronic 32[b], Nitronic 33[b], Nitronic 40[b], Nitronic 50[b], Nitronic 60[b]

(*Source:* Armco Inc., Specialty Steels Division, Copyright 1984, Armco Inc., Middletown, Ohio.)
[a] Nitronic, 17-4 PH, 15-5 Ph, 17-7 Ph, Ph 15-7 Mo, 21-6-9 are registered trademarks of Armco Inc. Middletown, Ohio. PH 13-8 Mo, 18 SR, 18-9 LW, 18-2 Mn, 18-3 Mn, 22-13-5, and 12 SR are trademarks of Armco Inc.
[b] Special Armco stainless steels.

TABLE 6.4
Stainless steels and their uses

Armco Stainless Steel	Composition (%)				Forms[a]						Major Characteristics	Metallurgical Structure
	Cr	Ni	C Maximum	Other Significant Elements	Bar	Rod and Wire	Sheet	Strip	Billets	Pipe and Tubing		
Nitronic 20	22.00 to 24.00	7.00 to 9.00	0.28 to 0.38	Mn 1.5–3.5 N 0.28–0.40	●	●			●		Designed for elevated temperature use where high mechanical strength and resistance to oxidation and sulfidation are important.	Austenitic, age-hardenable by heat treatment
Nitronic 32	16.50 to 19.00	0.50 to 2.50	0.15	N 0.20–0.45 Mn 11.00–14.00	●	●			●		Twice the yield strength of Type 304 and comparable corrosion resistance, nonmagnetic.	Austenitic, nonhardenable by heat treatment
Nitronic 33	17.00 to 19.00	2.25 to 3.75	0.08 maximum	N 0.20–0.40 Mn 11.5–14.5	●	●	●	●	●	●	Twice the yield strength of Type 304, comparable corrosion resistance and better stress corrosion cracking resistance, nonmagnetic.	Austenitic, nonhardenable by heat treatment
Nitronic 40	19.00 to 21.50	5.50 to 7.50	0.08	N 0.15–0.40 Mn 8.00–10.00	●	●	●	●	●		High strength with good resistance to oxidation and toughness at sub-zero temperatures.	Austenitic, nonhardenable by heat treatment
Nitronic 50	20.5 to 23.5	11.5 to 13.5	0.06	N 0.20–0.40 Mn 4.00–6.00 Mo 1.50–3.00	●	●			●		Approximately twice the yield strength of Type 316 and more corrosion resistant. Nonmagnetic.	Austenitic, nonhardenable by heat treatment
Nitronic 80	16.00 to 18.00	8.00 to 9.00	0.10	Mn 7.00–9.00 Si 3.50–4.50 N 0.08–0.18	●	●			●		Excellent wear and galling characteristics, twice the yield strength of Type 304 with comparable corrosion resistance.	Austenitic, nonhardenable by heat treatment
311 DQ	17.25	4.50	0.04	Mn 2.50 N 0.15 Cu 2.4			●	●			Improved drawability and strength over Types 301 and 304.	Austenitic, nonhardenable by heat treatment

(continued)

117

TABLE 6.4 *Continued*
Stainless steels and their uses

Armco Stainless Steel	Composition (%)				Forms[a]						Major Characteristics	Metallurgical Structure
	Cr	Ni	C Maximum	Other Significant Elements	Bar	Rod and Wire	Sheet	Strip	Billets	Pipe and Tubing		
Type 301	16.00 to 18.00	6.00 to 8.00	0.15	—		●	●	●	●	●	High work-hardening rate combines cold-worked high strength with good ductility.	Austenitic, nonhardenable by heat treatment
Type 302	17.00 to 19.00	8.00 to 10.00	0.15	—	●	●	●	●	●	●	Basic, general purpose type with good corrosion resistance and mechanical properties.	Austenitic, nonhardenable by heat treatment
Type 303	17.00 to 19.00	8.00 to 10.00	0.15	S 0.15 min	●	●			●		Machinability superior to that of most regular Type 303 grades.	Austenitic, nonhardenable by heat treatment
Type 304	18.00 to 20.00	8.00 to 10.00	0.08	—	●	●	●	●	●	●	Low carbon variation of Type 302, reduces carbide precipitation during welding.	Austenitic, nonhardenable by heat treatment
Type 304L	18.00 to 20.00	8.00 to 10.00	0.03	—	●	●	●	●	●	●	Extra-low carbon content minimizes harmful carbide precipitation due to welding.	Austenitic, nonhardenable by heat treatment
18-9 LW	17.00 to 19.00	8.00 to 10.00	0.10	Cu 3.00–4.00	●	●	●	●	●		Special Armco type with low work-hardening, excellent cold-heading characteristics.	Austenitic, nonhardenable by heat treatment
Type 308	19.00 to 21.00	10.00 to 12.00	0.08	—	●	●			●		High Cr and Ni produce good heat and corrosion resistance, used widely for welding rod.	Austenitic, nonhardenable by heat treatment
Type 308L	19.00 to 21.00	10.00 to 12.00	0.03	—	●	●			●		Extra-low carbon variation of Type 308 eliminates harmful carbide precipitation during welding.	Austenitic, nonhardenable by heat treatment
Type 309	22.00 to 24.00	12.00 to 15.00	0.20	—	●	●			●		High strength and resistance to scaling at high temperatures.	Austenitic, nonhardenable by heat treatment

Type	Cr	Ni	C	Other Elements	Characteristics	Structure
Type 309S	22.00 to 24.00	12.00 to 15.00	0.08	—	Similar to Type 309 but carbon lowered to minimize carbide precipitation and improve weldability. Welding wire.	Austenitic, nonhardenable by heat treatment
Type 310	24.00 to 26.00	19.00 to 22.00	0.25	—	Higher alloy content improves basic characteristics of Type 309.	Austenitic, nonhardenable by heat treatment
Type 310S	24.00 to 26.00	19.00 to 22.00	0.08	—	Similar to Type 310 but carbon lowered to minimize carbide precipitation and improve weldability. Welding wire.	Austenitic, nonhardenable by heat treatment
Type 316	16.00 to 18.00	10.00 to 14.00	0.08	Mo 2.00–3.00	Mo improves general corrosion and pitting resistance and high temperature strength over that of Type 302.	Austenitic, nonhardenable by heat treatment
Type 316L	16.00 to 18.00	10.00 to 14.00	0.03	Mo 2.00–3.00	Extra-low carbon version of Type 316, minimizes harmful carbide precipitation due to welding.	Austenitic, nonhardenable by heat treatment
Type 317	18.00 to 20.00	11.00 to 15.00	0.08	Mo 3.00–4.00	Higher alloy content improves basic advantages of Type 316.	Austenitic, nonhardenable by heat treatment
Type 317L	18.00 to 20.00	11.00 to 15.00	0.03	Mo 3.00–4.00	Extra-low carbon version of Type 317 for welded structures.	Austenitic, nonhardenable by heat treatment
Type 321	17.00 to 19.00	9.00 to 12.00	0.08	Ti × C, min	Stabilized to permit use in 800 to 1500°F range with reduced carbide precipitation.	Austenitic, nonhardenable by heat treatment
Type 347	17.00 to 19.00	9.00 to 13.00	0.08	Cb + Ta 10 × C, min	Characteristics similar to Type 321, stabilized by Cb and Ta.	Austenitic, nonhardenable by heat treatment
Type 348	17.00 to 19.00	9.00 to 13.00	0.08	Ta 0.10 max Co 0.20 max Cb + Ta 10 × C, min	Similar to Type 347 but Ta reduced for atomic energy applications.	Austenitic, nonhardenable by heat treatment
22-4-9	20.00 to 23.00	3.00 to 5.00	0.60	Mn 7.00–10.00 N 0.30–0.50	Good hot hardness at elevated temperatures.	Austenitic, age-hardenable by heat treatment

(continued)

119

TABLE 6.4 *Continued*
Stainless steels and their uses

| Armco Stainless Steel | Composition (%) | | | | Forms[a] | | | | | | MAJOR CHARACTERISTICS | METALLURGICAL STRUCTURE |
	Cr	Ni	C Maximum	Other Significant Elements	Bar	Rod and Wire	Sheet	Strip	Billets	Pipe and Tubing		
17-4 PH	15.00 to 17.50	3.00 to 5.00	0.07	Cu 3.00–5.00 Cb + Ta 0.15–0.45	●	●	●	●	●		Combines excellent corrosion resistance, high strength and hardness, low temperature hardening, and good fabricating characteristics.	Martensitic, precipitaion hardened
15-5 PH	14.00 to 15.50	3.50 to 5.50	0.07	Cu 2.50–4.50 Cb + Ta 0.15–0.45	●	●	●	●	●		Similar in properties and characteristics to Armco 17-4 PH but has superior transverse ductility and toughness.	Martensitic, precipitaion hardened
PH 13-8 Mo	12.25 to 13.25	7.50 to 8.50	0.05	Mo 2.00–2.50 Al 0.90–1.35	●	●			●		Excellent transverse properties, has superior strength and toughness.	Martensitic, precipitation hardened
17-7 PH	16.00 to 18.00	6.50 to 7.75	0.09	Al 0.75–1.25	●	●	●	●	●	●	Special ultra-high strength with good formability, excellent fabricating characteristics.	Semiaustenitic, precipitation hardened
PH 15-7 Mo	14.00 to 16.00	6.50 to 7.75	0.09	Mo 2.00–3.00 Al 0.75–1.50	●	●	●	●	●		Special type similar to Armco 17-7 PH but with higher strength.	Semiaustenitic, precipitation hardened
400	12.00 to 13.00	—	0.05	Al 0.50 max			●	●		●	Corrosion resistance comparable to Type 410 at lower cost. No Ti for better surface.	Ferritic, nonhardenable
Type 403	11.50 to 13.00	—	0.15	Si 0.50 max	●	●			●		Special high-quality variation of Type 410 for highly stressed parts.	Martensitic, hardenable

Type	Cr		C	Other	Description	Classification
409	10.00 to 12.00	—	0.06	Ti 0.50	Lowest cost of all stainless sheet and strip, provides economical corrosion and oxidation resistance for applications where surface appearance is not important.	Ferritic, nonhardenable
Type 410	11.50 to 13.50	—	0.15	—	Low cost general purpose stainless steel. Wide use where corrosion is not severe.	Martensitic, hardenable
Type 416	12.00 to 14.00	—	0.15	S 0.15 min	Has much better machinability than most regular Type 416 grades.	Martensitic, hardenable
Type 430	16.00 to 18.00	—	0.12	—	Most popular of chromium types. Combines good corrosion and heat resistance and mechanical properties.	Ferritic, nonhardenable
Type 430F	16.00 to 18.00	—	0.12	S 0.15 min	Free machining version of Type 430.	Ferritic, nonhardenable
18 SR	18.00	0.50	0.06	Si 1.0 Al 2.0	Excellent resistance to high-temperature scaling.	Ferritic, nonhardenable
12 SR	12.00	—	0.02	Si 0.50 Al 1.2 Ti 0.30 Cb 0.60	More oxidation resistant and more creep resistant than Type 409 stainless steel.	Ferritic, nonhardenable

(*Source*: Armco Inc., Specialty Steels Division, © 1984, Armco Inc., Middletown, Ohio.)

[a] Some Armco stainless steel grades may be subject to heat lot order accumulations. Plate items are available from several producers.

FIGURE 6.2
Gray cast iron showing graphite flakes, unetched (500 ×).

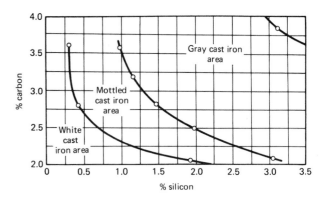

FIGURE 6.3
Composition limits for white, gray, and mottled cast irons (*Neely, Metallurgy,* 2d ed., © 1984 John Wiley & Sons, Inc.).

cipitate out to form flakes of graphite (Figure 6.2) when cast iron is cooled slowly, or cementite plus some ferrite if it is cooled rapidly. Also, graphitization is increased as the silicon content is increased (Figure 6.3). The iron-graphite equilibrium diagram (Figure 6.4) shows the various microstructures found in cast iron.

Gray Cast Iron

Gray cast iron is widely used to make castings and machinery parts since it is easily machined. Machine

bases, tables, and slideways are made of gray cast iron because it tends to remain dimensionally stable after an aging period and it makes an excellent bearing and sliding surface because of the graphite content. Cast iron has a lower melting point than steel and possesses good fluidity so it is easily cast into sand molds, even those having intricate designs. When low-cost but serviceable machine tool housings and frames are needed, the choice is almost always gray cast iron or one of its alloys because of its ability to deaden vibration and minimize tool chatter. Gray cast irons are often alloyed with elements such as nickel, chromium, vanadium, or copper to toughen and strengthen them. Unalloyed gray cast iron has two major

FIGURE 6.4
Iron-graphite equilibrium diagram (*Neely, Metallurgy,* 2d ed., © 1984 John Wiley & Sons, Inc.).

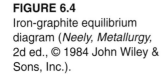

TABLE 6.5
Classes of gray iron

ASTM Number	Minimum Tensile Strength (psi)
20	20,000
25	25,000
30	30,000
35	35,000
40	40,000
45	45,000
50	50,000
60	60,000

(*Source:* John E. Neely, *Practical Metallurgy and Materials of Industry,* 2d ed., Wiley, New York, © 1984.)

drawbacks as an engineering material; it is quite brittle and it has a relatively low tensile strength, so it is not used where high stresses or impact loads are found. Cast irons may be ferritic or pearlitic in microstructure, which affects their strength. Gray cast irons are classified according to their tensile strength by ASTM numbers (Table 6.5). Although they can be welded, their high carbon content and their tendency to form extremely hard, brittle microstructures next to the weld make good welds difficult without preheating the base metal.

White Cast Iron

Unlike gray cast iron, white cast iron is virtually non-machinable because it is so hard. It contains massive cementite, which is the hardest substance in iron and steel. For this reason, it is mostly used for its abrasion resistance in such applications as wear plates in machinery. There are two ways white cast iron can be formed: by lowering the iron's silicon content and by rapid cooling, in which case it is called chilled cast iron. White cast iron cannot be welded.

Malleable Cast Iron

A cast iron that possesses all of the positive qualities of gray cast iron but has increased ductility, tensile strength, and toughness is called malleable cast iron. It is produced from white cast iron by a prolonged heat-treating process lasting for about 30 hours. This process essentially graphitizes the white cast iron to form tiny spheres of graphite instead of flakes as in gray cast iron (Figure 6.5). The iron matrix surrounds the graphite in this form, making it a

FIGURE 6.5
Micrograph of malleable cast iron (100 ×) (*Metals Handbook,* Vol. 7, 8th ed., American Society for Metals, 1972, p. 95. With permission).

more homogeneous metal, whereas graphite flakes (which are very soft and weak) cut across the iron matrix in every direction, causing it to be weak and brittle.

Nodular Cast Iron

Essentially the same purpose is achieved in nodular cast iron as in malleable cast iron; graphite spheres are formed, but without the extensive heat treatment. Nodular cast iron is known by several names: nodular iron, ductile iron, and spheroidal graphite iron. Like malleable cast iron, it has toughness, good castability, machinability, good wear resistance, weldability, low melting point, and hardenability. The graphite is formed into spheres in the ladle by adding certain elements such as magnesium and cerium to the melt just prior to casting. A vigorous, often spectacular, mixing reaction takes place that forms the graphite into tiny balls dispersed throughout the mix. Malleable and nodular cast iron castings are often used for machine parts where there is a higher tensile stress and moderate impact loading. These castings can be a less expensive alternative to steel castings or steel weldments.

NONFERROUS METALS

Nonferrous metals vary considerably in density (absolute weight per unit volume), color, melting points, and mechanical properties. Some, often called noble metals, such as gold and platinum, have a high resistance to corrosion. Many of the so-called space-age metals such as titanium and columbium were only relatively recently extracted from their ores in commercial quantities, and even aluminum, which we see everywhere today, was first extracted from ores in quantities only about 100 years ago. Most nonferrous metals and alloys are classified in a numerical system peculiar to that metal.

Aluminum

Aluminum is white or white-gray if the surface is oxidized. Unlike the oxidation or rust of iron, which contin-

FIGURE 6.6
Metal cladding. By cladding we use less of the scarce and expensive metals (Bureau of Mines).

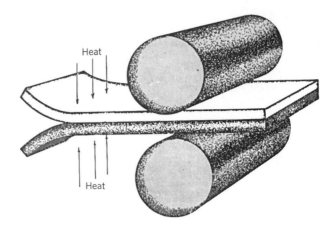

FIGURE 6.7
Roll bonding is one method of cladding metals. But if the heated surfaces oxidize, they will not bond (Bureau of Mines).

ues to form, the thin oxidized surface of aluminum prevents further corrosion. Aluminum weighs only 168.5 LB/ft³ compared to 487 LB/ft³ for steel, and pure aluminum melts at 1220°F (660°C).

Aluminum is classed as either wrought or cast. Wrought aluminum is cold worked and strain hardened by rolling, drawing, extruding, forming, or hot worked by forging. Molten aluminum is cast by gravity into sand and permanent molds or injected at high pressure into dies. Cast aluminum is generally less hard and strong than wrought products, yet some types can be hardened by heat treatments.

Pure aluminum resists corrosion better than aluminum alloys so, where corrosion is a problem, an alloy should not be used. However, rolled sheet stock of aluminum alloy is sometimes sandwiched between two very thin sheets of pure aluminum (Figures 6.6 and 6.7). Thus, the advantage of the high-strength alloy and the corrosion-resistant pure aluminum are combined in a single sheet. These clad aluminum sheets are used extensively in the aircraft industry.

There are several classification systems of specifications used to identify aluminum alloys: federal, military, the Society of Automotive Engineers (SAE), and the American Society for Testing and Materials (ASTM). The system most used by manufacturers is the one adopted by the Aluminum Association in 1954, which is shown in Table 6.6.

The first digit in Table 6.6 represents the alloy type. The second digit, here represented by the letter X, indicates any modifications made to the original alloy, and the last two digits identify either the specific alloy or impurity. The first entry in the series represents nearly pure aluminum. An 1120 aluminum contains no alloying element

TABLE 6.6
Aluminum and aluminum alloys

Code Number	Major Alloying Element
1xxx	None
2xxx	Copper
3xxx	Manganese
4xxx	Silicon
5xxx	Magnesium
6xxx	Magnesium and silicon
7xxx	Zinc
8xxx	Other elements
9xxx	Unused (not yet assigned)

(*Source:* John E. Neely, *Practical Metallurgy and Materials of Industry,* 2d ed., Wiley, New York, © 1984.)

and has 99.2 percent pure aluminum. This type cannot be hardened by heat treatment. In the 2000, 6000, and 7000 series, alloying elements make it possible to harden these alloys by solution heat treatment and aging. The alloys that cannot be heat treated are usually designated in the 1000, 3000, 4000, or 5000 series. However, all of these alloys are work hardenable and therefore have an additional designation for temper. In this case, temper denotes hardness from cold working.

Temper Designation Temper designations for those alloys that are not heat treatable sometimes follow the four-digit alloy series number.

—F As fabricated. No special control over strain hardening or temper designation.

—O Annealed, recrystallized wrought products only. Softest temper.

—H Strain hardened, wrought products only. Strength is increased by work hardening.

The symbol —H is always followed by two or more digits. The first digit, which can be 1, 2, or 3, denotes the final degree of strain (work) hardening.

—H1 Strain hardened only.

—H2 Strain hardened and partially annealed.

—H3 Strain hardened and annealed.

The second digit denotes the final degree of strain hardening.

2 = $\frac{1}{4}$ hard

4 = $\frac{1}{2}$ hard

6 = $\frac{3}{4}$ hard

8 = Full hard

9 = Extra hard

For example, 3003—H16 is an aluminum-manganese alloy, strain hardened to $\frac{3}{4}$ hard temper. A third digit is sometimes used to indicate a variation of a two-digit, —H temper.

Heat Treatment Classifications For aluminum alloys that can be hardened by solution heat treatment and precipitation hardening or aging, the symbol —T follows the four-digit number. Numerals 2 to 10 follow this letter to indicate the sequence of treatment.

—T2 Annealed (cast products only)

—T3 Solution heat treated and cold worked

—T4 Solution heat treated but naturally aged

—T5 Artificially aged only

—T6 Solution heat treated and artificially aged

—T7 Solution heat treated and stabilized

—T8 Solution heat treated, cold worked, and artificially aged

—T9 Solution heat treated, artificially aged, and cold worked

—T10 Artificially aged and then cold worked

For example, aluminum rivets may be made of 2024—T4 (an aluminum-copper alloy), which is solution heat treated, riveted into place (which is cold working), and then naturally aged.

Cast Aluminum Classification

Aluminum casting alloys are generally divided into three categories: sand casting alloys, permanent mold alloys, and die casting alloys. A classification system similar to that of wrought aluminum alloys is used (Table 6.7). The general purpose sand casting aluminum-copper alloy 212 has an ultimate strength of about 23,000 PSI. The die casting alloy 356—T6 is used for transmission cases and truck parts and has a tensile strength about 38,000 PSI.

Cadmium

Cadmium is used as an alloying element for low-temperature-melting metals such as solder, metals used to cast typesetting characters, for printing processes, bearing metals, and storage batteries. It has a blue-white color and is commonly used as a protective plating on steel parts such as screws, bolts, and washers. Cadmium compounds are extremely toxic and can cause illness when inhaled. Welding, heating, or machining these plated parts can produce toxic fumes. These fumes can be avoided with proper ventilation.

TABLE 6.7
Cast aluminum alloy designations

Code Number	Major Alloy Element
1xx.x	None, 99 percent aluminum
2xx.x	Copper
3xx.x	Silicon with Cu and/or Mg
4xx.x	Silicon
5xx.x	Magnesium
6xx.x	Zinc
7xx.x	Tin
8xx.x	Unused series
9xx.x	Other major alloys

(*Source:* John E. Neely, *Practical Metallurgy and Materials of Industry*, 2d ed., Wiley, New York, © 1984.)

Chromium

Chromium is a hard, slightly grayish metal that can take a brilliant polish and is often seen on automobile bumpers and trim as chromium plate on steel. It is very corrosion resistant and, when used as an alloying element in steel in quantities over 12 percent, it produces stainless steel; in lesser quantities, it produces a high-strength alloy steel. Chromium is too brittle to have any direct applications in its pure massive state.

Cobalt

This hard silver-white metal is widely used as a matrix or binder in which tungsten carbide particles are combined and formed into cutting tools by the powder metallurgy process. It is also used as an alloying element for high-speed and stellite cutting tools, for resistance wire, and for heat-resistant alloys used for jet engine blades. Dental and surgical alloys contain large percentages of cobalt and chromium; for example, 62 percent Co and 29 percent Cr in alloy A. They are not attacked by body fluids and do not cause irritation of the tissues as many other metals do. Cobalt is resistant to ordinary corrosion in air. Some iron-nickel-cobalt-chromium alloys are called superalloys because they are used where high strength and resistance to high temperature are required. Some of these alloys have the same expansion rate as glass and so are used for glass-to-metal joints and seals. Strong permanent magnets are made with cobalt and other metals. The familiar Alnico magnet material is an alloy of aluminum, nickel, cobalt, and iron. Cobalt and nickel are two of the few nonferrous metals that can be attracted to a magnet.

Columbium (Niobium) and Tantalum

Columbium, tantalum, and vanadium are all related chemically and they appear as a vertical group in the periodic table of the elements. However, columbium and tantalum are truly sister elements in more ways than one. Both occur in the same ores and have the same melting points. Unlike the other refractory (high-temperature) metals, they are both ductile and malleable at room temperature and therefore they can be easily formed. They are very immune to attack by strong acids, with the exceptions of fuming sulfuric and hydrofluoric acids. The anoxic oxide films they form are stable semiconductors and it is because of this property that they are used in capacitors and rectifiers. Columbium is used as a carbon stabilizer in stainless steels and as an alloying element for low-temperature service steels such as those used in Arctic pipelines. Tantalum is used in electronics and for surgical implants. Tantalum carbide, along with tungsten carbide and titanium carbide, is widely used for cutting tools in machining operations.

Copper and Copper-Based Alloys

Copper is a soft, reddish metal that has a number of valuable properties as a pure metal and in various combinations with other metals. Its high thermal and electrical conductivity, which is second only to that of silver, is of major significance. Copper has a high formability, corrosion resistance, and medium strength. To U.S. industry, the term copper refers to that element with less than .5 percent impurities or alloying elements. Copper-based alloys are those having more than 40 percent copper.

Deoxidized Copper   This type of copper differs chiefly in its lower electrical and thermal conductivity. It also has a somewhat higher ductility and is more readily formed and for this reason it is the most commonly used copper for the manufacture of tubular products such as those used in domestic and industrial plumbing.

Low Alloy Copper   Copper is often alloyed with very small percentages of other metals (from a fraction of 1 percent to approximately 2 percent) for the purpose of imparting such qualities as corrosion resistance, higher operating temperature, and increasing tensile strength and machinability. These additives include arsenic, silver, chromium, cadmium, tellurium, selenium, and beryllium. Copper cannot be hardened by heat treatments, but with the addition of about 2 percent beryllium it can be sufficiently hardened by precipitation and aging so that it can be used for making springs, flexible bellows, and tools. Because of their nonsparking quality, beryllium-copper tools are used in explosion-hazardous environments such as mines, powder factories, and some chemical plants. Since beryllium is quite expensive, an alloy containing only .4 percent Be with 2.6 percent Co was developed that is used as a substitute for some purposes, but the straight beryllium-copper alloy develops a higher strength and hardness by heat treatment than the cobalt-bearing alloy.

Oxygen-Free Copper This type is the purest of commercial coppers (99.92 percent minimum) and therefore is called a high-conductivity copper. It is used in electrical and electronic equipment, radiators, refrigeration coils, and distillers. Industrial copper is classified by a series of SEA numbers; for example, SEA No. 75 is a 99.90 percent deoxidized copper used for tubes.

Tough-Pitch Copper This copper contains a carefully controlled amount of oxygen, between .02 and .05 percent. Electrolytic refined tough-pitch copper is the most widely used type for electrical conductors and for building trim, roofing, and gutters. It lends itself to high tonnage production after being cast into wire, bars, and billets for further fabrication. Tough-pitch copper can withstand the ravages of time and weather because it does not harden with age and develop season cracks.

Brass

Brass is essentially an alloy of copper and zinc. As the relative percentages of copper and zinc vary, there is a corresponding variation in properties such as color, strength, ductility, and machinability. Brass colors usually range from white to yellow, and alloy brasses range from red to yellow. Alloy brasses rarely contain more than 4 percent of the alloying elements, either singly or collectively. These elements are manganese, nickel, lead, aluminum, tin, and silicon. However, in the plain brasses containing only copper and zinc, the zinc content can be as much as 45 percent. Gilding metal used for jewelry contains 95 percent copper and 5 percent zinc, whereas Muntz metal, which is used for sheet stock and brazing rod, contains 60 percent copper and 40 percent zinc. Lead is sometimes added to brass to increase its machinability. In general, because of their ductility and malleability, brasses are easily cold worked by any of the commercial methods such as drawing, stamping, spinning, and cold rolling.

Bronze

Bronze is an alloy of copper and tin and has been used for thousands of years. The tin-bronze today is known commercially as phosphor bronze because a small per-

centage of phosphorus is generally added as a deoxidizing agent in the casting of these tin-copper alloys. Phosphor bronze can range from 1.25 to 10 percent in tin content. Bronze colors generally range from red to yellow. Some bronzes have good resistance to corrosion near seawater. There are many types of bronzes that derive their names from an alloying element other than tin. Two of these are the silicon-bronzes and the aluminum-bronzes, which cover a wide range of metals. Phosphor-bronze can be cold worked and silicon-bronze can be hot and cold worked, but aluminum-bronzes are not severely cold worked. Most bronzes are easily cast into molds and they make good antifriction bearings. As with brasses, lead is often added to improve machinability. Both brasses and bronzes are designated by an SAE standard number system. For example, red brass casting metal is SAE standard No. 40, and SAE standard No. 64 is phosphor-bronze casting metal. These standard alloys used in the automotive industry can be found in reference books.

Die-Cast Metals

Die casting, which will be covered in Chapter 9, is a means of producing identical castings at a rapid rate by injecting molten, low melting point alloys into a mold with intense pressure. The mold then opens and the solidified part is ejected. Such small parts as carburetors, toys, and car door handles are mass produced by the process of die casting. Die-cast metals, also called pot metals, are classified into six major groups.

1. Tin-base alloys
2. Lead-base alloys
3. Zinc-base alloys
4. Aluminum-base alloys
5. Copper, bronze, or brass alloys
6. Magnesium-base alloys

The specific content of the alloying elements for any of the die-cast alloys may be found in materials reference books.

Indium

Indium is a very soft metal that can be scratched with the fingernail. It is silver-white, similar in appearance to platinum, and has a brilliant metallic luster. Indium adheres to other metals on contact and the molten metal clings to the surface of glass and wets it. Indium is used in bearing materials for wear reduction by diffusing it into metallic surfaces or as a coating of indium and graphite. Lead-indium alloys are used for solders and brazing materials. Indium in its pure form is not used as a manufacturing material; it is used only as it is alloyed with other metals.

Lead-Base Materials

Lead is a soft, heavy metal that is somewhat silvery when newly cut but gray when oxidized. It has been a very useful metal since ancient times when it was used for plumbing, utensils, and jewelry. Pure lead is so soft that it has limited usefulness, but when alloying constituents are added, its hardness is increased. Antimony in quantities up to 12 percent is added to harden and improve the mechanical properties of lead. Although lead has a high density, it has low tensile strength and low ductility, therefore it cannot be easily drawn into wire. It does have the property of exceptionally high malleability, which allows it to be rolled or compressed into thin sheets or foil. Lead shapes and wire cable sheathing can be easily extruded (pushed through a die). Lead has good corrosion resistance and is a good shielding material for gamma and X-ray radiation, but lead compounds are very toxic, and adequate protection is necessary in handling and manufacturing lead products.

Babbitt   A widely used antifriction lead alloy that is used for antifriction bearings is called babbitt. Babbitts can be lead, tin, or cadmium based. Lead-based babbitts contain up to 75 percent lead with antimony, tin, and some arsenic. Tin babbitts contain from 65 to 90 percent tin with antimony, lead, and some copper added. These are the higher grades and are more expensive than lead-based types. Cadmium-based babbitts resist higher temperatures than the other types. They contain from 1 to 15 percent nickel, a small amount of copper, and up to 2 percent silver.

Solders   Tin and lead are combined in various proportions to form low-melting-point solders that are used to join metals. The most common general purpose solders contain about 50 percent of each metal. They are sometimes called 50-50 solders. The canning industry uses solders with only 2 to 3 percent tin. When there is a need to lower the melting point so the metal to be joined will not warp, cadmium or bismuth is added to the lead-tin alloy.

Terne Plate   Terne metal is an alloy of tin and lead, the tin content being between 10 and 25 percent. Sheet steel coated with this alloy is called terne plate. This product has a high resistance to corrosion and is used for roofing and other architectural needs. Terne plate forms a good base for paints as used in the deep drawing of metal articles because the plating provides a good die lubricant. These coatings are applied to steel by the hot-dip process, or they are sprayed on or electrodeposited.

Magnesium

When pure, magnesium is a soft silver-white metal that resembles aluminum but is much lighter in weight (108.6 lb/ft^3 compared to 168.5 lb/ft^3 for aluminum and 487 lb/ft^3 for steel). When alloyed with other metals such as aluminum, zinc, or zirconium, magnesium has quite high strength-to-weight ratios, making it a useful metal for some aircraft components. Magnesium is also used in cast iron foundries as an additive in the ladle to produce nodular iron. Household goods, typewriters, and portable tools are some of the many items made of this very light metal. Magnesium when finely divided will burn in air with a brilliant white light. Magnesium alloys are designated by several numerical systems: Military, AMS, SAE, ASTM, and Federal, all of which may be found in reference handbooks. Magnesium can be shaped and processed by practically all the methods used to manufacture other metal products. It can be cast by sand, permanent mold, or diecast methods. It can be rolled, extruded, forged, and formed by bending, drawing, and other methods.

Manganese

This silver-white metal is seldom seen in its pure state since it is normally used as an alloying element for other metals. In steel production it is used as a deoxidizer and to control sulfur. All steel contains small percentages of manganese. Large percentages (12 percent or more) cause steel to become austenitic and no longer ferromagnetic. High manganese steel quickly work hardens and has excellent abrasion resistance as a result, so it is used extensively in making earth-moving machinery, rock crushers, and conveyors. Manganese is also used in some stainless steels.

Molybdenum

As a pure metal, molybdenum is used in high-temperature applications such as for filament supports in lamps and electron tubes. It is used as an alloying element in steel to promote deep hardening and to increase tensile strength and toughness.

Nickel and Nickel-Based Alloys

Nickel is a silvery white metal that is noted for its corrosion resistance. It is widely used for electroplating on other metals, especially steel, as a protection coating. It is also used as an alloying element in other metals and, like manganese, a large percentage (50 percent) of nickel in steel, or 8 percent Ni and 18 percent Cr as in austenitic stainless steel, causes it to become austenitic and lose its ferromagnetism.

Monel An alloy of 67 percent nickel and 28 percent copper plus some other impurity metals such as cobalt, manganese, and iron is called Monel. It is a tough alloy that is machinable, ductile, and corrosion resistant. Monel is widely used to make marine equipment such as pumps, valves, and fittings that are subjected to salt water, which is very corrosive to many metals.

Nichrome and Chromel These nickel-chromium-iron alloys are used as electric resistance elements in electric toasters and heaters. Inconel is similar to nichrome and chromel but is used to manufacture parts to be used in high-temperature applications.

Nickel-Silver This alloy could be considered one of the copper alloys since it contains a high percentage of copper. A common nickel-silver contains about 55 percent copper, 18 percent nickel, and 27 percent zinc. The color is silver-white, making it a substitute for silver tableware and other such products. Because of the nickel content, it has good corrosion resistance.

Precious Metals

Gold and silver have long been used for coinage and jewelry because they tend to resist deterioration by corrosion and are relatively rare. Both are fairly evenly distributed over the earth's surface, although not in sufficient amounts for economical recovery except in a few locations. The six metals of the platinum group are palladium (Pd), rhodium (Rh), iridium (Ir), osmium (Os), ruthenium (Ru), and platinum (Pt). Along with gold and silver these metals are generally called the precious metals group. Gold is used in dentistry and in the electrical and chemical industries. Silver has many uses in manufacturing and is well known in its use in photographic film. The specialty metals are all used for catalysts, as alloying elements, and in many industrial applications.

Tin

Tin is whiter than either silver or zinc. It has good corrosion resistance so it is used to plate steel. Vast quantities of tin plate are used in the food-processing industry. Tin is one of the constituents of bronze and is used in babbitts and solders. Because of the rapidly increasing price of tin in recent times, other less expensive metals or materials have been developed to replace it in some products.

Titanium

This silver-gray metal weighs about half as much as steel and, when alloyed with other metals, is as strong as steel. Like stainless steel, it is a relatively difficult metal to machine, but machining is accomplished with rigid setups, sharp tools, and proper coolants. The greater application for titanium and its alloys is in making jet engines and jet aircraft frames. However, many promising civilian, military, naval, and aerospace uses for titanium are being considered because of its high corrosion resistance strength, and because it weighs less than steel. At the present time, its high cost precludes many of these proposed applications.

Tungsten

Tungsten is one of the heaviest of the metals and has the highest known melting point of any metal (6098°F or 3370°C). It is not resistant to oxidation at high temperatures and, when used as lamp filaments, it must be in a vacuum or inert gas atmosphere. Tungsten is used for electrical contacts and for welding electrodes. It is used as an alloying element in tool steels and in cutting-tool alloys. The carbides of tungsten are the most widely used and most valuable cutting tools. Tungsten carbide powder combined with cobalt powder is compressed into tool shapes and sintered in a furnace in a process called powder metallurgy.

Uranium and Thorium

Both uranium and thorium are radioactive metals and alpha emitters. Competent authorities should be consulted regarding the handling of these metals and their ores. Uranium is a silver-white metal in its pure state and it can be machined like other metals. If finely divided, uranium will burn in air. Uranium is an important fuel for nuclear reactors. Thorium, a dark gray metal, is also used in nuclear reactors. Thorium oxide with 1 percent cerium oxide is used for gaslight mantles.

Vanadium

Vanadium is a silver-white, very hard metal that oxidizes when exposed to air. It is used almost exclusively as an alloying element in steel to impart improved impact resistance (toughness) and better elastic properties. Vanadium contributes to abrasion resistance and high temperature hardness in cutting tools and it promotes finer grain in steels, giving them toughness. Usually not more than 1 percent vanadium is alloyed with steels.

Zinc

Zinc is a white metal that is valued for its corrosion protection of steel. The familiar galvanized steel is coated with zinc by any one of several processes: hot dip, electrogalvanizing, sherardizing (heating steel in zinc dust), metal spraying, or zinc-dust paint. When zinc and iron are galvanically (electrically) in contact under corrosive conditions, the zinc becomes anodic and serves as a sacrificial metal, corroding in preference to the steel. The greater consumption of zinc is for galvanizing steel. Zinc-based die-cast metals also consume a high percentage of zinc production, as do brass and rolled zinc to a lesser extent.

Zirconium

Zirconium and titanium are similar in appearance and physical properties and, like the other reactive metals, zirconium will burn in air or explode when finely divided. In fact, it was once used as an explosive primer and as a flashlight powder for use in photography because of the brilliant white light it gives off when it explodes. Zirconium alloys are still used in flash bulbs. Perhaps one of its most notable uses is in nuclear reactors. It has good corrosion resistance and stability, and therefore has been used for surgical implants. Like titanium, it is somewhat difficult to machine.

MATERIALS IDENTIFICATION

We soon learn to identify natural materials such as wood, leather, stone, glass, and earthy materials, such as clay, by sight and touch. Similarly, we can learn to identify many manufacturing materials by sight. However so many varieties of these materials look exactly the same that other means must be used to identify them. For example, in the past when a distinction had to be made between low carbon and high carbon steel that look alike on the surface, identification was made by means of spark testing (Figure 6.8). The blacksmith simply touched the two samples on a grinding wheel and observed the sparks; the high carbon steel produced a brilliant display of lines and sparks and the low carbon steel showed few carrier lines and a few sparks. This method is still sometimes used today to compare unknown steels with known samples in steel yards and shops, although its accuracy is directly related to the operator's experience. Although it is still a useful tool, it is not a reliable method of identifying the alloy content of steel. Also, very few nonferrous metals produce a spark; exceptions are nickel, titanium, and zirconium. Nickel produces short carriers with no spark, similar to that of stainless steel, and titanium produces a brilliant white display of carrier lines, each having a spark at the end. The display of zirconium carrier and spark is identical to that of titanium (Figure 6.9).

Metals can often be identified by their reaction to certain acids and alkaline substances. A clean surface of a

FIGURE 6.8
Spark testing high carbon steel.

particular metal will become darkened or colored when a certain reagent is placed on it. For example, an aluminum sample can be quickly distinguished from a magnesium sample by placing a drop of a copper sulfate solution on each. The aluminum will not be affected but the magnesium will be blackened. Copper sulfate will turn ordinary steel a copper color, but it will not affect any type of stainless steel, nickel, Monel, or Inconel, all of which have a similar appearance. There are also many acid tests. Commercial spot testers are available in kits for the purpose of identifying metals. Magnetic testing is another means of identifying look-alike metals. Most ferrous metals, and those in which nickel or cobalt is the primary constituent, are attracted to a magnet, but some stainless steels, including the 300 series and the precipitation hard-

FIGURE 6.9
Spark testing titanium.

ening types, 17-4 Ph and 15-5 Ph, may or may not be magnetic. Austenitic stainless steel (300 series) is not attracted to a magnet unless it is cold worked; then it is attracted to the extent to which it is cold worked. Of course, none of these methods will identify all of the elements or how much of each is in a sample of metal, but this can be done by spectrographic analysis in the laboratory. Speedier results are obtainable with a portable unit (Figure 6.10). This device identifies and quantifies key elements (alloying and residual) plus carbon in metals. Fuess spectroscopes work on the same principle as the spark test except that the reflected light is refracted through a prism. Since each element produces characteristic emission lines, it is possible to identify which elements are present. It is also possible to compare the results to standard spectra as an aid in identifying alloys. Mobile spectrometers work on the Fuess principle and are not able to sort alloys but are preset for a particular matching test. They are calibrated to a known standard and can quickly identify that material with a match signal.

Using an X-ray analyzer is probably one of the best methods of sorting and verifying metals (Figures 6.11 and 6.12). Its greatest drawback is that it costs more than other

FIGURE 6.11
In as little as 5 sec the X-site portable X-ray analyzer can identify metals such as this bar stock. The abbreviated alloy name appears on the LED display within a few seconds (Photo courtesy of Kevex Corporation © 1982).

FIGURE 6.10
Spectromobile. A wheel-mounted spectrometer facilitates rapid, on-site identification and sorting of steel. It identifies and quantifies all key elements (alloying and residual) in metals and determines carbon content to within plus or minus .05 percent (Photograph courtesy of Spectrochemical Products Division, Baird Corporation).

FIGURE 6.12
Metal parts in a foundry are analyzed for content of various elements. This is done with an X-ray source that produces characteristic emissions from every element in the sample. A console with a microprocessor is connected to the hand-held unit that processes the data (Photo courtesy of Kevex Corporation © 1982).

systems. Scrap metals are among the most difficult metals to identify and sort. High-integrity steel products must have carefully sorted scrap. All of the previously mentioned methods may be applied to this task of sorting scrap metals. However, it seems that the more accurate a method is the more the system costs.

Plastics Identification

Because there is such a great variety of plastics and more are being developed every day, identification of plastics is very difficult by ordinary means. Certain solvents will dissolve certain types of plastics, but this would not be a reliable method of testing because many types of plastics react to the same solvent. However, as discussed earlier, when heat is applied to a thermoplastic material, it will melt and solidify again when cool and it can be reused, but heat applied to a thermosetting material will cause it to break down instead of melting and it cannot be reused. Of course, in either case, this would be a destructive test for the plastic object involved.

Some plastics with chemical odors can be identified by holding a sample in a flame. An odor of phenol indicates a phenolic material, and an acrid odor with a yellow flame indicates cellulose acetate or vinyl acetate. An odor of burnt wood indicates ethyl cellulose. Melamine has a fishy odor when burned and casein smells like heated milk. Polystyrenes give off an odor of domestic heating and cooking gas, whereas polyethylene and polyolefins produce little or no odor, burn with a smoky flame, and melt like a candle while burning. **Note:** Cellulose nitrate is very flammable, especially in thin films or sheets. Precautions must be taken to protect hands, face, and eyes as well as to prevent the danger of fire hazard when conducting any of these tests.

FLUID ANALYSIS

Industrial process oils are used in a broad range of applications in manufacturing processes. An overview of these petroleum products is seen in the Telura® numbering systems of the Exxon Company, USA (Table 6.8) The applications for these oils are in Table 6.9.

The Telura line of process oils consists of 19 grades suitable for a broad range of applications. The line consists of six general types of products.

1. **Aromatic**   Aromatic oils have good solvency characteristics. They are used to make proprietary chemical formulations and ink oils.

2. **Naphthenic**   Naphthenic oils are characterized by low pour points and good solvency properties. They

TABLE 6.8
Telura® numbering systems

The three-digit Telura grade number provides a key to the properties of the oil. The first digit designates the type of product, as follows:

1—Aromatic
2—Intermediate aromatic[a]
3—Naphthenic
4—Extracted naphthenic
5—Paraffinic
6—Extracted paraffinic
7—Special grades

The second digit is indicative of the approximate viscosity in centistokes (cSt) at 40°C, as follows:

0—<10	5—50 to 59
1—10 to 19	6—60 to 99
2—20 to 29	7—100 to 999
3—30 to 39	8—1000 to 1999
4—40 to 49	9—2000

The third digit is an arbitrary number used to discriminate between grades of similar viscosity.

(*Source:*   Reprinted with the permission of Exxon Corporation, "Telura® Industrial Process Oils," Exxon Corporation, 1984.)
[a]Products in this category are not currently available.

are used to make printing inks, shoe polish, rustproofing compounds, and dust suppressants, and they are used in textile conditioning, leather tanning, and steam turbine flushing. The colors of the low-viscosity Telura oils in this category are relatively light.

3. **Extracted Naphthenic**   These oils have outstanding color and color stability and good low-temperature characteristics. They are used in proprietary chemical formulations, dust suppressants, and rust-proofing compounds. Of the extracted naphthenic products, Telura 415 and 417 meet the requirements of Federal Food and Drug Administration (FDA) Regulation 21 CFR 178.3620(c) for mineral oils used in products that may have incidental food contact.

4. **Paraffinic**   Paraffinic oils are characterized by low aromatic content and light color. They are used in furniture polishes, ink oils, and proprietary chemical formulations.

5. **Extracted Paraffinic**   These oils offer light colors, low volatilities, relatively low pour points, and improved color stability. They are used in waste disposal system fluidizing oils, textile oils, and autoclave oils. Telura 607, 612, 613, and 619 are the oils in this

TABLE 6.9
Applications for Telura industrial process oils

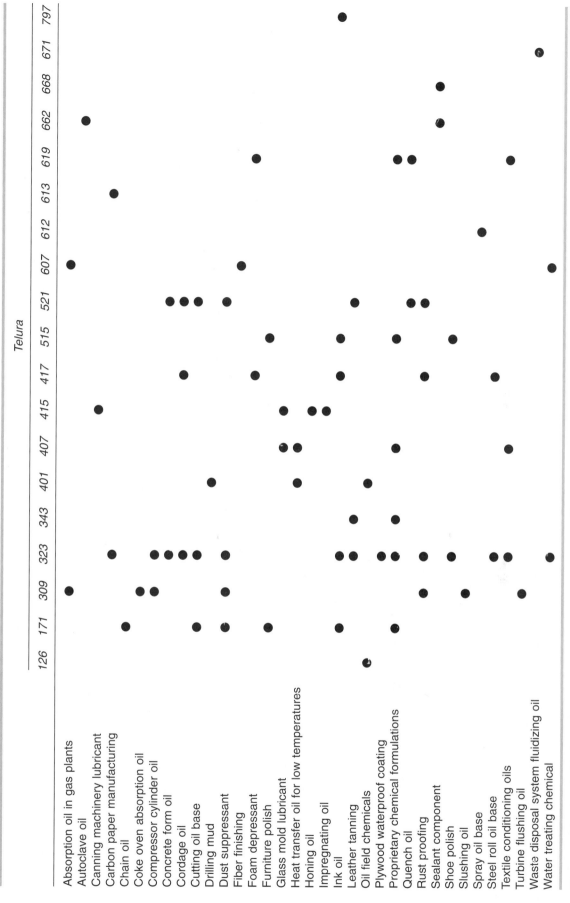

Telura

(*Source:* Reprinted with the permission of Exxon Corporation, "Telura® Industrial Process Oils," Exxon Corporation 1984.)

category that conform to the requirements of FDA Regulations 21 CFR 178.3620(c).

6. **Special Grade**   Telura 797 is a heavy, black residual oil used in newsprint ink formulations.

Oil Contamination

One of the main causes of wear and breakdown of machinery is contamination of lubricating or hydraulic oils by grit, water, and other unwanted materials. Regular inspection of industrial oils can save a great deal of breakdown time. This can be done by means of a fluid analysis kit (Figure 6.13). Examples of SAE oil contamination classification are also provided in the kit. Oil can also be contaminated with hazardous chemicals of which polychlorinated biphenyls (PCBs) are the most common, especially around electrical apparatus such as older transformers. PCBs are toxic and are thought to be a cause of cancer. Field tests for PCB spills or contaminated oil can easily be made with portable testers like the one shown in Figure 6.14.

Water

Aside from our municipal water systems, industrial water supply is of major importance. Water that has been used for industrial purposes is often contaminated with heavy metals and other materials that must be removed before this wastewater is returned to a river or other outflow.

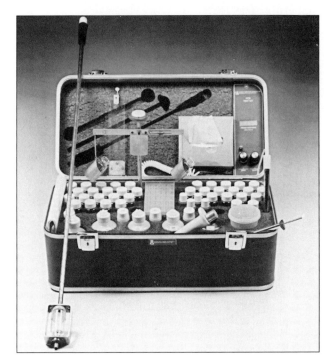

FIGURE 6.14
PCB field test kit for screening oil and soil. The soil collector is shown removed from the case and assembled (Courtesy of McGraw-Edison Company, Power Systems Division, Pittsburgh, PA).

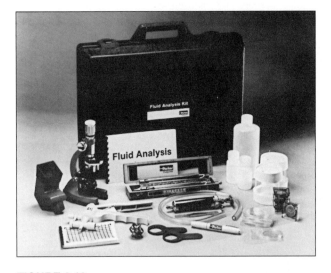

FIGURE 6.13
Fluid analysis kit. Oil samples in field, plant, and laboratory hydraulic systems can be checked easily for contaminants with this portable kit (Courtesy of Parker-Hannifin Corporation).

Various kinds of tests are made to assure the purity of wastewater; among them is an automatic metals monitor (Figure 6.15) that determines the level of cadmium, copper, lead, and zinc in wastewater.

Heavy metals may pose a serious threat to public health if they occur in sources of potable water such as rivers, lakes, and groundwater. The metals monitor, a completely automatic electrochemical installation, can detect and analyze all types of water without any human supervision.

Besides potable water, the instrument can monitor seawater, rainwater, and industrial wastewater. It can also be used to detect corrosion in power stations and waterworks, to find leaks in pipes, and to analyze the effluents of the metal finishing, plating, and chemical industries. Besides cadmium, copper, lead, and zinc, other metals can be detected by the monitor if the computer program and treatment of the sample are modified.

All of the control, monitoring, and calculating functions are performed by a microprocessor. The data are recorded four times an hour on a printer and an LED display. In addition, daily averages of the metal concentrations are printed once a day. If it is connected to a

FIGURE 6.15
Water analyzer for testing water for heavy metals
(Alewijnse Electrical Engineering, Environment Systems
Technology Division, Nijmagen, Holland).

telemetric recording circuit, the metals monitor will analyze concentrations ranging from .0005 to 1 mg/l.

MATERIALS APPLICATIONS

Material selection for a particular manufactured product is
an involved and sometimes complex process requiring a
broad understanding of the nature and behavior of materials. Many factors other than material strength and durability need to be considered. In view of domestic and foreign
competition, the cost of materials is of prime consideration, in addition to the manufacturing cost. Product warranties and customer service requirements are to be
considered as well as product liability and patent infringe-
ment. Many new materials for products are chosen
because of the necessity for weight reduction, especially

for energy savings in the field of transportation. However
there is an inherent risk in making a sudden change in
material for a product without extensive evaluation of its
long-term performance. Many hasty substitutions of materials have resulted in product liability cases. Some of the
factors involved in making a material choice are tensile
strength, compressive strength, torsional strength, yield
strength, fatigue strength, hardness, impact resistance,
wear resistance, ductility, machinability, formability, hardenability, weldability, corrosion resistance, availability,
and cost. Of course, many other factors relate to the specific selection, such as the prevailing conditions—temperature, atmosphere, and vibration—to which the products
will be subjected. Reference handbooks on materials are
helpful when making selections. Some of these are the
Metals Handbook and the *Source Book on Materials
Selection,* both published by the American Society of
Metals, and the *Cast Metals Handbook,* published by the
American Foundryman's Association.

Example: Galvanized versus Plastic-Coated Pipe

A West Coast manufacturer of irrigation equipment, OEM
(Original Equipment Manufacturer), had been using galvanized steel pipe to make pivot irrigators. These devices
slowly move a pipe around a center tower in a circular
pattern, spraying water that often contains liquid fertilizer
(Figure 6.16). Because the fertilizer is acidic, uncoated
steel pipe would corrode quickly. Since zinc is a sacrificial
metal, it soon begins to roughen inside, causing a turbulence that requires more energy to move the water through
the pipe (Figures 17a and 17b). A higher head (pressure) is
required to move the water through the galvanized sprinkler pipe than through a plastic pipe, for example, in order

FIGURE 6.16
Pivot irrigator covers 160 acres in a circular pattern
(Pierce Corporation, Eugene, OR).

FIGURE 6.17
After just a few years in the field, the section of galvanized pipe *(a)* shows the roughness and discoloration typical of a "worn-out" galvanized coating. The pipe with the electrostatically applied epoxy coating *(b)* still looks like new and is smooth to the touch even after several seasons of use (Pierce Corporation, Eugene, OR).

to maintain sufficient pressure at the outer sprinkler heads, requiring more electrical energy.

A search was made by the company engineers for a substitute pipe material to solve this energy loss problem. Although plastic had less friction than the other alternatives, it could not substitute for two reasons. It would tend to deteriorate from ultraviolet rays in sunlight and it has insufficient tensile strength to support the weight of water in a suspended pipe. Aluminum, besides being expensive, would not substantially improve the energy loss problem. The Hazen-Williams formula of the coefficient of friction, in pipe, in which higher numbers indicate lower friction, shows black steel pipe with a number of 100. Galvanized pipe and aluminum are both 120. Since weight was not a factor and higher strength was required, steel was still the preferred pipe material. The solution was to coat steel pipe with a suitable plastic material. Steel pipe was cleaned by shot-blasting and electrostatically sprayed inside and outside with an epoxy powder that was then heat dried. This epoxy coating is inert to the low acid in the fertilizer. It is very smooth and its Hazen-Williams number is 150, the same as that of plastic pipe. This investigation led to an amazing revelation. A substantial cost savings in electric power for the farmer's water pumps could be realized by reducing the friction in the pivot irrigation pipe. With the cost of electricity in the Midwest being 3 to 6¢/kwh, a ten-machine operation (each machine covers 160 acres) over a 10-year period would realize a savings of $30,000 to $50,000 or more. Thus, the switch to epoxy-coated steel pipe gives the machine purchaser, the farmer, a competitive edge; it also will

undoubtedly increase sales for that OEM company's product, the pivot irrigator.

Review Questions

1. Which is the most common classification system used in the manufacturing industry for machine and alloy steels, the SAE-AISI or the ASTM specifications?

2. What kind of steels are designated by the ASTM specifications?

3. Name the quenching medium used for W1 and O1 tool steels.

4. Which steel shafting is more expensive, cold rolled (CR) or ground and polished (G and P)?

5. Which steel bar having the same alloy content is more expensive, hot rolled (HR) or cold rolled?

6. State the greatest advantage of HSLA steels over plain carbon steels for use in modern automobiles.

7. One type of stainless steel that is often used for cutlery and surgical equipment can be hardened by heating and quenching. What number series is it classified under?

8. As gray cast iron solidifies, it is unable to hold more than 2 percent carbon in solution, yet it may contain 4 percent carbon. What happens to the remaining 2 percent?

9. Would you make a part from gray cast iron if it were to be subjected to high impact loads? Why?

10. What kind of cast iron might be acceptable for the part described in Question 9, particularly if the impact loading were moderate?

11. Why is a thin sheet of pure aluminum sometimes sandwiched (clad) on tough alloy aluminum?

12. In what two ways can aluminum be hardened?

13. What is the most widely used type of copper for electrical conductors?

14. Of all the means of identifying the alloy and carbon content of metals, which is the most accurate although the most expensive method?

15. If only low and high plain carbon steels are present, what simple method could be employed to distinguish between them?

16. A manufacturer of a consumer product wishes to make a material change from metal to plastic for the purpose of weight reduction and appearance. What should be done before the new product goes on the market?

Case Problems

Case 1: Separating Aluminum and Magnesium Bars

A manufacturing plant regularly received shipments of aluminum bars of various sizes on which machining and welding operations were performed. On one shipment the supplier had discovered the bars had been sent out unmarked and several magnesium bars had accidentally been mixed into the lot. There was no time to send for and receive a new shipment before running out of material. The two metals had to be separated, not only because magnesium was an unwanted material, but because it is a pyrophoric metal, tending to catch fire and burn fiercely when finely divided, as in machining chips, and from welding, when proper precautions are not taken. Therefore, it constituted an extreme hazard in this case where they were not normally using magnesium. The metal bars all looked identical and they varied so much in size that weight differences could not be used readily to identify them by unskilled help. A quick low-cost method was needed to separate these look-alike metals, one that an unskilled person could use. How would you solve this problem, considering what you have learned in this chapter?

Case 2: Cost of Steel Shipment

You receive a load of steel bars on a truck. The shipping order does not include the cost, which is on the invoice that will be mailed later. You need to know the approximate cost immediately in order to make a cost estimate for parts to be made from the steel bars. The load is made up of 20-foot-long steel bars: 10 flat bars (rectangular), $\frac{3}{4} \times 2\frac{1}{2}$ in., and 15 round bars $1\frac{1}{4}$ in. diameter. Metal shapes are sold by weight and different metals have different weights per unit volume. Steel weighs 487 lb/ft^3. The price of the steel in this shipment is .27¢/lb for the flat bars and .31¢/lb for the round bars. Weight in cubic inches of a steel bar is found by multiplying the cross-sectional area by the length in inches, multiplied by the weight of 1 in.^{3}s. There are 1728 in.^{3}s in 1 ft^{3}s. What is the total cost of the shipment?

PART II

MEASUREMENT AND QUALITY ASSURANCE

P art II introduces measurement and quality assurance with particular emphasis on tolerances, measuring processes, and calibration. Quality assurance is an integral part of today's manufacturing environment. No manufacturing process can be successfully designed and implemented without full consideration of how the quality of the product will be assured. This includes determining when, where, and how the product will be inspected as discussed in Chapter 7, Quality Assurance and Control.

Quality assurance must be maintained at every step in the manufacturing process, and quality is not assured unless all product specifications are met at each of those steps. In order to assure that the product meets all specifications, existing tolerances must receive full consideration, and tolerances establish the precision required of the manufacturing process. In designing a process to achieve this level of precision, the precision of both the measuring process and the manufacturing operation itself must be considered. The precision required of the measuring process affects the design of the measuring operation as well as the choice of inspection instrument. These issues are addressed in Chapter 8, Inspection and Measurement.

Product quality is of paramount importance in manufacturing. If quality is allowed to deteriorate, then a manufacturer will soon find sales dropping off followed by a possible business failure. Customers expect quality in the products they buy, and if a manufacturer expects to establish and maintain a name in the business, quality control and assurance functions must be established and maintained before, throughout, and after the production process. Generally speaking, **quality assurance** encompasses all activities aimed at maintaining quality, including quality control. Quality assurance can be divided into three major areas. These include the following.

1. Source and receiving inspection before manufacturing.

2. In-process quality control during manufacturing.

3. Quality assurance after manufacturing.

Quality control after manufacture includes warranties and product service extended to the users of the product.

CHAPTER 7

Quality Assurance and Control

Objectives

This chapter will enable you to:

1. Define the purpose and intent of quality control and quality assurance.

2. List the major responsibilities of quality control and quality assurance personnel in manufacturing.

3. Discuss aspects of nondimensional quality control.

4. Discuss quality control and assurance after manufacturing.

SOURCE AND RECEIVING INSPECTION BEFORE MANUFACTURING

Quality assurance often begins long before any actual manufacturing takes place. This may be done through source inspections conducted at the plants that supply materials, discrete parts, or subassemblies to manufacturer. The manufacturer's source inspector travels to the supplier factory and inspects raw material or premanufactured parts and assemblies. Source inspections present an opportunity for the manufacturer to sort out and reject raw materials or parts before they are shipped to the manufacturer's production facility.

The responsibility of the source inspector is to check materials and parts against design specifications and to reject the items if specifications are not met. Source inspections may include many of the same inspections that will be used during production. Included in these are:

1. Visual inspection

2. Metallurgical testing

3. Dimensional inspection

4. Destructive and nondestructive inspection

5. Performance inspection

Visual Inspections

Visual inspections examine a product or material for such specifications as color, texture, surface finish, or overall

FIGURE 7.1
Making a product visual inspection (Del Manufacturing).

FIGURE 7.2
Hardness testing is an example of metallurgical inspection (MTI Corporation).

appearance of an assembly to determine if there are any obvious deletions of major parts or hardware (Figure 7.1).

Metallurgical Testing

Metallurgical testing is often an important part of source inspection, especially if the primary raw material for manufacturing is stock metal such as bar stock or structural materials. Metals testing can involve all the major types of inspections including visual, chemical, spectrographic, and mechanical, which include hardness (Figure 7.2), tensile (Figure 7.3), shear compression, and spectrographic analysis for alloy content. Metallurgical testing can be either destructive or nondestructive.

Dimensional Inspection

Few areas of quality control are as important in manufactured products as dimensional requirements. Dimen-

sions are as important in source inspecting as they are in the manufacturing process. This is especially critical if the source supplies parts for an assembly. Dimensions are inspected at the source factory using standard measuring tools plus special fit, form, and function gages that may be required. Meeting dimensional specifications is critical to interchangeability of manufactured parts and to the successful assembly of many parts into complex assemblies such as autos, ships, aircraft, and other multi-part products.

FIGURE 7.3
Tensile testing a sample (Lockheed Co.).

specifications are known to be met in regard to the strength of materials, it is often not necessary to test further parts to destruction unless they are genuinely suspect.

Performance Inspection

Performance inspections involve checking the function of assemblies, especially those of complex mechanical systems, prior to installation in other products. Examples include electronic equipment subcomponents, aircraft and auto engines, pumps, valves, and other mechanical systems requiring performance evaluation prior to their shipment and final installation.

QUALITY CONTROL DURING MANUFACTURING

Quality control during manufacturing involves many of the same checks as in source inspection. A great deal of time and effort is invested in in-process quality control by the manufacturing industry. Major considerations for in-process quality control include the following:

1. Receiving inspection
2. First piece inspection
3. In-process inspection
4. One hundred percent inspection
5. Final inspection

Destructive and Nondestructive Inspection

In some cases it may be necessary for the source inspections to call for destructive or nondestructive tests on raw materials or parts and assemblies. This is particularly true when large amounts of stock raw materials are involved. For example it may be necessary to inspect castings for flaws by radiographic, magnetic particle, or dye penetrant techniques before they are shipped to the manufacturer for final machining. Specifications calling for burn-in time for electronics or endurance run tests for mechanical components are further examples of nondestructive tests.

It is sometimes necessary to test material and parts to destruction, but because of the costs and time involved destructive testing is avoided whenever possible. Examples include pressure tests to determine if safety factors are adequate in the design. Destructive tests are probably more frequent in the testing of prototype designs than in routine inspection of raw material or parts. Once design

Receiving Inspection

In receiving inspection, parts, material, and assemblies are checked against design specifications as they arrive at the manufacturer's production or assembly facility. If the items received are premanufactured parts requiring further assembly or processing, receiving inspection will inspect for correct dimensions and any other specifications required by the design.

First Piece Inspection

Any time a new part or product is made for the first time or when a production machine is set up for repeat production after having been used for another job, a first piece inspection is required. When the first item completes the manufacturing process it is inspected, and if it conforms to specifications further production is allowed to continue. If the first piece is not acceptable, the machine or in some cases the entire production line must be stopped and reset to correct any problems. First piece inspection is necessary before production can begin.

In-process Inspection

In-process inspection must be applied continuously during production. Each production station must have the necessary tools to accomplish the required inspection operations (Figure 7.4). Past practice of employing roving quality inspectors to randomly inspect parts throughout the factory has largely been replaced by in-process inspection performed by the machine operator or assembler at his or her workstation. In this manner the worker, often aided by statistical process control (SPC) measures, is made responsible for product quality.

One Hundred Percent Inspection

One hundred percent inspection refers to the inspection of every part or assembly produced. Though this might appear to be a way to solve all quality control problems, it is often impractical and altogether too expensive to

FIGURE 7.4
In-process inspections take place constantly during production (Del Manufacturing).

accomplish in routine manufacturing. This is especially true in high-volume production where it would not be feasible to inspect each and every item produced. Also, one hundred percent inspection does not always produce perfect quality. Inspection operations can be fallible and fail to detect defective products, especially in the case of highly repetitive manual inspection.

Regardless of its faults, one hundred percent inspection is often used for critical components. Critical aircraft parts may be one hundred percent inspected since the possible acceptance of an out-of-specification part could have serious consequences. Another place where such inspection may be applied is where nonconforming parts begin to show up continually in the production process, and the problem must be tracked down and immediately corrected. One hundred percent inspection may be necessary until production quality is restored, when normal in-process inspection may resume.

Final Inspection

Final inspection presents another opportunity to sort out parts or assemblies that do not meet specifications. Nonconforming parts are rejected to be scrapped or recycled for rework if possible. It is naturally desirable to avoid scrapping as many parts as possible since they often represent a large production expense.

NONDIMENSIONAL QUALITY CONTROL

Quality control and assurance practices must be observed for both dimensional and nondimensional specifications with equal dedication. This includes a wide variety of tests and inspections on all types of products, especially complex operating assemblies.

Nondimensional quality control inspection and testing are far-reaching and often involve not only state-of-the-art technology, but many truly remarkable engineering innovations. One example is the automotive dynamometer test that measures power output and emission levels of engines (Figure 7.5). Computerized equipment with the ability to analyze test data with great speed plays a vital part in the overall quality control and assurance effort. Oil samples from lift truck engines are spectrographically analyzed to determine chemical and molecular content (Figure 7.5). The samples are first vaporized (Figure 7.6) and then the light energy emissions from this process are computer analyzed (Figure 7.7) to determine the chemical makeup and content of foreign matter in the oil sample. The process can determine the type and amounts of metal

FIGURE 7.5
Automotive dynamometer testing (GM—Pontiac Division).

particles suspended in the oil and the data can be used to make a determination of where metallic abrasion is taking place within the engine during operation.

Aircraft and spacecraft airframes are tested to the same stresses that will be experienced in service. Spacecraft are enclosed in a large steel framework where hydraulic jacks exert forces of up to one million pounds to simulate what the vehicle will experience during launch,

FIGURE 7.6
Vaporizing an engine crank case oil sample for spectrographic analysis (Allis-Chalmers).

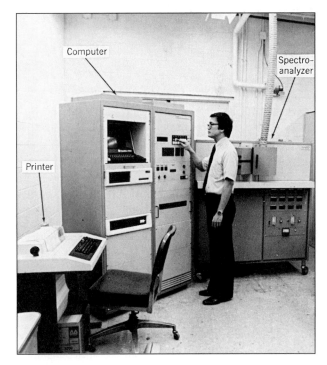

FIGURE 7.7
Computerized spectrographic analysis of the oil sample (Allis-Chalmers).

FIGURE 7.8
Space shuttle flight stress simulator subjects air frame to forces that it will encounter
in actual flight (Lockheed Co.).

spaceflight, and reentry into the earth's atmosphere (Figure 7.8). Another example of nondimensional quality control is computer analysis of microwave oven emissions (Figure 7.9). This type of inspection is vital to consumer protection.

QUALITY ASSURANCE AFTER MANUFACTURING

Quality control and assurance do not end at the manufacturer's shipping dock. They continue well after manufacture in the form of warranties providing factory or factory-authorized service, spare parts, field service, service contracts, reliability studies, failure reports and recommendations, and repair and reconditioning product support.

Purchasers of almost any product expect and are entitled to a warranty or guarantee that will be in force for a period of time after the product arrives in the hands of the end user. Warranties vary in length ranging from a few months to lifetime full replacement. Some warranties cover full repair and replacement for a short period of time and cover only specific components for an extended time. Extended warranties may be purchased at extra cost in many cases. Related to this are service contracts for which the customer pays a flat extra cost, sometimes a

FIGURE 7.9
Computer analysis of microwave oven emissions (Whirlpool Corp.).

percentage of the purchase price, in exchange for repair service for the duration of the contract.

Critical plant equipment that fails in service must often be restored to functional capability as quickly as possible, and efficient service can be critical to the continuing function of plants and businesses. Many manufacturers provide factory service on their products. This capability may be extended over wide geographical areas by using factory-authorized service through other businesses. Spare parts can often be purchased from either the original maker or a factory-authorized service center.

Many manufacturers, especially those of high-technology products, are concerned with reliability and failure rates of their products in service. Product failure and reliability analysis is useful in determining where improvements or correction of deficiencies in the production or design need to be made. Reliability analysis can greatly improve a manufacturer's product quality. The compilation of reliability statistics can be used effectively in product advertising, thus enhancing the manufacturer's standing in the business.

PRODUCT TESTING

Product testing is vital to successful manufacturing, and manufacturing industries often place a large emphasis on product testing. Testing may take place at the prototype stage before an actual product is manufactured for sale. Product testing also may be carried out in controlled markets where a few preproduction models are placed in selected hands for the purpose of testing and evaluation. In the case of aircraft, for example, a license to manufacture will not be granted until design and safety specifications have been met. The manufacturer must assume the costs of manufacturing and testing preproduction prototypes.

Endurance testing of automobiles is another significant product testing activity. Production models are test driven through many different environments to simulate what the vehicles will encounter in actual service. By continuous testing, several years of typical service of the vehicle can be compressed into a few months. The data collected can be invaluable in determining design problems and in predicting the reliability and effective service life of the product.

Review Questions

1. What is the difference between quality control and quality assurance?
2. What are the three major areas of quality assurance?
3. What are the responsibilities of the source inspector?
4. What items does the source inspector look for?
5. Give two examples of metallurgical testing.
6. How would a given metal be checked for alloy content?
7. What is involved in a dimensional inspection?
8. Why are dimensional inspections important in manufacturing?
9. What is the difference between destructive and nondestructive testing?
10. Where might destructive testing be applied?
11. What is a performance inspection?
12. What are the major points of in-process quality control?
13. How are statistics used in quality control?
14. What is the purpose of a first piece or first article inspection?

CHAPTER 8

Inspection and Measurement

Objectives

This chapter will enable you to:

1. Describe why, how, and with what tools a product is inspected.

2. Describe a gage that could be used for comparison inspection.

3. Identify common tools for dimensional measurement.

4. Select an effective measuring instrument for a given application.

5. Describe the relationship between tolerance, precision, and resolution and use this to select inspection methods and equipment.

6. Define calibration, relate the importance of measurement standards, and describe how instrument traceability in measurement operations is assured.

7. Explain how the effect of temperature changes differs for steel and aluminum workpieces.

Manufacturing involves the modification of raw materials for the purpose of meeting product specifications established when the product is designed. The manufactured product must conform to the intended specifications so that it may perform its intended function effectively. Many types of specifications exist. The weight, length, area, or volume of an object may be specified. For any surface of an object the shape, smoothness, or ability to reflect light may be specified. Sound pitch or volume level, speed, acceleration, and many other factors also may be specified. This chapter will deal with specifications of size that are commonly critical to manufactured products, such as length, width, height, diameter, depth, and thickness. The chapter will also include form specifications involving angles, shape, or surface qualities.

Any product must be inspected to judge compliance with the established specifications. This step is necessary because a product that does not conform to the design specifications may fail to perform its intended function and therefore be of little value. Should this occur the manufacturer will have wasted the material, employee time, use of machinery, and the efforts of the company to support the manufacturing operation. More importantly, a nonconforming product may fail in use, leading to customer dissatisfaction and possible injuries for which the manufacturer may be liable.

Products are inspected in many different ways. A product may be visually inspected to ensure general conformance to specifications. A feature of the product such as length may be compared to an inspection tool to judge whether it meets a specification, or a feature may be measured with an instrument and the resulting reading compared to the specification to judge conformance. All these methods—visual inspection, comparison inspection, and measurement—are commonly used in manufacturing.

Computers and other electronic devices have found wide application in precision measurement. They have made high precision measurement routine and reliable, permitting precision manufacturing to flourish and produce many of the high-quality products that consumers enjoy. Examples of electronic measuring devices include high resolution digital readouts for machine tools (Figure 8.1), the three-axis coordinate measuring machine (Figure 8.2), and many types of digital electronic instruments and gages used for precision measurement (Figure 8.3).

INSPECTION

Since manufacturing operations involve varying degrees of precision and create products of many different sizes and shapes with both internal and external features, many

FIGURE 8.1
High discrimination electronic digital readout systems may be retrofitted to existing machine tools, greatly enhancing their positioning and measurement accuracy (MTI Corporation).

FIGURE 8.2
The three-axis coordinate measuring machine used in auto body dimensional inspection (American Motors Corporation).

FIGURE 8.3
Electronic comparators and digital height master (The L.S. Starrett Company).

different inspection instruments are needed. These range from simple plugs, rings, and templates to semiprecision tools such as steel rules and protractors and to highly precise measuring instruments such as micrometers, height gages, and coordinate measuring machines.

Measurement in manufacturing processes is commonly confined to size specifications. Frequently used expressions of size include width, depth, outside and inside diameter, center-to-center distance of part features such as holes, and wall thickness of tubes and pipes (Figure 8.4). In addition, form specifications (Figure 8.5) such as angles, perpendicularity (squareness), flatness, straightness, concentricity (or eccentricity), and surface roughness are encountered in manufacturing operations, although these require special measuring devices.

Tolerances

An understanding of **tolerances** is vital to almost any successful manufacturing operation. A tolerance is a product specification that defines the acceptable range of sizes through which the part will fit and function as intended. Tolerance is required because it is not possible to manufacture any part to an exact size; however, it is possible to closely approach an exact size, depending on design requirements and the capability of the machinery used for manufacturing. It also is possible to measure parts reliably to assure that the product conforms to the tolerance. For each product component and assembly the designer must designate acceptable tolerances that will accommodate variations in the manufacturing process.

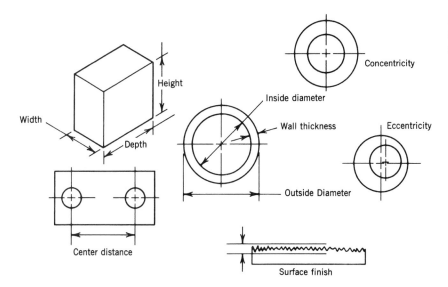

FIGURE 8.4
Common size and form specifications.

Tolerances generally fall into two important categories, size and form. Several methods of specifying size tolerances may be used (Figure 8.6). Most manufacturing blueprints utilize "limits" tolerancing in which the largest and smallest sizes are specified. This creates a range of permissible manufacturing error, or tolerance. Other common methods for designating tolerances state the theoretical exact size plus or minus a permissible amount of deviation. "Bilateral" tolerancing states the exact size with both plus and minus permitted deviation. "Unilateral" tolerancing states the exact size and includes only a plus or minus deviation, not both.

The designer, of course, would prefer that the parts are manufactured as near the exact theoretical size as possible; however, the tolerance is necessary to accommodate variations that exist in even the most carefully controlled manufacturing processes. For example, manufacturing precision parts to close (small) tolerances of a few millionths of an inch generally requires more care and attention than looser (larger) tolerance production. Close tolerance manufacturing requires higher technology in manufacturing equipment and inspection tools. All this

extra effort results in higher manufacturing costs when close tolerances are specified by the designer.

Manufacturing drawings should contain specifications for form and position dimensions and tolerances as well as specifications for size dimensions and tolerances. Tolerances of form include squareness, parallelism, straightness, flatness, circularity, angularity, and concentricity, and position tolerances include the location of holes or other features (Figure 8.7). Both size and form specifications are vital for the manufacture of a part that will fit and function as intended.

Consider the example illustrated in Figure 8.8. Flange A must bolt to mating part B. A size dimension is provided on the drawing showing the diameter of the bolt holes. A hole size sufficient to contain the bolt is naturally required; however, if the position of the holes is not maintained within tolerance the parts may not properly assemble, even if the diameter of the holes conforms to the size tolerance. Therefore the manufacturer must maintain enough control of the manufacturing process to conform to both the hole position tolerance and the hole size tolerance.

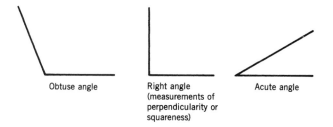

FIGURE 8.5
Angular quantities must be measured.

PRECISION, RESOLUTION, AND TOLERANCE

Precision is a complex issue that is critical to any measuring process. Precision involves the resolution of the measuring instrument, the reliability and repeatability of the instrument, the condition of the surfaces of the workpiece, the skill of the worker operating the instrument, the temperature, and any other environmental factor that

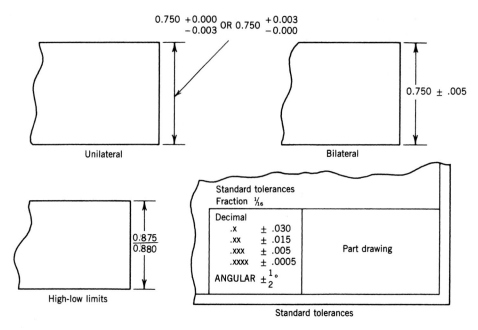

FIGURE 8.6
Types of tolerances and standard tolerances.

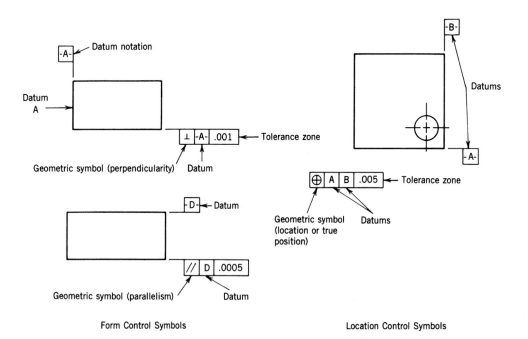

FIGURE 8.7
Examples of form and position tolerances.

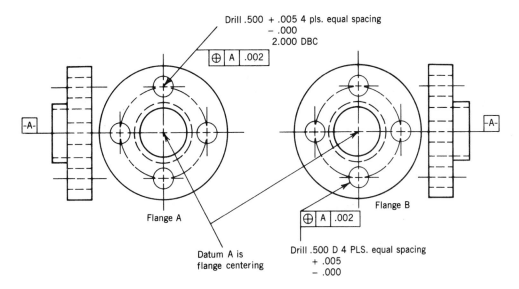

FIGURE 8.8
Accurate part feature locations are essential to successful assembly.

may affect the reading. Precision can be defined as a range of readings within which the process will measure repeatedly. Determining the precision of a measuring operation is difficult and requires a statistical study, but it can be estimated by measuring a feature on a part several times and figuring the difference between the highest and lowest readings. For example if a part was measured one hundred times and the readings varied from a low of .513 in. to a high of .519 in., then the precision of this measuring operation could be estimated as .519 − .513 = .006 in. If the instrument or any environmental factors are changed, the precision of the measuring operation may also change.

The resolution of a measuring instrument is easy to determine. It is the smallest deviation that the instrument can detect. This is equal to the smallest graduation on the scale or readout of the instrument. The precision of a measuring operation could never be finer than the resolution of the instrument utilized. An instrument with .001 in. resolution could never measure to .0005 in. precision or any precision finer than .001. Do not be fooled, however, into believing that the precision is always equal to the resolution of the instrument. Due to many factors influencing the measuring process, the precision is often more coarse (larger) than the resolution of the instrument.

The precision of a measuring operation has a very close relationship to the tolerance being measured. The commonly used rule requires that the precision must be no larger than one-tenth of the tolerance range. This is a critical rule that must be considered in selecting a measuring instrument and designing a measuring process.

Example: The length dimension of a part is shown on the blueprint as .905 to .915 in. (.910 ± .005 in.).

The tolerance range is .915 − .905 = .010 in.

The required precision is .010 ÷ 10 = .001 in.

The maximum permitted size variation in the manufacturing process is .010 in., and the maximum permitted variation in the measuring process is .001 in. This assures that most of the manufacturing variation permitted according to the blueprint will be reserved for the manufacturing process that creates the product feature. The production process may vary .010 − .001 = .009 in., but the measuring process may vary only .001 in.

COMPARISON INSPECTION

Many production inspection operations are performed for the purpose of determining conformance to tolerances, and an actual reading of the part size is not needed. Here the techniques of comparison inspection can be applied. In comparison inspection dimensions are checked against tolerances by simply comparing the part to an inspection tool (gage) that is known to be the correct size. If the gage fits, then the part dimension is in tolerance. If the gage does not fit, then the part dimension is out of tolerance. In many situations it is not necessary to know how far the part is out of tolerance, since any amount out of tolerance renders the part unacceptable. However, there may be a possibility of reworking rejected parts to restore them to

FIGURE 8.9
Fixed GO and NO-GO gages for checking tolerances.

FIGURE 8.11
Ring gage for checking threads.

design specifications. This is done in a secondary rework operation.

In most cases fixed gages are used for comparison inspection. These inspection tools do not display a reading; they simply fit or do not fit. Acceptance or rejection of the part feature is decided purely on this basis. This concept of measurement is often referred to as GO and NO GO or GO/NO GO. Gages used in this type of inspection are known as GO and NO GO gages (Figure 8.9).

Typical GO/NO GO gages are plug gages (Figure 8.10) used for comparison inspection of internal diameters and ring gages for external diameters (Figure 8.11). An alternative gage for quick GO/NO GO comparison inspection of outside diameters is the snap gage (Figure 8.12). Some examples of GO/NO GO gages are thread plug gages and thread ring gages. The thread plug gage is widely used in checking tolerances of internal threads. One end of the gage, marked the GO end, will screw into the thread if the thread is in tolerance. The other end of the gage, marked NO GO, will not screw into the thread if the thread conforms to the tolerance. The worker has a comparison tool that can be quickly applied to determine whether thread dimensions are within tolerances.

Screw-pitch gages (Figure 8.13) are used to determine thread pitches (distance between threads), and radius gages (Figure 8.14) are used to check the form of arcs of specific radii on manufactured parts.

There are several advantages in this type of inspection. Operation of the gage is extremely simple, so variations introduced by differences in worker skill are minimized. The worker is not required to read or interpret any numeric data, eliminating another source of error. Although precision inspection tools are expensive and often represent a large portion of production tooling cost, the gages used for comparison inspection are simple, durable, and relatively inexpensive when compared to other types of inspection equipment.

A disadvantage of fixed gaging for comparison inspection is that a separate gage is required for each different size to be inspected. This may cause a large expense if gages must be purchased for many different dimensions. The primary disadvantage of comparison inspection, however is that very little information about the part is gathered other than in-tolerance (accept) or out-of-tolerance (reject). A measurement of feature size would

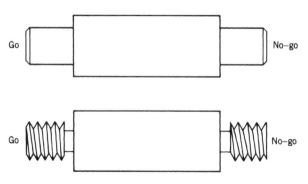

FIGURE 8.10
Plug and thread gages.

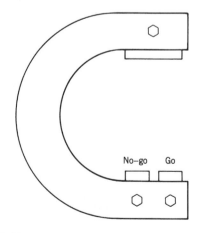

FIGURE 8.12
Snap gage. Such tools are useful for making a quick check of dimensions.

FIGURE 8.13
The screw-pitch gage is used to determine thread pitches (Compliments of Mitutoyo/MTI Corp.).

be helpful for controlling the manufacturing process that created the feature, but no size reading is obtained in a comparison inspection operation.

MEASUREMENT OF DIMENSIONS IN MANUFACTURING OPERATIONS

Measurement involves the use of a measuring instrument to obtain a numerical reading of the length, width, thickness, roundness, flatness, etc. of a feature of a product. The reading can be compared to the tolerance to judge conformance, and the measurement may be used to adjust the manufacturing process and to assure that it continues to produce conforming products. The selection and application of the wide range of tools and equipment

FIGURE 8.14
Radius gages.

used in dimensional measurement are one of the largest and most interesting manufacturing support activities. As the age of high technology has progressed, production of parts to tolerances of a few millionths of an inch has become commonplace in today's high-precision manufacturing environment, and the requirement for precision measurement has grown in lockstep.

Measurement Units

The industrial world, for the most part, has adopted the metric system of measurement, known as the Système International d'Unites or the International Metric System (SI). This system is based on the meter (m), which is 39.37 inches long. In manufacturing almost all metric linear dimensions are measured in millimeters (mm). The United States is committed to conversion to the metric system from the English system of feet and inches; however, this conversion is being gradually accepted in the United States. Many U.S. manufacturing plants and machine shops continue to use the English system of measurement, so machinists, manufacturing technologists, and engineers often refer to conversion tables or use conversion factors such as the following:

1 in. = 25.4 millimeters (mm)
1 mm = .03937 in.

Measuring Tools

Gages and instruments used for measurement of dimensions in manufacturing operations may be separated into three major categories: direct-reading instruments, deviation-type gages, and transfer-type instruments.

Direct-reading instruments have a graduated scale, and measurements are read directly from that scale. Measuring instruments are only as reliable as the user's skill, and accuracy can be affected by the following:

1. Line matching. This requires good eyesight.
2. Parallax in reading. An angled view of matching lines or a measured object can cause an erroneous reading.
3. Deformation of the workpiece or instrument. For example, putting too much pressure on an instrument when measuring a part may stretch or spring the instrument and distort the reading.
4. Temperature effects. A workpiece or measuring tool can expand and contract with temperature changes, thus altering the reading.

Deviation gages have a graduated scale, but the displayed reading does not include the entire part size. The reading represents the difference between the size of a

FIGURE 8.15
In production measurement, only the tolerance portion of a dimension must be measured.

Gage

Low limit adjusted to .740

High limit adjusted to .760

Exact size

.750 ± .010

Tolerance zone

Low limit .740

High limit .760

Only this range needs to be measured

master and the measured feature. For example, if the dimension .750 ± .010 in. (Figure 8.15) requires inspection and the production process yields parts that are all close to this size, it is not necessary to inspect the .750 in. portion of the dimension each time. Only the tolerance of ±.010 needs inspecting, so it is not necessary to use a measuring tool that has the capability to measure over the full range of the .750 in. dimension. Although it would be acceptable to measure this dimension with such a tool, a production gage designed and adjusted to check only the deviation in size from a .750 in. master would be much faster, more reliable, and easier to use.

Transfer measuring tools do not have a graduated scale but are used to gage and transfer the dimension to a direct-reading instrument that produces a reading.

Semiprecision Measuring Tools

Measuring tools such as machinists' combination sets and machinists' steel rules should be considered to be semi-

precision tools. The steel rule (Figure 8.16) can be read with average eyesight to an accuracy of $\frac{1}{64}$ in., and a metric rule, .5 mm. English rules are made in four fractional divisions, $\frac{1}{8}$, $\frac{1}{16}$, $\frac{1}{32}$, and $\frac{1}{64}$ in. in addition to $\frac{1}{100}$ and $\frac{1}{10}$ in. decimal divisions. Metric rules normally have 1 centimeter (cm), 1 mm, and $\frac{1}{2}$ mm divisions. Machinist's rules are often used with a square head or combination set (Figure 8.17). Squareness and angularity are measured with the square head and protractor of the combination set. The precision of the protractor is only the nearest degree, not minute or second portions of a degree. Square heads vary in precision but often are manufactured within a few minutes of perpendicular.

Spring calipers (Figure 8.18) are sometimes used to transfer measurements to steel rules and are reliable for that use to about $\frac{1}{64}$ in. Thickness gages (Figure 8.19) consist of hardened steel leaves made to specific thicknesses that are marked on each leaf in either inch or metric measure. They are commonly used to determine the dimension of small spaces between product features and

FIGURE 8.16
Steel rules.

FIGURE 8.17
Combination set. A square head, protractor, or centerhead may be mounted on a graduated blade for various uses (Compliments of Mitutoyo/MTI Corp.).

FIGURE 8.19
Thickness gages, also called feeler gages, have leaves of various thicknesses (Compliments of Mitutoyo/MTI Corp.).

are used in conjunction with a straightedge to determine the flatness of a surface. Rule-depth gages (Figure 8.20) are used to determine the depths of grooves and shoulders by the transfer method.

Tools for Precision Measurement

Precision measurement involves direct-reading instruments such as micrometers, calipers, optical comparators, coordinate measuring machines, and surface roughness analyzers. Deviation gages also are used for highly pre-

FIGURE 8.18
Spring calipers for *(a)* inside and *(b)* outside measurements.

FIGURE 8.20
The rule-depth gage is a semiprecision measuring tool, used here to measure the depth of a counterbore.

FIGURE 8.21
Small (0 to 1 in.) outside micrometer in use.

FIGURE 8.23
This depth micrometer reads in thousandths of an inch on the thimble and in a digital readout counter. It has interchangeable rods so the range can be extended (Compliments of Mitutoyo/MTI Corp.).

cise measurement; they include indicators, precision height gages, air gages, and electronic gages. Tools such as small hole gages and telescoping gages are required for precision transfer measurements.

Micrometers (commonly called mikes by machinists) are perhaps the most widely used precision measuring tools in manufacturing. Outside (caliper-type) micrometers measure lengths and outside diameters and typically have a measuring range of 1 in. These can be purchased in sizes ranging from 0 to 1 in. (0 to 25 mm) (Figure 8.21) to very large instruments that can measure dimensions up to several feet (Figure 8.22). Depth micrometers are used to make precision measurements of the depths of slots and shoulders (Figure 8.23). Screw thread micrometers

FIGURE 8.22
Large outside micrometer (The L. S. Starrett Company).

FIGURE 8.24
Screw thread micrometer that uses different interchangeable anvils (Compliments of Mitutoyo/MTI Corp.).

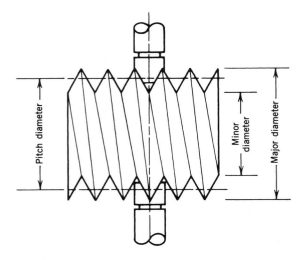

FIGURE 8.25
Measurement for thread pitch diameters is made on the flank of the threads.

FIGURE 8.26
Inside micrometer set with interchangeable rods that give it a wide range of measurement (Compliments of Mitutoyo/MTI Corp.).

(Figure 8.24) have interchangeable anvils for different thread pitches and a 60 degree point on one anvil, making possible the precision measurement of 60 degree screw thread **pitch diameters** (Figure 8.25). Inside micrometers (Figure 8.26) have the same barrel and graduated thimble as outside or caliper-type micrometers but are used to measure internal surfaces such as bores. Micrometers typically measure in .001 in. or .01 mm increments, but digital micrometers and micrometers equipped with a special vernier scale can measure in .0001 in. or 1 micron (μm) increments.

Vernier scales are used on vernier calipers (Figures 8.27 to 8.30), on some micrometers, and on many other types of measuring instruments. The purpose of the vernier is to mechanically magnify tiny variations in size so that they can be read by the human eye. The vernier divides instrument graduations into smaller increments by matching them against a vernier scale. More commonly used in manufacturing are the dial-type and digital readout calipers, which are quicker and easier to read. Calipers are very versatile instruments able to measure outside diame-

FIGURE 8.27
Vernier calipers; the upper one measures in inches and the lower one measures in millimeters (Compliments of Mitutoyo/MTI Corp.).

FIGURE 8.28
Vernier caliper measuring outside diameter. This vernier caliper measures both inches and metric.

FIGURE 8.30
Vernier caliper measuring depths.

ters, inside diameters, and depths, but they cannot measure as precisely as micrometers. Vernier or dial calipers are capable of discriminating to .001 in. or .02 mm, and digital calipers read .0005 in. or .01 mm increments.

Deviation gages used in precision measurement include mechanical dial indicators (Figure 8.31) and test indicators in addition to electronic digital indicators. Indicators are precision instruments that detect minute movements of a plunger or pivoting needle. Dial (Figures 8.31 and 8.32), and test (Figure 8.33) or digital indicators can detect deviations of .001 or .0001 in. or equivalent metric measures. Indicators have many uses, from measuring size dimensions such as height on a surface plate to

checking form specifications such as runout (eccentricity). Indicators are often fastened to a surface gage (Figure 8.34) and used with a precision height master (Figures 8.35 and 8.36) or mounted as dial indicator comparator stands (Figures 8.37 and 8.38 on p. 162). Many types of

FIGURE 8.29
Vernier caliper measuring inside diameter.

FIGURE 8.31
Dial indicator (Compliments of Mitutoyo/MTI Corp.).

FIGURE 8.32
Dial indicators are often mounted on magnetic stands so that they can be fastened where necessary to machine parts for taking measurements (Compliments of Mitutoyo/MTI Corp.).

FIGURE 8.33
Test dial indicator (Compliments of Mitutoyo/MTI Corp.).

FIGURE 8.34
Test dial indicators such as this one are often used for transferring precision measurements from workpiece to precision height gage or gage block stacks. To do this the indicator must be mounted on a surface gage or a standard height gage. These measurements usually take place on a granite surface plate (The L. S. Starrett Company).

production widely used for measurement (Figure 8.39 on p. 163). The electronic comparator, the electronic counterpart of the dial indicator comparator, can measure in millionths of an inch (Figure 8.40 on p. 164).

Air gages (Figure 8.41 on p. 164) and electronic gages are more precise deviation-type gages that are used to measure inside diameters and other features. Air gages (Figure 8.42 on p. 164) and electronic gages are used where many parts of the same dimension must be very precisely measured.

Precision transfer-type instruments include telescoping gages (Figure 8.43 on p. 165), small-hole gages for measuring hole diameters (Figure 8.44 on p. 165), and adjustable parallels to gage slot widths that are difficult to measure (Figure 8.45 on p. 165). All of these gages adjust to the size of the feature and then are measured using a micrometer to produce a reading.

Form Measurement

The importance of precision measurement of size dimensions is well understood, but measurement of form and location specifications is equally important. An example is

FIGURE 8.35
The precision height master is capable of measuring to
.0001 in. (The L. S. Starrett Company).

FIGURE 8.36
Precision height master (MTI Corporation).

two parts mated by a four-hole bolt pattern. Regardless of the size of the holes, if one or more holes is located incorrectly or the holes are not cylindrical (incorrect form) the parts may not fit together properly. Form measurements often have been shortcut or ignored in the past due to their painstaking, time-consuming, and error prone nature. The primary tools for form and location specification measurements have been the indicator, surface gage, flat surface plate, and 90 degree angle plate. The complex setups and procedures required in using these tools to measure perpendicularity, flatness, straightness, roundness, cylindricity, concentricity, feature position, and other specifications were costly and required consummate skill. The advent of the coordinate measuring machine and other electronic measuring instruments has greatly simplified and increased the precision and reliability of the measurement of form and location specifications. The corresponding improvements in quality and conformance in manufacturing have created a breakthrough in performance, reliability, longevity, and aesthetics in today's products.

The universal bevel protractor is a precision tool used for measuring angles (Figure 8.46 on p. 165). It is more accurate than the combination set protractor, which can measure only in degrees. The bevel protractor has a vernier scale that makes possible a reading within the range of 5 minutes of arc.

The optical comparator (also called a profile projector) (Figure 8.47 on p. 166), where a magnified image of the part being inspected is projected onto a screen, precisely measures size dimensions that are not accessible to other instruments and is very useful for measuring angles and inspecting form specifications (Figure 8.48 on p. 166).

The toolmaker's microscope has a two-axis stage upon which parts are moved and viewed under high magnification (Figure 8.49 on p. 167). Electronic tracking of the movement of a stylus as it traces the workpiece surface also can be used to measure the two-dimensional (profile) geometry of machined surfaces (Figure 8.50 on p. 167). Other form measurement procedures include inspecting and graphing part geometry such as roundness (Figure 8.51 on p. 167).

FIGURE 8.37
Mechanical dial indicator comparator (MTI Corporation).

The coordinate measuring machine (CMM) consists of a probe mounted on sliding ways that permit free motion in three-dimensional space (Figure 8.52 on p. 168). When the probe touches the workpiece the position of each of the three axes is communicated to an attached computer, which calculates the position of the probe. By calculating the relationship between multiple point locations on the part that has been probed, features can be measured quickly, easily, and reliably. Feature size may be measured using coordinate measuring machines, but the most important capability of the CMM lies in measuring form dimensions, such as flatness, roundness, feature position, parallelism, and many other geometric features, without the complexity and potential errors of difficult surface plate setups. For example, the measurement probe can be used to quickly touch multiple points on the part being inspected. Using the position data for these points, the computer can calculate a location or the deviation from a geometrically perfect feature. The process is extremely fast and reliable, and the computer can automatically generate reports of measurement results.

The smoothness of the surface of a machined part is called surface finish (surface roughness). When a cross section of a machined surface is magnified, the surface finish looks like a mountain range, a series of peaks with valleys in between, each valley created by the tip of the

FIGURE 8.38
Dial indicator comparator stand. Small parts can be quickly measured with this tool (Compliments of Mitutoyo/MTI Corp.).

FIGURE 8.39
Dial indicator gage for fast and accurate measurement in a production situation (MTI Corporation).

FIGURE 8.40
Digital electronic comparator (MTI Corporation).

cutting tool as it removes material. Surface roughness is measured in microns (metric) or microinches (millionths of an inch) from a smooth 32 microinches to a coarse 500 microinches. Ground or polished finishes are smoother than 32 microinches, and a surface with roughness of 4 microinches has a mirrorlike polished finish. Surface finish is vital to many parts such as gears, bearings, and hydraulic sealing surfaces. Surface roughness can be measured by comparison inspection or by direct measurement.

FIGURE 8.41
Air gages measure dimensions as a function of air flow and pressure (Federal Products Corporation).

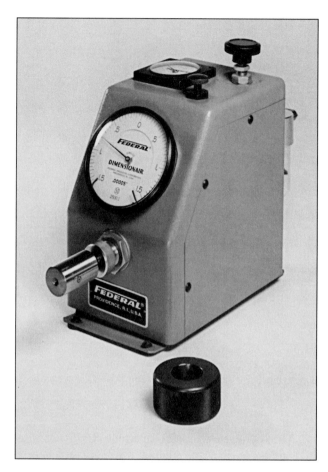

FIGURE 8.42
The air gage measures clearances between bores and nominal shaft sizes. The setting ring is used for adjustment purposes (Federal Products Corporation).

In comparison inspection the surface roughness can be estimated by matching a workpiece surface with blocks that display a series of increasingly smooth surfaces (Figure 8.53 on p. 168). The profilometer surface measuring instrument is used for more precise measurement of surface finish (surface roughness) (Figure 8.54 on p. 168 and Figure 8.55 on p. 169). The profilometer directly measures surface finish as it traces, records, and displays the depth of the peaks and valleys.

Effects of Temperature Changes

All materials, especially metals, expand or contract with temperature changes, and this size change can affect measurements. Whether the workpiece is made of steel, aluminum, or other materials any precision measurement is meaningless if the measuring temperature is not known, so a standard measuring temperature of 68°F has been established worldwide. Any measurement made at any other temperature has some amount of error due to tem-

FIGURE 8.43
Telescoping gages.

FIGURE 8.45
Adjustable parallel is measured with a micrometer.

perature effects. Care must be taken when the workpiece and measuring instrument are made of two different materials, because temperature variations will affect the two materials differently.

Extreme precision measurements are performed in temperature-controlled (68°F) gage laboratories with fil-

tered air and controlled humidity. Workpieces and instruments must be allowed to "soak" or stabilize to gage room temperature for a period of time before measurement to assure that the standard measuring temperature has been reached throughout. As a rule of thumb for steel workpieces, if the tolerance is finer than .002 in. then normal workplace temperatures can significantly affect common measurements. Since aluminum expands more than steel with temperature changes, measurements of aluminum workpieces with tolerances finer than .004 in. are subject to the same temperature concerns. In these circumstances measurement must be performed in an environment where the temperature remains within 10 degrees of the standard 68°F measuring temperature, or better yet in a temperature-controlled environment. For large dimensions and

FIGURE 8.44
Small-hole gage being measured with a micrometer.

FIGURE 8.46
Universal bevel protractor (Compliments of Mitutoyo/MTI Corp.).

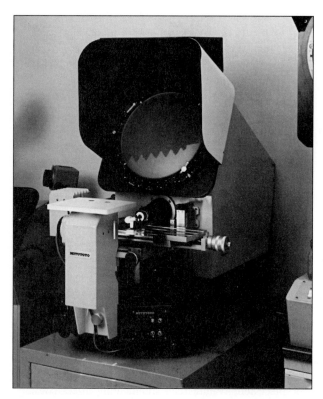

FIGURE 8.47
The optical comparator projects a greatly magnified image
of the part onto a large screen (MTI Corporation).

where the temperature differs greatly from 68°F, even
more care should be taken to assure that temperature does
not affect the validity of measurements.

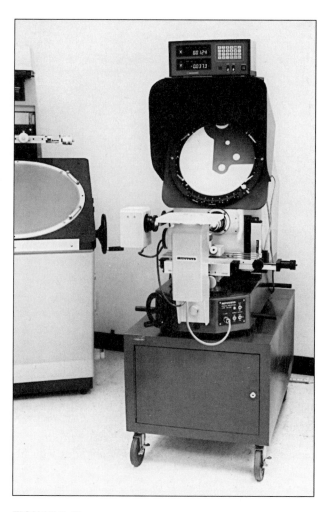

FIGURE 8.48
Optical comparator or profile projector can be used for
precision measurement or comparing curves and radii
(Compliments of Mitutoyo/MTI Corp.).

CALIBRATION, STANDARDS, INTERCHANGEABILITY, AND TRACEABILITY

Manufacturing enterprises depend upon the concept of
interchangeable parts for modern interchangeable manufacturing. Few complex assemblies are made by single
manufacturers. It is more likely that many different makers will participate in the process, each one contributing a
part or parts of the final product. It is easy to see how
assembly problems could occur if specifications regarding
size as well as geometry of assembly parts were not
inspected with worldwide consistency to assure conformance. Mechanical parts would not fit, nor could replacement parts be obtained with any assurance that they would
fit or have the same dimensions as the original.

The **interchangeability** of manufactured parts
depends entirely upon workpiece dimensions conforming
to design tolerances, and different parts that fit into an
assembly are frequently made by many different manufacturers. To make this possible, instruments and gages
used by all participating manufacturers must measure consistently, and this is accomplished through frequent calibration. Calibration of measuring tools is extremely
important to the control of product quality.

Calibration is the process of comparing measuring
tools against standards of known size. The process is a continuous one in almost any manufacturing industry. In a typical manufacturing plant the calibration of inspection tools
is the responsibility of the gage laboratory. Calibration

FIGURE 8.49
Toolmaker's microscope, for a magnified look at
dimensions (Nikon Inc., Instrument Division).

FIGURE 8.51
Measuring and graphing roundness, a form inspection
(MTI Corporation).

equipment is very precise, sensitive, delicate, and expensive and must be used in controlled environments.

Inspection tools used in the factory are sent on a regular basis to an on-site gage laboratory where precision calibration procedures are carried out in a temperature controlled (68°F) environment. The tools are allowed to stabilize to gage room temperature, they are compared to appropriate standards, adjustments are made to bring the tools to conformity, and then the tools are tagged with a

FIGURE 8.50
Measuring profile geometry electronically (MTI
Corporation).

sticker indicating the date of calibration. It is the responsibility of the inspector to ensure that the calibration is current for any inspection tools used on the production floor.

Calibration procedures are by no means unique to the measurement of dimensions. Calibration is equally important in many other areas such as electronics where frequency standards must be maintained. Calibration of equipment for reading pressures, flow, weight, colors, performance, and other specifications is required for manufactured products.

Gage blocks (Figure 8.56) are among the most common and versatile measurement standards used in manufacturing. Gage blocks are used in production measurement, machine setup, and calibration of inspection tools (Figure 8.57). They are used together with sine bars (Figure 8.58) to make angles as precise as 1 to 5 seconds of arc.

A gage block set (Figure 8.59) consists of a series of blocks of various heights that are made from a hard, wear-resistant material, usually hardened steel. Each block has been manufactured under controlled conditions and is dimensionally very close to its nominal size. A gage block marked .500 in. will be, at the standard

FIGURE 8.52
Three-axis coordinate
measurement machine with
computer readouts and
printer (Sheffield
Measurement Division).

precision measuring temperature of 68°F, within a few
millionths of an inch of its marked size. These sets may
be purchased in several grades of precision: the most
inexpensive gage blocks are within .000020 in. of nominal size, and laboratory grade gage blocks will be within
.000002 in. of nominal. Form tolerances including flatness and parallelism of the working surfaces of the block
are also very close.

Nearly perfect dimensions make the gage block a
very effective standard, and gage block sets are designed

FIGURE 8.53
Roughness comparison standards.

FIGURE 8.54
Surface finish, a measurement of length in millionths of
inches (MTI Corporation).

FIGURE 8.57
Gage blocks assembled into stacks make versatile and highly accurate measuring tools (MTI Corporation).

FIGURE 8.55
Surface texture analyzing instruments (Compliments of Mitutoyo/MTI Corp.).

so that several gage blocks can be stacked to make a master of the desired length. If these gage blocks are properly "rung" together with no contaminants or air between blocks, then the stack will be extremely close to the calculated size.

In today's manufacturing environment, parts manufactured on opposite sides of the world may be required to fit together and function effectively in a product assembly. This can be accomplished only if the size standards used for calibration are precisely correlated to one worldwide length standard. This is called "instrument **traceability.**" Master measurement standards are commonly established and maintained by government agencies such as the National Institute of Standards and Technology (NIST), formerly called the National Bureau of Standards (NBS), in the United States. The NIST maintains a length standard that is consistent within atomic precision with the

FIGURE 8.56
Gage blocks, one of the most common length measurement standards in manufacturing (MTI Corporation).

FIGURE 8.58
Precision sine bars (Compliments of Mitutoyo/MTI Corp.).

FIGURE 8.59
Gage block sets (Compliments of Mitutoyo/MTI Corp.).

worldwide standard. U.S. manufacturing facilities contract with regional services for calibration of gage blocks and other length standards, and these regional calibration services maintain length standards obtained from the NIST. Other regions throughout the world maintain similar systems. All inspection tools used in the world can be consistent because through this procedure they are all traceable to one worldwide standard of length.

Review Questions

1. What are tolerances and why are they important in manufacturing?

2. Name two types of tolerance and explain them.

3. What are geometric dimensions and tolerances and why are they important in manufacturing?

4. Is there any difference between tolerances of size and tolerances of geometry?

5. Cite several examples of geometric tolerances of form.

6. Cite an example of geometric tolerances of location.

7. Which is more important, tolerance of form or tolerance of size? Why?

8. What is GO and NO GO gaging?

9. Cite several examples of tools for production gaging.

10. What is the meaning and purpose of calibration?

11. What are used as common shop measurement standards?

12. Who is finally responsible for national measurement standards?

13. What is meant by calibration traceability?

14. How are computer and electronics used in measurement?

15. What is the purpose of final and 100 percent inspection?

16. What are some examples of nondimensional quality control?

17. What quality assurance factors occur after manufacturing and shipping?

18. Why is product testing important to a manufacturer?

19. If you had only an inch-reading micrometer and obtained a measurement on a part of .315, you might suspect that it is a metric measurement since most inch dimensions are usually found in decimals of fractions; the nearest fraction is $5/16$ which equals .3125 in. Using the conversion factor 1 in. = 25.4 mm, what would the metric equivalent of your inch measurement be?

20. Is a machinist's steel rule considered to be a precision measuring tool, or a semiprecision measuring tool?

21. How can a precision measurement be transferred from an inside diameter of a bore to an outside direct reading instrument such as a micrometer caliper?

22. If an inch micrometer can discriminate to .0001 in. and a metric micrometer to .01 mm, which instrument can measure to the smallest or finest increment? Use the conversion factor 1 in. = 25.4 mm or 1 mm = .03937 in.

23. Since angular measure is in degrees, minutes, and seconds, which instrument, the bevel protractor or the combination set protractor, has the most precision?

24. How can surface roughness be measured?

25. When sizes between two tolerance dimensions need to be checked quickly by a machine operator at a production machine, what type of measuring tool is used?

26. Gage blocks can have an accuracy of (a) ten thousandths, (b) hundred thousandths, or (c) millionths of an inch.

Case Problems

Case 1: Calibration

A machine operator, using standard production gages, checks parts as they come out of the production process. Parts appear to be in tolerance, but are later rejected in final inspection as being out of tolerance. What could be the cause of this problem and how might it be corrected? Relate your answer to the purpose and procedures of calibration.

Case 2: Geometric Tolerances

A manufacturer of electric motors acts as a vendor to several different customers making use of several different motor models. The company ships several motors to a new customer with which it has not previously done business. A fit problem develops when the new customer attempts to attach motors to its equipment. The customer company complains that the motor manufacturer's product is not within advertised tolerances. However, the motor manufacturer indicates that the motors in question when used by other customers have caused no assembly problems.

What could cause this problem? What steps might be taken to ensure that there will be no assembly problems in the future? Relate your answer to tolerances of both size and geometry.

Case 3: Translating Shop Terminology into Precise Dimensions

A millwright telephones your shop to request a number of roller chain sprockets, some to be bored to $2\frac{7}{8}$ in. and some to $1\frac{7}{16}$ in., "for a sliding fit." To do what the millwright wishes, you would have to make the bore $1\frac{1}{2}$ to 2 thousandths of an inch over these standard shafting sizes. Convert these crude data into precision decimal inch dimensions first and then to metric, because your shop is now on metric measure.

PART III

SURVEY OF MANUFACTURING PROCESSES

P art III, Chapters 9 through 19, covers specific manufacturing processes. Conventional manufacturing processes including casting, cold and hot rolling, forming, forging, plastics molding, and joining of materials are discussed in detail. Newer processes and methods are presented where they apply.

Materials such as metals, plastics, and ceramics are made into useful articles and consumer products by many and varied means. Metals are cast into molds in a number of different ways to form intricate small shapes or massive machine parts. Metals are also rolled, pressed, and hammered into dies or forced through dies to make special shapes. For example, iron and steel are heated to a high temperature so they can be easily formed by forging (hammering and squeezing). Although forging was once only a hot-metal operation, today cold upset forging is practiced even on steel. At intermediate temperatures, a metallurgically superior product can be produced for some purposes. For example, in warm forming, tough materials like SAE 52100 steel are routinely manufactured into high-quality parts with lower production costs than cold forming.

A large segment of manufacturing is devoted to the processing of sheet metal in cold working operations such as blanking and stamping. Home appliances, automobiles, and many other products depend on sheet metal processing.

Machinery of all kinds that requires precision parts depends on the machining and machine tool industry, which could be considered the foundation of modern manufacturing. The tool and die industry, though little known and understood, is the backbone of modern industry.

Virtually every manufacturing process is in some way dependent upon tool and die shops. Stamping, pressing, injection molding, and die casting are just a few of the processes for which tool and die services are absolutely necessary.

Manufacturing of plastic and composite materials is steadily growing and is replacing many products formerly made of metal, leather, and wood. However, many of these manufacturing industries are interdependent. For example, plastic injection molds require special tool and die machining processes; the manufacture of the molding machine is a machine tool process.

CHAPTER 9

Processing of Metals: Casting

Objectives

This chapter will enable you to:

1. Explain the use of patterns for sand casting.

2. Describe the green sand molding process.

3. Describe the shell molding and investment casting processes.

4. Describe centrifugal casting, permanent mold, and die casting processes.

5. Describe several common types of melting furnaces.

6. Show how good casting design can prevent many casting problems.

Casting is one of the oldest methods of manufacturing metals. Prehistoric humans made tools by pouring molten metal into open molds made of stone or baked clay. Cast objects over 4000 years old have been found dating from ancient Assyrian, Egyptian, and Chinese cultures.

The process of **casting** metals is done by pouring or forcing molten metal into a **mold** cavity having a desired shape. When the metal has solidified, the casting is removed from the mold. Virtually any shape can be produced by this method, in some cases with such precision that subsequent machining is not required.

When designing a metal part to be manufactured, an engineer must choose a method of production. The part can be made by one or more processes, including machining from solid metal, welding fabrication, powder metallurgy, pressing and cold forming, hot forging, or casting. The main advantage of casting over other manufacturing processes is that parts with intricate internal cavities/passages (e.g., faucets, exhaust manifolds, etc.) can be made. This geometric feature would be difficult or even impossible to produce with any other manufacturing process. Besides, the casting process results in parts with smooth, flowing designs, either for practical or decorative purposes. Also, the metal can be placed only where it is required. Thus, the economy of using less metal for a part (especially when a very expensive metal is used), the eye-pleasing appearance (such as in machinery housings), and the possibility of producing intricate shapes are the factors that set casting apart from other manufacturing processes.

Sand casting is typically not a rapid method of production, whereas die casting is a relatively rapid process. Other casting processes lend themselves to production, but none should be considered to have high production rates as compared to punch press work or powder metallurgy.

Casting processes involve a large segment of the metals industry. These range from the tiniest precision parts to huge castings for machinery sections weighing many tons (Figure 9.1). Some metals that are too difficult to machine, such as those used for aircraft turbine impeller blades, can be cast to a precision shape not requiring any subsequent machining. Other softer metals, such as aluminum, are used to form articles such as carburetors for automobiles. It would take many hours of machining to make a complicated carburetor part from solid metal in a machine shop, but it takes only seconds in a die casting machine.

Several general requirements are needed for the production of good castings, regardless of the method used:

1. A method of melting the metal to the correct temperature.

2. A mold cavity of the desired shape, with sufficient strength to contain the metal without distorting or hav-

FIGURE 9.1
Centrifugal casting, 131,000 lb, 5 ft. diameter 35½ ft. long, to be used for a suction roll shell in a paper machine. Here it is being machined to size (Copyright 1976, 1978, 1984, Sandusky Foundry & Machine Co., Sandusky, Ohio).

ing too much restraint on the molten metal as it solidifies. Also, the mold must be designed to avoid internal porosity and cracking.

3. Molds must be arranged so that when molten metal is introduced into the mold, air and gases can escape so the casting will be free from gas-related defects.

4. Any mold material (cores) in internal cavities must have a provision for its removal. Finishing operations are often required to remove any excess material from the casting.

PATTERNS

The first requirement in making a casting is to design and make a pattern (Figure 9.2). **Patterns** are usually made of wood when only a few castings are needed. For larger quantities and when wear (sand abrasion) becomes a problem, patterns are made of metal, such as aluminum or bronze. Hard, tough plastics are also used for patterns. Patterns include several types—single-piece, split-piece, loose-piece, match-plate, and cope and drag. The type of pattern used in the mold making depends on the required production. For example, small production is economical with a single-piece pattern; however, large production requires use of mechanized molding with application of match-plate pattern. Previously made casting can also be used as a pattern if additional shrinkage is not a factor.

Shrinkage Allowance   When metal solidifies from the liquid state, it shrinks. Thus, in order to reproduce casting of a desired dimension, a shrinkage allowance must be added to the pattern. Each metal has a different shrinkage. The following are shrinkages in in./ft for some metals:

Cast iron	1/8
Steel	1/4
Brass	3/16
Aluminum	5/32
Magnesium	5/32

Draft   Patterns must have draft or taper to permit removal from the mold (Figure 9.3). Draft is usually

FIGURE 9.2
Metal match-plate patterns (Eugene Aluminum & Brass Foundry).

FIGURE 9.3
All sand patterns must have draft so they can be removed from the mold.

about 2 to 3 degrees. Without draft on the pattern, parts of the sand mold would be broken as the pattern is pulled out.

Other Allowances   If there are areas on the casting that will have machined surfaces, the pattern needs to provide machining allowance that will leave additional metal for removal. The amount of machining allowance depends on the roughness and accuracy of the finished casting. Enough material must be left for subsequent machining in order to cut under sand inclusions on the surface of the casting.

SAND CASTING

In sand casting, a specially prepared sand that is mixed with different binders and additives is used as a mold material. The appropriately conditioned sand is then compacted around a pattern that has the shape of a desired casting. Sand has the advantage of being highly refractory (can resist high temperatures without melting), so metals like cast iron and steel can be cast easily in sand molds. Although there are many different molding materials, sand casting accounts for the greatest tonnage of all castings produced.

Molding Sands

The basic characteristics of sand are: it is (1) easily molded and capable of holding accurate detail, (2) reusable, and (3) inexpensive. Sand molding processes can be classified into the following groups: green sand molding, heat-cured resin binder processes, cold box resin binder processes, no-bake resin binder processes, and silicate and phosphate bonds. The most common type of sand molding process is green sand molding. Green refers to the fact that the molding sand contains moisture. Other

sand molding processes, because of their higher strength, are often used for cores that form holes and hollow spaces in green sand molds.

The most common molding sand is a natural silica sand (SiO_2); however, other sand types such as zircon ($ZrSiO_4$), olivine, and other sands are also used in special applications. The sands can be classified into two groups: natural and synthetic. The natural sand is the sand in its natural form. The synthetic sands are natural sands that have been washed, screened, classified, and blended to meet the requirements of a particular application. The synthetic sands are most commonly used in foundries.

General Characteristics of Molding Sands

Molding sand must have several characteristics:

1. **Cohesiveness.** The ability to be packed (rammed) in a mold and retain its shape. This property is achieved by adding various sand additives (e.g., clay, water, resins).
2. **Refractoriness.** The ability to withstand high temperatures.
3. **Permeability.** Porosity that allows gases to escape through the mold.
4. **Collapsibility.** The ability to allow freedom for the solidifying, shrinking metal to move without fracturing, and to allow the cast part to be removed easily from the mold. This property is also achieved by adding appropriate sand additives (e.g., corn flour, dextrin, etc.).

Preparation of Green Sand Mix

The green sand mixture (sand, clay, and water) is placed in a muller (Figure 9.4), where the ingredients are thoroughly mixed in order to obtain the proper consistency. A typical muller consists of a large tub in which an arm with rollers swings around, forcing the rollers over the sand.

To achieve required molding properties, the sand grains must be of the right size, while clay and water must be added in the right proportion. To obtain consistently good castings, the mold must be of appropriate hardness, strength, and permeability. Some standard tests are commonly applied in the foundries to control process variables (e.g., grain size and grain distribution, moisture content, green strength, hardness, etc.). These tests are performed both during sand preparation as well as during mold making in order to assure high-quality molds and castings.

FIGURE 9.4
Three types of mulling machines used for the conditioning of molding sand. *(a)*
Continuous muller. *(b)* Conventional batch muller. (*Metals Handbook,* Vol. 5, 8th ed.,
American Society for Metals, 1970, p. 163. With permission.)

Molding

A series of steps in manually producing castings can be
seen in Figures 9.5 through 9.10. Although many kinds of
patterns are used, such as split patterns, single piece, and
loose piece types, patterns for green sand molding are
usually made in two halves and are called match plate patterns, or in the case of larger castings cope and drag patterns. Cope refers to the top half (Figure 9.11 on p. 180)
and drag refers to the bottom half (Figure 9.12 on p. 180).

The cope and drag pattern is placed in a flask that is
made in two halves (cope and drag) (Figure 9.13 on p.
180). First, the drag half of the flask is filled with sand. The
sand is then rammed into place and afterward struck off

FIGURE 9.5
Casting sand is being pushed around on a jolt-squeeze
machine (Eugene Aluminum & Brass Foundry).

FIGURE 9.6
Drag (bottom half) of sand mold after it has been formed
around the match-plate pattern (Eugene Aluminum &
Brass Foundry).

FIGURE 9.7
Core being placed in mold (Eugene Aluminum & Brass Foundry).

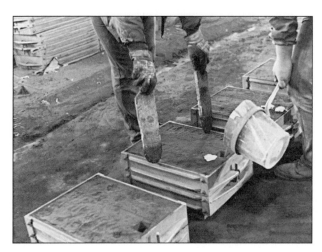

FIGURE 9.9
Molten aluminum being poured into the mold (Eugene Aluminum & Brass Foundry).

(leveled off) even with the top of the flask with a straight edge (Figure 9.14 on p. 180). The drag half of the flask and pattern is also rammed in the same manner (Figure 9.15 on p. 180). The pattern also has an extension called a core print if a core is used. This space in the sand mold provides a support for the ends of the sand core. One or more vertical holes are provided for pouring the metal. This hole, called a sprue, has a pouring cup on the top. The role of the pouring cup is to make pouring easy and to keep the sprue constantly filled with metal. The sprue is on the bottom connected to the runner, which is connected to the mold cavity through the gate. When the molten metal is poured, it fills the mold cavity and other holes in the top of the mold, called risers. The risers are used as reservoirs to feed liquid

metal to the mold cavity as it shrinks while solidifying. They also provide for escaping gases and allow impurities to float to the top of the riser and out of the casting. The cope and drag pattern is made in segments that can be separated, such as the riser and downsprue which are removed from the pouring side of the cope half of the flask, and the pattern is removed from the parting line side after turning it over. Finally, the patterns are removed from the mold sections. Cores are installed, the two halves of the flask are assembled, and the mold is ready for casting. The resulting casting is similar to the pattern (Figure 9.16 on p. 181).

The sand mold and pattern, as well as two halves of a sand mold are prevented from sticking together by dusting them with a commercially prepared parting powder. The

FIGURE 9.8
Cope (top half) being placed on drag flask (Eugene Aluminum & Brass Foundry).

FIGURE 9.10
Solidified casting broken out of molds (Eugene Aluminum & Brass Foundry).

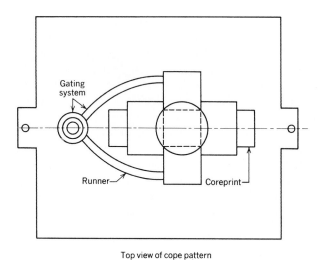

Top view of cope pattern

Side view of cope pattern

FIGURE 9.11
Top view and side view of cope match-plate pattern.
The down sprue and riser patterns are removed from
one side and the cope pattern is removed from the
other side (White, Neely, Kibbe, Meyer, *Machine
Tools and Machining Practices,* Vol. I, © 1977 John Wiley
& Sons, Inc.).

inside of a sand mold is often coated with mold wash (e.g.,
graphite solution) to prevent the sand from sticking to the
casting. It also results in a smooth surface of the casting.

Molten metal is poured into the mold, where it solidifies. After a specified cooling time the casting is removed
from the mold. There are several ways to remove a casting. The most common method is vibration, where the
mold is broken by placing it in a shake-out, i.e., a vibrating machine, that brakes the sand sticking to the casting,
as well as the cores. The sand from the broken-up mold
and from the shakeout is transferred to another machine

FIGURE 9.12
Drag match-plate pattern (White, Neely, Kibbe, Meyer,
Machine Tools and Machining Practices, Vol. I, © 1977
John Wiley & Sons, Inc.).

FIGURE 9.13
Sectional view of cope and drag with core in place on
the mold board. At this point it is ready to make the
casting (White, Neely, Kibbe, Meyer, *Machine Tools
and Machining Practices,* Vol. I, © 1977 John Wiley &
Sons, Inc.).

FIGURE 9.14
Sectional view of cope pattern in flask with the sand
rammed in place and struck off.

FIGURE 9.15
Sectional view of drag pattern in a flask with the sand
rammed in place and struck off (White, Neely, Kibbe,
Meyer, *Machine Tools and Machining Practices,* Vol. I, ©
1977 John Wiley & Sons, Inc.).

that reconditions it and prepares it for reuse. The sand
conditioning consists of crushing, screening, magnetically
separating, and impinging of sand grains against a wear
plate at high velocity. The fines (dust) are vacuumed away
and the clean sand is transported to hoppers to be used
again.

Cores

Cores are parts used to produce a desired cavity in a casting. To reduce the manufacturing cost, the cores should be
made as a part of the mold whenever possible. However,
due to the complexity of desired hollow sections, the
cores are most commonly made separately. Cores are
made by compacting a sand mixture into the core box

FIGURE 9.16
Completed casting (White, Neely, Kibbe, Meyer, *Machine Tools and Machining Practices,* Vol. I, © 1977 John Wiley & Sons, Inc.).

FIGURE 9.17
Sectional view of core box with the core (White, Neely, Kibbe, Meyer, *Machine Tools and Machining Practices,* Vol. I, © 1977 John Wiley & Sons, Inc.).

FIGURE 9.18
Two halves of core fastened together after being removed from core box and baked in oven (White, Neely, Kibbe, Meyer, *Machine Tools and Machining Practices,* Vol. I, © 1977 John Wiley & Sons, Inc.).

(Figures 9.17 and 9.18), and subsequently curing the mixture. The compaction can be done manually or mechanically by using a core blowing machine. Today, most cores are made with one of many organic binder processes. These processes can be classified as: no-bake binder processes, heat-cured binder processes, and cold box binder processes. In the no-bake and cold box processes the binder is cured at room temperature, while in the heat-cured binder process the binder is cured at elevated temperatures. Some cores are made hollow; thus, they use less sand and are also lighter to handle.

The core is placed in the sand mold, in the core print, and the mold is now ready for pouring (Figure 9.19).

The Advantages and Disadvantages of Sand Casting

The greatest advantage of sand casting is that almost any metal can be poured in the sand mold and there is almost no limit on size, shape, or weight of the part. Sand casting provides the most direct route from pattern to casting. Tooling costs are low and the gravity casting process is economical. Among the limitations involved in sand casting is the need for machining in order to finish the castings, especially large ones having rough surfaces. In addition, it is not practical in the green sand process to cast parts with long and thin sections.

Sand castings are extensively used for machine tool housings, bases, slideways, and other parts, and they are also extensively used in the automotive industry for making engine blocks.

Castings must be defect free, and their quality must be consistent from casting to casting. Statistical process control (SPC), quality controls for castings, are rapidly

FIGURE 9.19
Large sand mold being gravity cast with molten cast iron (© 1983 Pottstown Machine Company, Pottstown, PA).

becoming a survival issue for foundries since scrap losses must be kept to an absolute minimum because of the strong foreign competition in this industry.

EVAPORATIVE CASTING PROCESS

In the evaporative casting process, the pattern, sprue, and riser are made of foamed polystyrene. They can be made as a single piece, or they can be made separately and then glued together. The completed polystyrene pattern is then coated with a thin layer of refractory material. The coated pattern is placed in the flask and dry sand is compacted/vibrated around the pattern. When the metal is poured, the heat vaporizes the polystyrene pattern almost instantaneously, leaving the mold shape intact as it is being filled with metal (Figure 9.20). This process can be used for casting parts of any size and shape since the patterns do not need to be removed prior to casting, thus eliminating the need for draft on the pattern (Figures 9.21 and 9.22). Also, problems related to mold shift are eliminated since the entire pattern is assembled before covering with sand, allowing for corrections prior to pouring. Another advantage of the process is the low cost of sand preparation since no additives or binders are used for sand conditioning. The process is economical only for prototyping or large production, i.e., only if the patterns themselves can be mass produced in a die.

SHELL PROCESS

The shell molding process is a type of sand casting process that provides a finer detail and smoother finish because the sand is finer and it is combined with plastic resin to make a

FIGURE 9.20
Evaporative casting process. This photo depicts the way the intense heat of molten aluminum vaporizes a foam pattern, leaving behind a high-quality casting (Photo courtesy of Ford Motor Company).

FIGURE 9.21
Polystyrene (styrofoam™) pattern for a bronze casting. The pattern is made in sections (Eugene Aluminum & Brass Foundry).

FIGURE 9.22
Bronze casting called *Coast Spirit,* permanently located in front of the library at the University of Victoria, British Columbia (Eugene Aluminum & Brass Foundry).

The advantage of the shell process is the high quality of the surface finish of the casting. In addition, the mold is very strong, allowing casting of parts with fine details. Also, molds have long shelf life, thus allowing production flexibility. A disadvantage of the process is the relatively high cost of the core boxes. However, shell molding is an economical process for high quantity production due to savings in the machining time of castings, low labor costs associated with the process, and high productivity. Many shapes can be made by this process. In the case of the complex shapes, individual segments are manufactured separately and assembled together into the mold. Shell mold casting has good repeatability and most ferrous and nonferrous metals can be cast by this method.

PERMANENT MOLD CASTING

The greatest disadvantage of sand casting is that a new mold must be made for each casting. In addition, some inherent dimensional inaccuracies are present in the sand casting. These disadvantages gave rise to the development of a permanent mold. Despite the name, permanent molds can be reused at most for several thousand pours, after which they lose their true shape and must be scrapped. Most permanent molds are made of gray cast iron or steel. Graphite molds are often used for casting higher temperature metals. The molds are made by machining processes and are hand finished or polished. A refractory wash is applied to the mold prior to casting in order to prolong its life. When cores are needed, they can be made from metal, in which case they are reused, or they can be made from sand, in which case they cannot be reused. The mold halves may be hinged (Figure 9.24) or mounted on a casting machine (Figure 9.25) so they can be opened and closed quickly and accurately.

In permanent mold casting, the metal is poured from a ladle, using gravity to fill the mold. Cast iron and nonferrous metals are cast in this manner. The molds are heated prior to a run and the temperature is maintained by the molten metal. This is necessary to avoid chilling the metal too quickly, which can produce laps or cold-shuts in the casting (cracks where two adjacent portions of metal do not solidify together). Usually only simple shapes are cast by this process because of the rigidity of the mold and shrinking of the metal when it solidifies. Since these molds are not permeable, they must be vented at the plane where the two mold halves come together. In addition to gravity pour, low-pressure and vacuum-pouring techniques are used in conjunction with the permanent mold process.

Good surface finish, good dimensional accuracy, and low porosity can be obtained with permanent molds.

smooth mold surface. The process can be automated for mass production. This method requires the use of metal patterns that are heated to temperatures of up to 450°F (232°C), and coated with a silicon release agent. The pattern is then placed on the dump box, which contains sand coated with phenolic resin (Figure 9.23). The assembly is then turned upside down causing the sand to fall and cover the heated pattern. Some of the sand adheres to the pattern and hardens, forming a shell around it. The thickness of the shell depends on the length of time the pattern is in contact with the sand. After the prescribed curing time, the dump box is inverted to remove the loose sand. The pattern and adhering sand are placed in an oven and heated to a temperature of 600° F (316° C) for 1 or 2 min. The half-shell is then removed from the pattern and the two shells are glued or clamped together to form a complete mold. This thin mold is strong enough for casting small or thin parts, but heavier castings require the use of a backup material such as shot or sand in a pouring jacket.

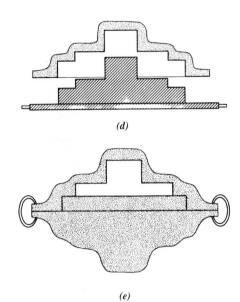

FIGURE 9.23

Shell molds. *(a)* Heated metal pattern in dump box with sand. *(b)* Dump box is turned, leaving some sand adhering to the pattern. *(c)* Pattern is placed in an oven for 1 or 2 min to harden it. *(d)* Half-shell mold is stripped from pattern. *(e)* Mold halves are clamped together with core. The top half-shell is shown as a sectional view (White, Neely, Kibbe, Meyer, *Machine Tools and Machining Practices,* Vol. I, © 1977 John Wiley & Sons, Inc.).

FIGURE 9.24

Book-type manually operated permanent mold-casting machine, used principally with molds having shallow cavities (*Metals Handbook,* Vol. 5, 8th ed., American Society for Metals, 1970, p. 266. With permission).

FIGURE 9.25
Manually operated permanent mold-casting machine with straight-line retraction, required for deep-cavity molds (*Metals Handbook,* Vol. 5, 8th ed., American Society for Metals, 1970, p. 266. With permission).

Repeated use of molds and a rapid production rate with low scrap loss make this casting process ideal for moderate production runs of a few thousand pieces. Some disadvantages are high mold cost, limited intricacy of casting shape, and unsuitability for high-temperature metals such as steel.

SLUSH CASTING

Slush casting is used with permanent molds to make a shell of metal in the mold. The molten metal is poured into the mold and allowed to solidify to a certain wall thickness against the mold, and the remainder of the molten metal is dumped out, thus producing a shell. Toys, lamp bases, and ornamental objects are made with this process.

CENTRIFUGAL CASTING

Centrifugal casting is a process in which molten metal is poured into a rapidly revolving mold. The liquid metal is forced to conform to the shape of the mold by centrifugal forces many times the force of gravity. In this process the mold rotates about either a horizontal or a vertical axis. No core is needed to make an inner surface since this process naturally produces a hollow shape such as pipe (Figures 9.26 and 9.27). The thickness of the mold controls the cooling rate and therefore the grain structure of the cast part. Also, if a hard outer wear surface is needed with an inner machinable soft metal, two dissimilar metals can be used—the outer one to harden when solidified and the inner one to remain relatively soft. For example, a thin wall of stainless steel can be poured first and then less expensive low carbon steel can be poured, giving the pipe a corrosion-resistant jacket at low cost. Lighter elements, impurities, and slag collect on the inner wall of a centrifugal casting and can be removed later by machining.

Vertical centrifugal and semicentrifugal processes involve the rotation of a mold about its vertical axis. The metal is poured into a central reservoir that allows the liquid metal to flow outward into the spinning mold by centrifugal force. This process provides a denser structure in the metal than ordinary gravity casting does. It is ideal for steel wheels, brake drums, and cylinder barrels. A similar process, called centrifuging, rotates a group of molds or an entire mold located a certain distance from the center of rotation. A pouring basin allows the metal to flow outward into the molds. This method produces superior quality castings.

Centrifugal casting machine

Machine during casting

Machine during stripping

FIGURE 9.26
Machine for centrifugal casting of iron pipe in a rotating water-jacketed mold (*Metals Handbook,* Vol. 5, 8th ed., American Society for Metals, 1970, p. 267. With permission).

FIGURE 9.27
Large centrifugally cast tube (Copyright 1976, 1978, 1984, Sandusky Foundry & Machine Co., Sandusky, Ohio).

INVESTMENT CASTING

One of the oldest methods of casting metals is the investment casting (i.e., lost wax) process. The pattern is made of wax coated (invested) with a thick layer of refractory material (Figure 9.28). After the refractory material has hardened the wax is melted and poured out of the mold and reused for another pattern. The mold is then preheated close to pouring temperature and molten metal is then poured. This method is used for dentistry, arts and crafts, and for any type of very accurate near net-shape casting. When it is used for industrial purposes, significant production rates are possible.

FIGURE 9.28

Steps in the production of a casting by the solid investment molding process, using a wax pattern. *(a)* Wax sprue pattern. *(b)* Pattern assembly (wax patterns attached to wax sprue). *(c)* Pattern assembly in flask after mold slurry has been poured. (Precoating of pattern assembly with slurry is required for metals with pouring temperatures above 2000°F.) *(d)* Mold after pouring. *(e)* Solidified casting after mold has been broken away. *(f)* One of four castings after removal from sprue. (*Metals Handbook,* Vol. 5, 8th ed., American Society for Metals, 1970, p. 239. With permission).

Plaster of paris can be used for the mold material in the lost wax process. The mold must be broken and destroyed to remove the cast part. Plaster of paris molds can be used only for the lower temperature metals such as aluminum, zinc, tin, and some bronzes.

Metals having higher pouring temperatures, such as cast iron and steel, can also be mass precision cast. For a production casting process, a method of mass producing the wax patterns is needed. The patterns are mass produced by injecting the molten wax into a split mold in the same way plastic objects are rapidly made in injection molding machines (Figure 9.29). The wax patterns are assembled in groups on a sprue in clusters (Figure 9.30). The wax assembly is dipped into an investment, i.e., slurry

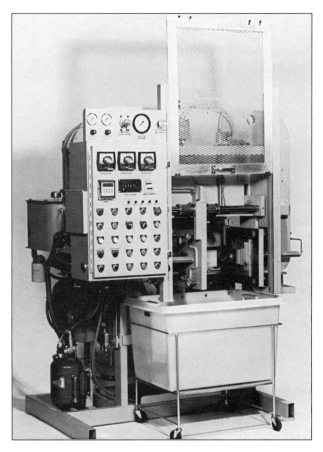

FIGURE 9.29
Automatic wax pattern molder. Large quantities of
duplicate wax patterns can be produced on this machine
(Courtesy of Mueller-Phipps International, Inc.,
Poughkeepsie, N.Y.).

of refractory material. It is dipped several times to
develop a fairly thick shell. The shell is then heated in an
oven to melt out the wax, which is later reused. Prior to
casting the mold is heated in a furnace close to pouring
temperature and then may be placed in a flask where a
heavy mold material is poured around it for additional
support. The molten metal can be poured into the mold by
gravity, pressure, or vacuum.

This method can be automated to some extent and is
widely used in the automotive and aerospace industries.
Some of the advantages of investment casting are:

1. Unusual and nonsymmetrical shapes that would not
allow withdrawal of an ordinary pattern are easily produced.

2. Smooth surfaces and high accuracy can be obtained.
Machining can be reduced or eliminated.

3. Nonmachinable alloys can be cast with high precision.

Most parts made by this process are limited to a small
size and there are other limitations, such as in the use of
holes and cavities. The process is more involved than some
other casting processes, such as permanent molding and
die casting, and is therefore more expensive per casting.

SHAW PROCESS

In this process, a rubbery jelling agent and a slurry of
refractory aggregate is poured over the pattern. This rubbery mold hardens sufficiently to be easily stripped off the
pattern and will return to the exact shape of the pattern. The
mold is ignited to burn off the volatile elements, and it is
then placed in a furnace and brought to a high temperature.
The mold is then ready for pouring. The advantage of this
process is good permeability and good collapsibility, which
allow for production of delicate and intricate shapes with
fine detail and higher quality castings.

DIE CASTING

Die casting is similar to permanent molding in that a
metal mold made in two halves is used. The difference is
that the metal is not gravity poured into the mold (die),
but instead the metal is injected under high pressures
ranging from 1,000 to 100,000 PSI. This requires massive
machines that are generally operated hydraulically to
exert the hundreds of tons of force necessary to hold the
two halves of the die together when the molten metal is
being injected (Figure 9.31 on p. 190).

The dies are usually made of alloy or tool steel and
are quite expensive to make. Some have one or two identical mold cavities for larger parts and others may have
several different cavities (Figure 9.32 on p. 190). Some
dies are more complicated and have sections that move in
several directions. Grooves or overflows around the cavity
on the parting face permit gases to escape. The overflows
of excess metal must be trimmed off by a secondary operation after the casting is removed from the mold. This
trimming is done with trimming dies that also remove the
sprues and runners. The mold must also have provisions
for water cooling so that a constant operating temperature
can be maintained. Knock-out pins provide for ejection of
the part when the die is opened (Figure 9.33 on p. 191).
When cores are used, they are made of metal and are usually drawn out before the die is opened. Cores are
retracted, either in a straight line or in a circular motion.

It is easy to see that these complex dies are quite
expensive. Their costs can range from $5,000 to more than

FIGURE 9.30

Steps in the production of a casting by the investment shell molding process, using a wax pattern. *(a)* Wax sprue pattern. *(b)* Pattern assembly (wax patterns attached to wax sprue). *(c)* Pattern assembly dipped in ceramic slurry. *(d)* Pattern assembly stuccoed in fluidized bed. Dipping *(c)* and stuccoing *(d)* are repeated until required wall thickness of mold is produced. *(e)* Completed mold after wax pattern has been melted out (mold shown in pouring position). *(f)* Mold after pouring. *(g)* Solidified casting after mold has been broken away. *(h)* One of four castings after removal from sprue. (*Metals Handbook,* Vol. 5, 8th ed., American Society for Metals, 1970, p. 239. With permission.)

FIGURE 9.31
Die casting machine (TVT Die Casting & Mfg. Inc.)

Single cavity die   Multiple cavity die

Combination die   Unit die

FIGURE 9.32
Die casting molds showing four different types of die cavities (Courtesy of American Die Casting Institute).

$10,000, depending on complexity and size. Obviously, this process is not suitable for small quantity production or for making very large parts. The large investment for dies and machines will pay off only when very large quantities (20,000 to millions of parts per year) are required.

Die casting is a means of producing castings of lower temperature alloys at a relatively high rate. These castings are usually thin walled, smooth, and highly accurate (Figures 9.34 and 9.35). The process is highly adaptable to the manufacture of small parts, such as automobile door handles, carburetors, wiper motors, kitchen appliance parts, and thousands of small items we use every day.

The quality of die castings is high because of a rapid cooling rate that produces fine grains in the metal. The surfaces tend to be harder than the interior as a result of the chilling actions of the metal die. Porosity is sometimes a problem as a result of entrapped air, but with proper venting this can be overcome.

Zinc-based alloys are the most commonly used in die casting. Other metals used in die casting are alloys using copper, aluminum, and magnesium. Zinc alloys have the lowest melting point, about 700°F (371°C), and so have a less destructive effect on dies. Aluminum and magnesium alloys melt at about 1100°F (593°C) and copper alloys melt at about 1700°F (927°C). Therefore, dies using these alloys have a shorter life because of thermal shock which causes crazing (microcracking) on the die surfaces.

Die Casting Machines

The two basic systems and machines used for die casting are the cold-chamber machines (Figure 9.36 on p. 192) and

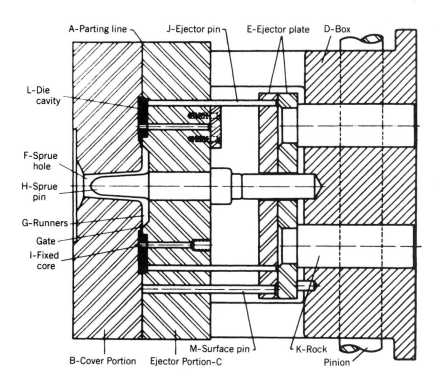

A-Parting line   J-Ejector pin   E-Ejector plate   D-Box

L-Die cavity

F-Sprue hole

H-Sprue pin

G-Runners

Gate

I-Fixed core

B-Cover Portion   Ejector Portion-C   M-Surface pin   K-Rock   Pinion

FIGURE 9.33
Typical die (Courtesy of American Die Casting Institute).

the hot-chamber machines (Figure 9.37). Cold-chamber machines have a separate melting and holding furnace. For each casting cycle, the molten metal is fed into the cold chamber and then is forced into the die by the plunger. These machines are often used for casting metals that melt at higher temperature because the chamber remains relatively cool and there is less tendency for spalling or cracking than in the hot-chamber machine. Although the cycle time is somewhat slower than that of the hot-chamber machine, the productivity is still high, as much as 100 cycles per hour.

A typical hot-chamber machine with a submerged gooseneck injector and plunger is illustrated in Figure 9.37. This type has an oil-fired or gas-fired furnace with a cast iron pot for melting and holding the metal. The

FIGURE 9.34
Part of a die showing mold cavity at center, raw material on the left, and product at the right (TVT Die Casting & Mfg., Inc.).

FIGURE 9.35
Some of the great variety of parts made by die casting (TVT Die Casting & Mfg., Inc.).

FIGURE 9.36
Cold-chamber machine. Diagram illustrates die, cold chamber, and horizontal ram or plunger in charging position. (Courtesy of American Die Casting Institute).

plunger and cylinder are submerged in the molten metal that is forced through the gooseneck and nozzle into the die at each cycle. This type of die casting machine can be used only for the low melting point zinc-based and tin-based alloys. This machine is much faster in operation than the cold-chamber machine.

Several problems and difficulties have plagued the die casting process, especially when attempts were made to speed up or fully automate die casting systems. The major problems are flash (Figure 9.38), sticking, and soldering (molten metal adhering to the die). Many kinds of automatic unloading and loading devices have been developed and die sprayers have been used to control these problems with mixed results. Other problems such as surface defects, dimensional inaccuracies, and internal porosity have been addressed also. Increasing the injection plunger velocity and reducing the total filling time reduced these problems.

FURNACES AND HOT METAL HANDLING

The cupola furnace (Figure 9.39) has been of primary importance in the melting of cast iron for foundry work. Coke (produced from coal) is used for producing heat in the cupola furnace. A typical cupola furnace is a circular steel shell lined with a refractory material such as firebrick. It is equipped with a blower, air duct, and wind box with tuyeres for admitting air into the cupola. A sand bottom keeps the molten metal from burning through and is sloped so that the iron may flow out when the tap hole is opened. The process begins by starting a wood fire on the sand bottom and placing coke on top. When the coke bed is thoroughly ignited and is about 2 to 4 ft thick, iron

FIGURE 9.37
Hot-chamber machine. Diagram illustrates plunger that is submerged in molten metal, gooseneck, nozzle, and dies. Modern machines are hydraulically operated and are equipped with automatic cycling controls and safety devices (Courtesy of American Die Casting Institute).

FIGURE 9.38
Die cast parts (clutch arm) showing flash on the left and
flash removed on the right (Consolidated Metco, Inc.).

scraps or pigs are added and alternated with coke plus
small amounts of limestone for a flux. The blower is
turned on and the iron begins to melt at the top of the coke
bed. The molten iron is drawn off into a ladle, and suitable
metallurgical tests are made. Metallurgical control of the
cast iron is made at the furnace before the ladle is moved
to the casting floor. These tests reveal the amount of carbon and other conditions. Certain alloy-rich materials or
silicon to control graphitization are added to the ladle.
Also, if nodular iron is desired, elements such as magnesium or cerium are added to the ladle to cause the graphite
to form in the shape of nodules (spheres) instead of flakes.
Nodular iron is much tougher and stronger than gray cast
iron, which contains graphite in the form of flakes. When
the tests are completed and the additions are made in the
ladle, the molten metal is ready to be poured into the prepared molds.

FIGURE 9.39
The cupola furnace (Neely, *Metallurgy*, 2d ed., © 1984 John Wiley & Sons, Inc.).

FIGURE 9.40
Molten metal being poured from an induction furnace into a ladle. The furnace is tilted to make the pour (Inductotherm Corp.).

When fossil fuels such as coke are burned, considerable smoke and fumes are released into the atmosphere. In areas where air pollution is a critical problem, these furnaces are fitted with dust collectors and other air pollution equipment; where cheap electricity is available, the cupola furnaces are replaced with electric furnaces. Electric furnaces for melting cast iron can be either electric arc, electric induction, or electric resistance types (Figures 9.40 to 9.42).

POURING PRACTICE

Ladles are usually constructed of steel and lined with firebrick or other refractory material (Figure 9.43). The interior of the ladle is heated or kept hot while in use so as not to cool the molten metal. In small foundries, a hand-held, shank-type ladle is often used to pour small quantities (Figure 9.44). This type requires two persons to use it. The teapot ladle can contain more metal and is supported on an overhead monorail or crane. The handwheel is turned to tilt the ladle and make the pour. This and the bottom pour types keep the slag and oxidized metal from going into the mold. In some operations the molds are placed on a pouring floor, whereas other operations are automatic or semiautomatic, the molds being carried along a conveyor to the ladles where a measured amount of molten metal is poured into the mold.

CASTING CLEANUP

After the solidified castings are removed from the mold, they are cleaned in a shake-out and small parts are often

FIGURE 9.41
Cross section of a 60-cycle coreless furnace with a rammed lining (Courtesy of Ajax Magnethermic Corp., Warren, Ohio).

FIGURE 9.42
Front view of 250KW, 1KHz phasor and left swing furnace (Courtesy of Ajax Magnethermic Corp., Warren, OH).

FIGURE 9.43
Tea pot ladle.

FIGURE 9.44
Shank-type ladle.

put in a tumbler. The risers and gates must be knocked off (often this can be done on cast iron because of its brittleness) or cut off by means of sawing or use of an abrasive cut-off wheel.

CASTING DESIGN AND PROBLEMS

Since many factors influence the quality and performance of a casting, careful attention must be given to several design requirements if the best results are to be obtained. The more complex castings require more considerations than the simple ones. The draftsman or engineer should understand the problems involved or confer with the foundrymen for suggested alterations. Sufficient allowance of metal should always be provided when machining operations are to be performed. Size of the casting, and surface roughness, are involved in this decision. The location of the parting plane is also very important since the pattern must be extracted at this plane without disrupting the sand mold. Other factors to be considered are:

1. Weight of the casting and mold strength
2. Effective gating and sufficient riser
3. Number of cores and their placement
4. Required dimensional accuracy
5. Radii, thickness of sections, and amount of shrinkage.

Cracking can occur at the sharp corners and where thick sections join with thin sections (Figure 9.45). Proper radii can reduce this problem. Also, the cooling rates are greater for the thin section than for the thick one; this also causes cracking at the juncture. Cooling rates may be increased in a thick section by using "chills," metal sections embedded in the mold to absorb heat (Figure 9.46).

FIGURE 9.45
Casting problem: thin to thick section. The spokes, being smaller, cool more rapidly than the thick rim, causing them to break.

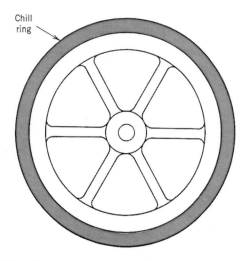

FIGURE 9.46
Chill in a mold around the rim of a wheel. This equalizes the cooling rate between the thick rim and the thin spoke. Odd numbers of spokes are often used to reduce cracking.

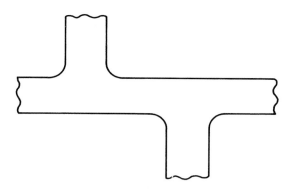

FIGURE 9.47
Staggered ribs.

The shape of casting should be designed to have as uniform a thickness as possible, but in many products this is not possible. Parts that have ribs or spokes are subject to this kind of cracking, but often the ribs can be staggered (Figure 9.47). When parts intersect, such as in a 90 degree turn, interior corners (fillets) should be provided, but if they are too generous, hot spots may develop and cause shrinkage defects (Figure 9.48). Holes provided by cores can be used to eliminate these weak points in a casting. If risers are inadequate to supply molten metal when the casting is shrinking during solidification, shrinkage cavities can result (Figure 9.49).

FIGURE 9.48
Shrinkage areas, hot spots cause voids.

FIGURE 9.49
Shrinkage cavities.

REVIEW QUESTIONS

1. When precision surfaces are required on a casting, which one of the several types of castings is most likely to need a machining operation subsequent to casting? Which two types are least likely to need subsequent machining?

2. What are patterns for sand casting made of? What material is used to make patterns for shell molds? Which kind of pattern volatilizes when the metal is poured?

3. What characteristic of sand casting patterns is most necessary for removal of patterns?

4. What is shrink rate and what does it have to do with patternmaking?

5. What are cores and of what materials are they made?

6. If you had to choose between a steel casting and a gray cast iron casting for a crane hook, which would you choose? Why?

7. A gear housing for a tractor is to be sand cast. Which metal would you choose for this casting, gray cast iron or steel? Why?

8. You are required to manufacture 1000 simple, small parts that are to be cast in aluminum. Would using a permanent mold be less expensive per part than preparing 1000 sand molds? There is no time limit. Explain your reason.

9. What is slush casting and what is it used for?

10. Name a relatively rapid method of casting pipe and tubing.

11. In which type of casting is the pattern removed by heating and melting it? List three advantages in using this casting process.

12. Name the most rapid method of producing small cast parts. Is this system practical to use for only 20 parts? For 100? Explain the reasons for your answer.

13. Which fuel does the cupola furnace use? What metal is usually melted in it? Name one other type of furnace used for the same purpose.

14. How does the molten metal get from the furnace to the mold?

15. What happens to the sand castings after they have solidified in the mold and cooled sufficiently to handle?

16. A cast iron casting has failed in use by breaking at a T joint where one leg meets the other at the joint. On inspection, a large shrinkage cavity is found in the joint, which obviously caused the failure. How could this have been prevented?

17. In a sand casting, the top edge near the riser is full of impurities and porosity; also, a large shrinkage cavity is present. What could be the cause and how can it be corrected?

18. What is the major difference between a hot-chamber and a cold-chamber die casting machine? Which one is used mostly for low-temperature melting alloys?

Case Problems

Case 1: Hydraulic Cylinder Head Problem

A design for a metal cylinder calls for a metal cylinder head (Figure 9.50) that must withstand considerable forces. Zinc-based alloys, aluminum and magnesium cannot be used because they are not strong enough for the size limitations of the design. Cast iron, bronze, or steel may be used since they all have sufficiently high tensile strengths. These metals vary somewhat in cost and in ease of casting. The least expensive material and process to be used to manufacture these cylinder heads of about 100 parts per 8 hours, for a total run of 1000 heads, must be selected. Only 20 days are given to produce them. The options are:

1. Machine all holes, cavities, and surfaces from solid bars of steel. These come in 20-ft-long bars that are sawn off in suitable lengths for machining.

2. Make sand castings of steel and finish by machining.

3. Make sand castings of cast iron and finish some surfaces by machining.

4. Cast the parts in cast iron by investment casting.

5. Use permanent molds and cast them in bronze.

6. Use the process of die casting with bronze and eliminate machining altogether.

Which method would produce the least costly parts and get the job done on time?

Case 2: Manufacturing a Small Bronze Gear

A small bronze gear must be made at a high production rate. Machining is ruled out as being too slow, and although extrusion (forcing metal through a die) of these gears could be a competitive option, it is also ruled out because no extrusion machine is available to the manufacturer. The order requires 10,000 gears to be delivered within 30 days. It will take about 18 to 20 days to make either sand casting patterns, permanent molds, or die casting dies. This leaves 10 to 12 days for production.

This means that about 1000 parts per day must be manufactured, allowing two days for packaging and shipping. The manufacturer will receive payment of $200,000 for the complete order. Which casting method should be used?

FIGURE 9.50
Metal cylinder head. See Case 1.

In earlier times, wrought iron and steel were forged by shaping a piece of red hot metal on an anvil by pounding on it with a hammer. The hammering process not only produced a desired shape, it also improved the grain structure of the metal, resulting in increased strength. When the blacksmith needed to make a round bar from a square one, a half-round die block, called a swage, was inserted in a square hole in the anvil and an upper half-round die was held in the hand by a handle. The square bar was heated red hot in a forge and placed in the lower die while the upper one was hammered down on the bar. This process was repeated all along the bar, thus producing a length of round stock. This is roughly the process used for hot forging various shapes today, using open and closed dies and powerful presses. Today, steel bars of various shapes and dimensions are usually rolled in a steel mill. However, larger sections and other more complex and irregular shapes are often forged. Modern forging processes account for a large part of basic metalworking. Often a forged or rolled shape is subjected to a secondary process of metalworking, such as stamping or machining.

CHAPTER **10**

Processing of Metals: Hot Working

Objectives

This chapter will enable you to:

1. Explain how molten steel is formed into industrial shapes.
2. Show the significance of recrystallization and grain structure of hot rolling and forging.
3. Describe several methods of hot forming metals and explain their advantages.
4. Describe two methods of hot forming pipe and tubing.

HOT ROLLING

In Chapter 2 you learned how raw materials, iron ore, coke, and limestone are used to extract pig iron and how steel is produced in furnaces. The molten steel must then be formed into useful shapes. The process of shaping steel begins by teeming or pouring molten steel into ingot molds. The molds are then moved to the stripper where, after the metal solidifies, the ingot is removed. Before the ingot can be rolled, it must be heated uniformly to about 2200°F (1204°C) throughout in order to allow uniform flow of material during the rolling operation. The soaking is done in soaking pits (Figure 10.1).

Rolling Mills

The soaked ingot is sufficiently plastic to be taken to the rolling mills to be shaped into one of the intermediate forms called blooms, slabs, and billets (Figure 10.2). Blooms are forms of semifinished steels that have square cross section 6 × 6 in. or larger. A slab is rolled from an ingot or a bloom and has a rectangular cross section of width 10 in. or more and thickness 1.5 in. or more. A billet is rolled from a bloom and is square with dimensions 1.5 in. on a side or larger. Blooms and slabs are most frequently rolled on a reversing two-high rolling mill (Figures 10.3 to 10.5 on the following pages).

Blooms are further processed in structural rolling mills (Figure 10.6 on p. 201) into railroad rails or different shapes such as I beams, angles, wide flange beams, zees, tees, and H-piles. They are made of plain low carbon

FIGURE 10.1
Powerful tongs lift an ingot from the soaking pit where it was thoroughly heated to the rolling temperature (American Iron & Steel Institute).

steel which can be drilled, heated, or welded without such complications as work hardening, embrittlement, or cracking. Blooms are sometimes further reduced in size to billets on billet mills (Figure 10.7) before being processed into finished products such as bar and rod stock. Bar and rod (from which wire is made) can be made of either low carbon steel or higher carbon and alloy steel.

As the ingot is carried through the massive rolls, the hot plastic steel is squeezed to a smaller thickness. The rolls are brought closer together on each pass, causing the ingot to get longer and thinner as it passes between the rolls (Figure 10.8 on p. 202). Rolled shapes become quite long and must be cut to standard customer-length (Figure 10.9 on p. 203). Different configurations of the rolling mills include two-high, three-high, four-high, cluster mill,

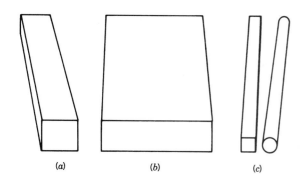

(a)   (b)   (c)

FIGURE 10.2
(a) Blooms, *(b)* slabs, and *(c)* billets are formed from ingots.

FIGURE 10.3
Reversing two-high roughing mill (American Iron & Steel Institute).

FIGURE 10.4
Operator at console of reversing two-high roughing mill (Courtesy of Bethlehem Steel Corporation).

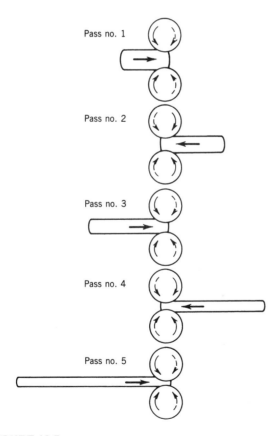

Pass no. 1

Pass no. 2

Pass no. 3

Pass no. 4

Pass no. 5

FIGURE 10.5
Diagram of sequences in rolling hot ingot into a slab on a reversing two-high rolling mill (American Iron & Steel Institute).

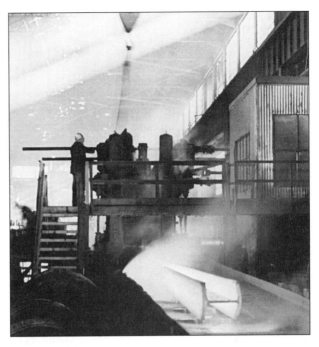

FIGURE 10.6
Structural steel and rails are rolled from blooms. Standard shapes are produced on mills equipped with grooved rolls. Wide flange sections are rolled on mills that have ungrooved horizontal and vertical rolls (American Iron & Steel Institute).

FIGURE 10.7
Steel billet emerges from 35-in. blooming mill. The billet has been rolled from an ingot. The product of this mill will eventually be made into bars. Dials on the two-stand mill indicate size of billet being rolled. Alloy and tool steel ingots are processed in this particular mill (Courtesy of Bethlehem Steel Corporation).

FIGURE 10.8
Diagrams demonstrating
various arrangements of rolls
and rolling procedures used
in steel mills. Operating under
extreme heat and pressure,
rolls must frequently be
changed, and a supply of new
or refinished mill rolls is
always maintained from the
mill's own machine shop
(American Iron & Steel
Institute).

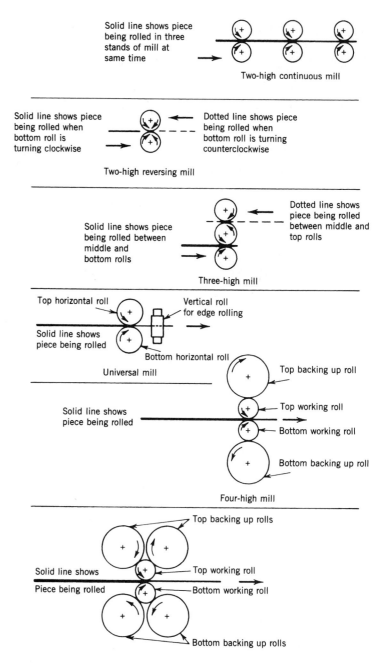

and tandem rolling mill. The two-high rolling mill can be used for reducing the stock by reverse action. However, because of the large inertia of the rolls it requires a complex mechanism for reversing direction of rotation. The three-high rolling mill has advantage over the two-high mill in that the direction of rotation of rolls does not have to be reversed. In one pass, stock is passed between the top two rolls, then the stock is lowered and passed between the lower two rolls, after which it is lifted and the operation is repeated (Figure 10.8). The only disadvantage of this configuration is that it requires an expensive handling system. Four-high (Figures 10.8 and 10.10) and

cluster mills (Figure 10.8) use rolls of small diameter to reduce the dimension of the stock, since smaller diameter rolls require less energy and force for the same effect as do large rolls. However, since small rolls are insufficiently rigid, they must be backed by larger rolls. Finishing mills for plates, strip, and sheets are often tandem rolling mills with a number of mill stands in series, each adding to the final shape. The tandem rolling mill requires a complex system for adjusting rotation speeds of the rolls in order to prevent uneven thickness or even fracture of the stock.

All hot rolled steel forms a black or gray mill scale on the surface that is caused by oxidation of the red hot steel

FIGURE 10.9
Hot saw cuts finished rolled shapes to customer length after delivery from the 48-in. finishing mill (Courtesy of Bethlehem Steel Corporation).

in the presence of oxygen in the air. The oxide scale helps protect the steel from corrosion after it has cooled.

STRAND CASTING

A second procedure of steel processing that bypasses ingot teeming, stripping, soaking, and rolling in roughing mills is called strand casting (Figure 10.11). It is also called continuous casting, because the molten metal is continuously supplied to a tundish or molten metal reservoir (Figure 10.12 on p. 205) where it is fed through a water-cooled copper mold from which it emerges as a continuous strand of steel in a desired cross section as slabs or billets. Sprays of water under high pressure cool and harden the metal still further. The strand of metal is cut into suitable lengths by a traveling torch as the steel moves along on rolls toward the mill stands (Figure 10.13 on p. 205). In some strand casters, the descending column of metal is cut to the desired length while it is still in the vertical position, then the cut length falls forward onto the rolls and is carried away. At this time the metal is at a rolling or forging heat, just under the temperature of solidification. Both ingot and strand casting procedures are primarily designed for high-volume production. The continuous strand casting unit at Bethlehem's Burns Harbor, Indiana, plant can convert 300 tons of molten steel to solid slabs in about 45 minutes.

During the strand casting process the steel strand cools from the surface toward the center, forming large, columnar, dendritic grains (Figure 10.14 on p. 206). These large grains are characterized by low strength and high elongation, which in most cases is considered undesirable. In addition, internal voids can be created due to the shrinkage of the material. The shrinkage voids result in reduced cross sectional area and even more importantly they act as stress risers, which significantly reduce the strength of the structure. These undesirable grains and internal shrinkage voids can be removed or altered by the process of hot working: rolling, drawing, extruding, or forging. Whether the slabs or billets are made by casting ingots or by the continuous-strand casting process, the subsequent rolling processes in mill stands remain the same.

FIGURE 10.10
Plate enters the four-high finishing stand in the 110-in. sheared plate mill. The plate is rolled to the desired length and thickness by being passed back and forth through the finishing stand rolls. The mill has AGC (automatic gage control) and is completely computerized (Courtesy of Bethlehem Steel Corporation).

Hot Metal Ladle

Hot steel is transported rapidly by ladle from electric and open hearth furnaces to the casting unit and is fed into the tundish.

In the water-cooled mold the steel begins to solidify. A solid shell is formed.

Roller Aprons and secondary cooling

The refractory-lined tundish controls the flow and distribution of metal into the molds.

Operator's Console

Gantry Service crane

Solidifying steel enters the secondary cooling zone. Cooling is accomplished by direct water spray. Roller aprons are arranged to guide and support the strands and simultaneously take up the ferrostatic pressure exerted by the liquid metal core upon the strand shell.

Roller Leveler

Here rolls withdraw and level the strands.

Traveling Slab Cut-off Torch

Slabs are cut into predetermined lengths and removed by roller tables.

Slab Run Out Table

FIGURE 10.11

Diagram of continuous (strand) casting (American Iron & Steel Institute).

FIGURE 10.12
Molten steel is poured from a ladle into the tundish, which regulates the flow of liquid steel into two water-cooled copper molds for the slabs to be continuously cast (Courtesy of Bethlehem Steel Corporation).

RECRYSTALLIZATION

Steel ingot, with its typically coarse columnar grain structure, is quite unsatisfactory for applications where high strength is required. A part made directly from steel ingot can easily fail under impact load. The columnar grains in a cast ingot must be recrystallized to give steel the required strength. This is done by a hot working process, such as forging or rolling (Figure 10.15).

Factors that influence the grain size resulting from hot deformation are: (1) initial grain size, (2) amount of deformation, (3) finishing temperature, and (4) rate of cooling.

During hot or cold working process, grains are deformed and their internal energy is increased. Heating worked metal to an appropriate temperature will cause recrystallization of the grains. During recrystallization numerous nucleii are formed within the larger deformed

FIGURE 10.13
Perfectly formed slabs of steel are cut to required lengths by an automatic flame torch after passing through the continuous slab caster (Courtesy of Bethlehem Steel Corporation).

FIGURE 10.14
Large dendritic crystals form in solidifying steel, which must be reformed into smaller uniform grains by the rolling process.

and the size of new grains is a function of the time and temperature. If the structure is left for an extended period of time at high temperature, the recrystallization process will be followed by a grain growth stage at which smaller grains will continue to grow and eventually start merging into larger grains. The process of recrystallization and grain growth can occur simultaneously with hot working operation, and sometimes it is intentionally performed, as an intermediate annealing process. This allows the material to be worked to a great degree without fracturing.

The forming processes result in a condition called anisotropy, i.e., metal has nonuniform mechanical properties (e.g., strength, elongation) in different directions. This is due to the grain deformation and their longitudinal orientation. As a result, the material is more ductile, i.e., can be deformed without cracking or splitting, to a greater extent along one direction than the other (Figure 10.16). For example, the metal can be bent to a smaller radius in the direction of rolling than it can at 90 degrees from direction of rolling. The anisotropy of cold or hot formed metals is very important in the welding and cold-forming processes.

grains. Keeping the structure at high temperature will cause the nucleii to grow, causing large grains to be transformed into smaller grains. In order to initiate this process it is required that at least minimum critical work be imparted on the material. The rate at which transformation occurs depends on the temperature to which the part is being heated. The amount of transformed grains

FORGING PROCESSES

In ancient times, copper and other nonferrous metals, which are sufficiently soft and ductile, were shaped while cold by hammering. Iron and steel, which are very strong but not ductile enough to be hammered into a shape while cold unless the piece is very small or thin, were heated before forming. Heating metals to forging temperatures greatly increases their plasticity and workability. Therefore, hot

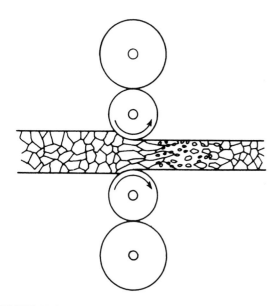

FIGURE 10.15
Recrystallization in hot rolling (Neely, *Metallurgy*, 2d ed. © 1984 John Wiley & Sons, Inc.).

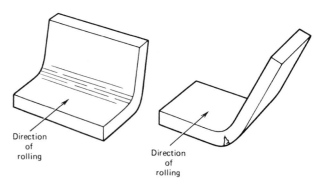

FIGURE 10.16
This illustrates the fibrous quality of rolled steel, called anisotropy. Since the fibers are in the direction of rolling, the metal is stronger along one axis than in the other (Neely, *Metallurgy*, 2d ed., © 1984 John Wiley & Sons, Inc.).

FIGURE 10.17
Woodcut showing a large wrought iron ship's anchor being forged in eighteenth century France. The arm and the shaft of the anchor are being forge welded together in this view (Dover Publications).

Vol. VII, Marine, Forge des Ancres, Pl. X.

forging was and still is a useful method of forming steel articles, both large and small.

In ancient Egypt, Greece, Persia, India, China, and Japan forging was used to produce armor, swords, and agricultural tools. In the thirteenth century, the tilt hammer came into use in which water power was used to raise or tilt a lever arm to which a weight was attached. When it fell, a heavy blow was delivered to the hot iron. In the mid-seventeenth century at the Saugus Iron Works in Massachusetts, a 500 pound weight on the end of a wooden beam was used as a forging hammer in which a cogged wheel driven by a waterwheel periodically raised and dropped the weight onto an anvil. Thus it was possible to make large forgings such as heavy ship's anchors (Figure 10.17). In 1783, Henry Cort of England built a rolling mill that had grooved rolls, making shaped bars of uniform cross section possible.

Forging gives metal good static, fatigue, and impact strength. This is why most good tools are forged and critical parts such as automobile wheel spindles and axles are made as forgings. Forged parts are often machined to provide bearing surfaces, splines, key seats, and so forth, but the essential strength of the forging remains. When parts are made by machining from the solid material, the grain flow is not always in the most advantageous direction and does not flow around shoulders and corners the way forgings do (Figure 10.18). Thus, machined parts are not as strong or resistant to failure as forgings.

Forging Machines

Today, large forgings (Figure 10.19) are produced in massive hydraulic presses. Smaller articles are forged by a variety of methods and machines. **Open-die drop forging,** impression die drop forging, press forging, **swaging,**

and **upset forging** are the most commonly used types of forging operations in manufacturing. When choosing a particular forging operation, several things must be considered: force and energy required, size of the forging, repeatability, speed and workability, and metallurgical structure. In addition, the amount of time hot metal is in contact with the die is important to die life and is also a consideration when choosing a forging machine for a particular forging operation.

Drop Hammer The drop hammer (Figure 10.20) is a development based on the old tilt hammer. Today, drop hammers are usually steam or air assisted to increase the force of the hammer blow and to raise the weight after each blow. Large articles such as machinery shafts that are to be finished by machining are usually open-die forged, using repeated blows while turning and moving the hot metal with tongs or a manipulator.

During this process the metal is squeezed or hammered to increase its length and decrease its cross section, or it is upset, i.e., pressed to increase its diameter and

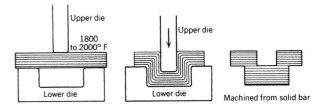

FIGURE 10.18
Grain flow in a solid bar as it is being forged, compared to a machined solid bar (White, Neely, Kibbe, Meyer, *Machine Tools and Machining Practices,* Vol. I, © 1977 John Wiley & Sons, Inc.).

FIGURE 10.19
A large forging to be used in a large aircraft. Subsequent machining operations will be performed on it (Lockheed Co.).

force. When upper and lower dies are used, it is called impression die forging. This method increases production rates and ensures repeatability. Often two or more progressive dies are used, each contributing to the final shape (Figure 10.21).

Mechanical Presses These machines make use of a heavy flywheel that stores energy for the forging stroke. Mechanical presses (Figures 10.22 and 10.23) of this type have an eccentric shaft or crank that moves the ram down and back to the starting position when a clutch is actuated. The greatest pressure is exerted at the bottom of the stroke. Inherent problem with these mechanisms is the possibility of jamming the ram near the bottom dead center, but most mechanical presses have a stall release that frees the stuck ram and upper die when this happens. Compared to drop forging machines, these machines operate more quietly, allow more accurate control, and produce higher force and energy. For this reason, the mechanical press is often chosen when a part must be made to closer tolerances. During the past decades, there has been a gradual trend toward the

FIGURE 10.20
Drop forging hammer. Steam or air pressure is used to raise the hammer and force it down on the heated metal.

shorten the length. Drop hammers perform repetitive blows to bring the workpiece to size, whereas mechanical and hydraulic presses can often make a forging in one or two strokes. However, both types of forging machines are adaptable to automated forging processes.

In open-die forging, drop hammer operators require a skill in giving the hammer just the right amount of

FIGURE 10.21
Progressive dies used with a mechanical press. Here a round bar at forging heat will be moved into the die by means of mechanical fingers. At each stroke kickouts move the partially formed forging to the next die. The bar is converted to two connecting rod forgings and the last die operation is that of removing the flash (Illustration courtesy of National Machinery Co.).

FIGURE 10.23
Mechanical press with a cutaway view showing the driving mechanism. Maximum die used in a press depends on shut height, which is the vertical opening with the ram down and adjustment up (Photo courtesy of the Minister Machine Company).

FIGURE 10.22
Automatic mechanical presses are high production forging machines (Illustration courtesy of National Machinery Co. Maxipres® is a registered trademark of National Machinery Co.).

FIGURE 10.24
Many years ago parts were made one at a time by an operator, often under dangerous conditions, unlike automatic press systems today (Graebener Press Systems, Inc., Providence, RI).

mechanical press forging. Fewer blows are required per forging and less operator skill is needed with these machines. Mechanical presses squeeze rather than impact the hot metal with a shorter forging-to-die contact time, thus extending die life.

A modern version of an old principle (Figure 10.24) is a high-energy type of mechanical forging press (Figure 10.25) that utilizes a vertical screw (Figure 10.26) instead of a crank or eccentric shaft to transfer the flywheel energy to the ram. This arrangement provides uniform force throughout the stroke, unlike crank presses in which the major tonnage is confined to the end of the stroke. Like the drop hammer, the screw press does not have a bottom dead center. A continuously rotating flywheel imparts motion to a screw when a clutch is engaged. It appears that the screw press combines two of the best features of the drop hammer and the crank press, high impact and a squeezing action. Even with hard blow forging, the press frame and mechanical components are not subject to high stresses and the die is in contact with the hot metal a very short time. Often only one or two blows are needed to complete a single forging (Figure 10.27 on p. 212). High-energy forging presses provide high flexibility since they can be used for either open or closed die forging operations.

Hydraulic Presses   Hydraulic presses (Figure 10.28 on p. 212) are the slowest types, but their advantage is that they can exert very high forces required for large forgings. Instead of mechanical linkages and cranks, the press ram is powered with one or more large hydraulic cylinders. Oil is forced at high pressure into the cylinders and the ram moves at a constant rate. Smaller cylinders return the ram to its upper position. Large hydraulic presses can exert thousands of tons of pressure, making them ideal for some of the largest forgings (Figure 10.29 on p. 212).

FIGURE 10.25
High-energy forging press (Photo courtesy of Siempelkamp Corporation).

FIGURE 10.26
Drawing of the parts of a high-energy screw press (Photo courtesy of Siempelkamp Corporation).

Clutch cylinder

Clutch piston

Flywheel

Thrust bearing

Return cylinder

Lubrication oil

Press frame

Screw drive

Ram

Ram bolster plate

Table bolster plate

Ejector

FORGING DIES

Forging dies can be classified in three general groups: (1) open dies, (2) impression dies, and (3) flashless dies. In open die forging, the workpiece is compressed between two flat or almost flat dies that do not hinder lateral flow of the material. In impression die forging, the material is squeezed between two dies where lateral motion is significantly restrained. A small amount of metal is allowed to flow out of the die. This excess material is called flash, and it is removed after forging. Flashless die forging involves compression of material in which the lateral flow of material is completely restricted, and no flash is formed during

operation. The process requires that the exact amount of material be placed into a die. If too much material is placed, forces exerted on the die can be so high that the die can be damaged. Alternately, if not enough material is placed, then an incomplete filling of the die will result.

Design of forging dies can be very simple (open dies) or very complex (impression dies), and it is similar to that used for die casting. In the case of the impression dies, there is a parting line where the two halves of the die meet. To allow flow of excess material (flash) out of the die, a gap between the two dies (gutter) is provided, which is connected to the die cavity through a channel (land). The excess metal—**flash** (Figure 10.30)—must be removed in a

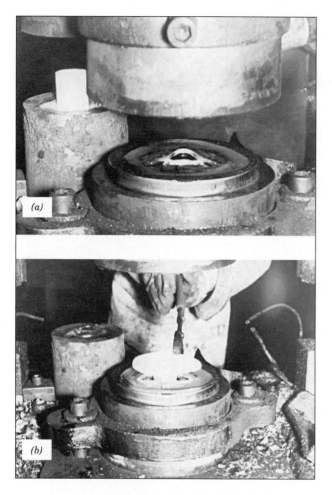

FIGURE 10.27
Preform in one-blow forging on a high energy press. *(a)* Preforging or descaling. *(b)* One-blow finish forging (Photo courtesy of Siempelkamp Corporation).

FIGURE 10.28
Large open-die hydraulic forging press with manipulator positioning the forging ("Forging Presses," Schirmer-Plate-Siempelkamp, Hydraulische Pressen GmbH, Krefeld, Germany).

FIGURE 10.29
This huge forging press is used to dish heavy plate for pressure tank heads ("Pressing for All Applications in Metalforming," Schirmer-Plate-Siempelkamp, Hydraulische Pressen GmbH, Krefeld, Germany).

FIGURE 10.30
Forged connecting rods showing flash that must be removed, either in the automatic die set or as a secondary operation in a separate die (Illustration courtesy of National Machinery Co.).

FIGURE 10.31
Die showing parting line and gutters for flash overflow.

FIGURE 10.32
Typical automatic progressive die set for use with a 2000-ton press. Guide pins at the rear align the upper half of the die. Parts at the front show the progressions in making the part; the last one to the right shows the sheared-off flash (Illustration courtesy of National Machinery Co.).

FIGURE 10.33
Hand-fed crankshaft progressive die used on a 3000-ton press (Illustration courtesy of National Machinery Co.).

secondary operation, usually in a small shear press. There must be adequate draft so the part can be removed from the upper and lower die halves without sticking. Generous corner fillets and corner radii are necessary to allow for a smooth flow of material (Figure 10.31).

Normally, only simple designs can be made in forging dies (Figure 10.32). Sometimes a series of dies is necessary to preform the material so it will conform to the shape of the next set of dies until the final shape is made (Figure 10.33 to 10.35). Computer aided design (CAD) is quite useful in designing these multiple dies. From the starting shape of used stock, each step altering its shape prior to the finished shape can be displayed on the screen. The determination can be made as to the amount of deformation and its required force for each step so it will be within the capacity of the equipment with the least strain on the dies. Cast and powder metal preforms are often used in automated forging operations.

A forging die must be able to withstand extremely high temperatures of the red hot metal. Special hot-work tool steels are used for this purpose to prevent spalling and cracking of the dies.

UPSET FORGING

The term *hot upset forging* refers to a deformation operation in which a cylindrical stock is increased in diameter and reduced in length. Its prime application is in the fastener industry, for manufacturing nails, bolts, and similar hardware products (Figure 10.36).

Hot forming machines are massive and complex automatic feeding, shearing, forging, and punching systems capable of very high output (Figure 10.37). The process of nut making takes four steps on a three-die hot former (Figure 10.38 on p. 215). The heated metal stock is fed into a stop where it is sheared off (Figure 10.39 on p. 215). The

FIGURE 10.34
Hot crankshaft forging in a die after the last blow. It will now be removed from the die and the flash will be sheared off. (Illustration courtesy of National Machinery Co.).

FIGURE 10.36
These are typical shapes made in hot formers—short, round, symmetrical, and with a hole (Illustration courtesy of National Machinery Co.).

FIGURE 10.35
Stages in the formation of a crankshaft by hot forging from bottom to top (Illustration courtesy of National Machinery Co.).

FIGURE 10.37
A hot former is an automatic feeding, shearing, forging, and punching system capable of very high output. This machine makes 150 parts per minute (Illustration courtesy of National Machinery Co.).

FIGURE 10.38
The tooling arrangement for a three-die hot former progressing from right to left shows the steps in nut making—upset, form, and punch out the hole (Illustration courtesy of National Machinery Co.).

Finished nut blank

Punchout   Form   Upset

₵ Stock

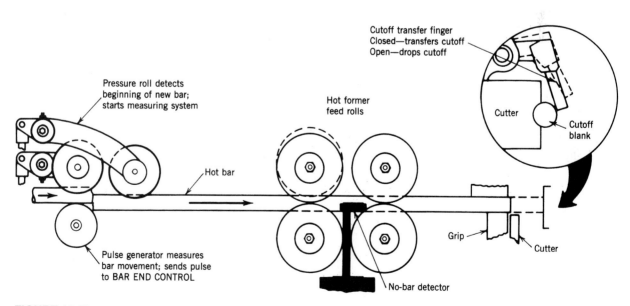

Cutoff transfer finger
Closed—transfers cutoff
Open—drops cutoff

Cutter

Cutoff blank

Pressure roll detects beginning of new bar; starts measuring system

Hot former feed rolls

Hot bar

Pulse generator measures bar movement; sends pulse to BAR END CONTROL

No-bar detector

Grip

Cutter

FIGURE 10.39
Hot shearing. A computer measures bar end movement into the hot former, then signals feed rolls and cutoff transfer to cycle at preset times (Illustration courtesy of National Machinery Co.).

FIGURE 10.40
Press system for the coining of bicycle bearing races.
Parts are oriented and conveyed up the incline on the left,
then fed down the tube escapement device that admits
them individually into the open die. The blanks continue
through a five-station progressive die with the operation
steps shown in Figure 10.34 (Graebener Press Systems,
Inc., Providence, RI).

first die upsets the blank; the second die forms the blank to
a hexagonal shape and partly forms the hole; and the third
die punches out the hole and the blank nut is ejected. One
of the advantages of hot forming is that a large percentage
of the hot blank can be displaced to make various shapes.
Bolt heads, flanged axles, and similar shapes are made in a
process called **heading,** which is similar to the nut making
process. Heading can be done by either a hot forging or a
cold upset process. The trend seems to be toward the cold
upset forging process and away from hot work upsetting.
However, shapes such as bearing caps and front wheel hubs
are hot formed, whereas more symmetrical shapes such as
wrist pins and gear blanks are cold formed. Heading and
the cold upset process are fully covered in Chapter 11,
Processing of Metals: Cold Working.

Another type of very high production forming system
that is in sharp contrast to the older methods (as shown in

FIGURE 10.41
Principle of the Graebener knuckle-joint press system
(Graebener Press Systems, Inc., Providence, RI).

Figure 10.24) is the knuckle-joint press principle (Figures
10.40 and 10.41). Machines of this sort use progressive
dies in an advanced automated system. Small parts such
as socket wrenches and bicycle bearing races (Figure
10.42) are produced with such presses. Hot, cold, or warm
forming can be done on knuckle-joint presses, depending
on the requirements of a design and toughness of the
metal. Open-end and box-end wrenches are perfect examples of parts that require a forging process. For example,
an end wrench made from steel strip by blanking would
be a very poor product, tending to bend or break in use.
Forgings are much stronger. Combination box-end and

FIGURE 10.42
Sequence in producing bicycle
bearing races: *(a)* Material.
(b) Blanking. *(c)* Upsetting.
(d) Forming. *(e)* Finishing.
(Graebener Press Systems, Inc.,
Providence, RI).

(a)   *(b)*   *(c)*   *(d)*   *(e)*

FIGURE 10.43
Interlinked press line for the production of ⅞-in. combination wrenches: (left) Line 1—GK 360 and (right) Line 2—GK 800 (Graebener Press Systems, Inc., Providence, RI).

FIGURE 10.45
Wrenches emerging from automatic machine (Graebener Press Systems, Inc., Providence, RI).

open-end wrenches are produced on two presses that are linked by a conveyer (Figure 10.43). Together they produce about 1800 pieces per hour. Since the material is a chrome-vanadium alloy, it can only be hot forged. Combination wrenches are produced in a sequence of operations (Figure 10.44). Wrenches are carried out of the machine after the final finishing operation (Figure 10.45). The wrenches collect in containers (Figure 10.46) for transfer to other operations.

FIGURE 10.44
Sequence of operation for ⅞-in. combination wrenches: (a) On a GK 800: (1) Setting. (2) Straightening of head. (3) Forming. (4) Trimming. (b) On a GK 360: (1) Piercing. (2) Flat coining. (3) Edge coining. (Graebener Press Systems, Inc., Providence, RI).

SWAGING

Swaging, like nut making and bolt heading, can be either a hot- or a cold-forming operation. However, considerably more force and more massive machines are needed for cold than for hot forming. Therefore, larger, heavier parts are usually hot formed. Swaging involves reducing the diameter of a rod or tube (Figure 10.47).

Swaging operation is often performed in a rotary swaging machine. Typical swaging types are shown in Figures 10.48 to 10.50 on pp. 218 and 219. The two-die swaging machine delivers blows at a rate of 1800 to 4000 per minute by action of the backers, thus alternately striking the work rolls. During the process, dies also rotate around the part as they hammer the work.

In a modification of the swaging process called radial forging, instead of dies rotating around a rod, the rod rotates between dies as it feeds into hammering dies. Nevertheless, the result is the same.

HOT EXTRUSION

Extrusion is a process of forcing metal through a die, similar to squeezing toothpaste out of a tube. Nonferrous metals may be extruded cold or warm, whereas ferrous metals usually need to be at a forging temperature to make them plastic enough to be extruded. Square and round tubular products, structural shapes, and round, square, or hexagonal solid shapes are some examples of extruded pieces. There is almost no limit to the kinds of intricate cross-

FIGURE 10.46
The finished product—⅞-in. combination wrenches
(Graebener Press Systems, Inc., Providence, RI).

sectional shapes that can be made by this process. Extrusion produces surfaces that are clean and smooth with accurate shapes that can be held within close tolerances. The great advantage of the extrusion process is that it allows manufacture of intricate shapes. For example, inexpensive bronze gears of various size and shapes can be extruded in lengths of 20 to 40 ft, and later cut off to desired lengths. However, in cases where a product can be rolled instead of extruded, the choice should be a rolled product if large amounts are needed because rolling is a less expensive method for shaping large quantities of metal. For short runs, extrusions may be more economical since tooling costs are much lower.

The two basic types of extrusion process are (1) direct extrusion and (2) indirect extrusion. In direct extrusion metal is placed into a container, a ram is pushed, and the work billet is forced to flow through one or more openings in a die. Alternately, in indirect extrusion a die is mounted on the ram. As the ram is pushed against the work billet, material flows through the die opening and around the ram. Tubular products can be made by both processes. However, direct extrusion requires a mandrel that allows the metal to flow around it (Figure 10.51).

HOT DRAWING

Drawing is a process in which the cross section of a bar, rod, or wire is reduced by pulling it through a die opening. The process is similar to extrusion, except that in this process the material is pulled through a die, whereas in extrusion the material is pushed through. Cold drawing uses relatively thin metal and changes its thickness only slightly, whereas hot drawing deforms the metal to a very great extent. Hot drawing is used to make thick-walled parts of simple cylindrical shapes (Figures 10.52 to 10.54 on the following pages). Heavy-duty hydraulic cylinders, artillery shells, and oxygen tanks are made by the hot drawing process.

HOT SPINNING

Although most metal spinning operations are carried out on cold metal, very large, tough metals are spun while hot. Domed heads for pressure vessels are often made by hot spinning. This process consists of shaping flat or preformed metal disks over a rotating form. Pressure applied by the spinning tool causes the hot metal to flow over the form. Since simple tooling is needed for

FIGURE 10.47
Tubing being necked down by
the swaging process
(Courtesy: Aluminum
Company of America).

Hardened steel rolls

Backers

Die

Die slot

Spindle

Straight expanding | Double flanging | Double flaring

Flaring | Curling out | Flanging

FIGURE 10.48
Types of expanded tube ends (Aluminum Company of America).

Curl in

Reducing

FIGURE 10.49
Types of reduced tube ends (Courtesy: Aluminum Company of America).

External bead | Internal bead | Closed bead

FIGURE 10.50
Types of beaded tubes (Courtesy: Aluminum Company of America).

(a)

(b)

FIGURE 10.51
Forward extrusion. *(a)* Use of the spider mandrel to produce hollow extruded shapes. *(b)* Mandrel fastened on the ram producing a hollow shape.

FIGURE 10.52
Hot drawing press, called a push bench, for the production of thick-walled hollow bodies and pressure vessels ("Presses for All Applications of Metalforming," Schirmer-Plate-Siempelkamp, Hydraulische Pressen GmbH, Krefeld, Germany).

FIGURE 10.53
After the mandrel has forced the hot metal through the draw ring, a stripping device removes the forging as the mandrel is drawn back ("Presses for All Applications of Metalforming," Schirmer-Plate-Siempelkamp, Hydraulische Pressen GmbH, Krefeld, Germany).

FIGURE 10.55
Piercing of solid billet to make seamless tubing (White, Neely, Kibbe, Meyer, *Machine Tools and Machining Practices,* Vol. I, © 1977 John Wiley & Sons, Inc.).

spinning, it is a less expensive method of forming circular and cylindrical shapes than the use of drawing or stamping dies.

SEAMLESS TUBING

Seamless tube is hot formed by the process of roll piercing (Figure 10.55). A heated, cylindrical billet is passed through rotating conical rolls whose axes are in different planes. This action results in high tensile stresses forming a crack at the center of the billet that flows over a conical mandrel that shapes and sizes the thick-walled tube. A following operation in a rolling mill with grooved rolls and a mandrel further elongates the tube and reduces its wall thickness. Seamless tube is often cold drawn as a finishing operation.

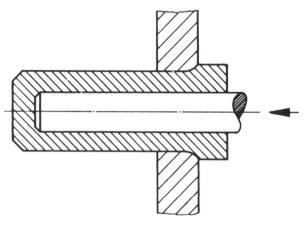

FIGURE 10.54
Diagram showing the principle of producing thick-walled steel bodies ("Presses for All Applications of Metalforming," Schirmer-Plate-Siempelkamp, Hydraulische Pressen GmbH, Krefeld, Germany).

PIPE WELDING (ROLL FORMING)

Hot-formed pipe is made of **skelp,** a low-carbon semifinished steel strip. Essentially, the hot strip is drawn through a series of rolls that gradually curl the flat strip into a cylinder (Figure 10.56). As the pipe is formed, the edges are forced together, and at a white heat they make a weld bond. This is called butt welding. Since the skelp is unwound from a continuous coil, the pipe must be cut to lengths with a flying saw, one that follows the moving pipe as it cuts it to length. Small diameter pipe, up to 3 in., is butt welded. For larger diameters it is lap welded to

FIGURE 10.56
Small-diameter pipe formed
by continuous butt welding.
Skelp is formed and welded
into pipe as it passes through
sets of rolls. A flying saw cuts
the pipe into lengths
(Courtesy of Bethlehem
Steel Corporation).

give it greater strength. Hot-formed pressure-welded pipe is much less expensive than seamless tubing and is widely used for underground water systems, household plumbing, and for many industrial uses. However, because it is pressure welded (butt welded), it is not acceptable for high-pressure applications such as in hydraulic machinery where the seamless pipe is used.

Review Questions

1. By what means are white hot ingots formed into slabs, blooms, and billets?

2. What modern method of steel manufacture bypasses the ingot stage and produces slabs or blooms directly?

3. Large dendritic crystals that weaken the metal form in cast metals such as ingots and strand casting. How are these removed from the metal?

4. What is anisotropy in metals?

5. Why are very large forgings heated although smaller forming operations are sometimes carried out cold? Explain.

6. What kind of forging machine would you choose to use for a very large open-die operation? Which would you choose for making a run of 10,000 high-precision parts?

7. Hot metal tends to stick to forging dies. What design characteristic helps to prevent this from happening?

8. What is "flash" and how is it removed?

9. If you were required to design a machine part, such as an automobile axle, that will be subjected to high shock loads, would you choose a forging? A casting? Or would you machine it from solid bar stock? Why?

10. How are bolt and nut blanks formed?

11. What kinds of shapes can be made by hot extrusion?

12. A tank end having a hemispherical shape must be made of ¾ in. thick flat steel. Name two methods by which this operation can be accomplished.

13. Briefly explain how seamless tubing is made and how common steel water pipe is made.

Case Problems

Case 1: Failure of a Tractor Spindle

A front wheel axle spindle on a tractor became damaged from a bearing failure. A replacement part would take 6 weeks to get if it were ordered from the factory and the farmer needed the tractor for his spring planting. A new spindle was made at a local machine shop from solid steel of the same alloy as the original. After 2 weeks of use, it broke in half and the wheel fell off. Considering what you have learned in this chapter, what do you think caused the failure of the machined spindle?

Case 2: Hot Pressing

Open die forging on flat anvils usually requires a pressure of 3 to 5 tons per square inch on the tool faces for mild steel at a forging temperature. Harder alloy steels may require two or three times that amount, and up to 15 tons per square inch are needed for the very hardest steels. Pressures may need to be doubled if swages or dies are used.

A 3-in. square billet, 10 ft long, of hot rolled mild steel is to be open die forged into a lifting arm for a hoist. Upper and lower swaging dies will be used to form half the length of the bar into an elliptical shape from the square. Each of the lower and upper swages has a surface area of 20 in.^{2}. A 300-ton hydraulic forging press is available. Will it be sufficient for the job? Approximately how much pressure will this forging operation require?

CHAPTER 11

Processing of Metals: Cold Working

Objectives

This chapter will enable you to:

1. Explain the effects of cold working on metals.
2. Describe how hot-rolled steel is prepared and cold finished in steel mills.
3. List a number of cold-forming operations and explain their principles, advantages, and uses.
4. Account for the difference between cold-rolled and machined parts and select one process over another for a particular product.
5. Make a choice among cold-forming operations for a particular product.

In ancient times, cold working was accomplished by hammering on soft metals such as gold, silver, and copper for jewelry and other ornamentation. Iron wire was made in the Middle Ages by pulling iron rods through progressively smaller holes in a steel plate. Later, the improved properties of cold-drawn products were discovered and a new industry was born. After being hot formed in a steel mill, steel shapes such as bars, sheets, and tubes could be descaled and further shaped by the process of cold forming. Today, cold-formed products range from very fine hypodermic needles to huge pipeline tubes and from tiny hair-size filaments to propeller shafts for ships. Almost any conceivable shape can now be made by one or more processes of cold forming. Greater strength-to-weight ratios, better finishes, and closer dimensional tolerances are some of the advantages a cold-worked metal has over a hot-worked metal.

FACTORS IN COLD WORKING

One of the most valuable characteristics of metals is known as **plasticity,** the ability of metals to be deformed permanently in any direction without cracking or splitting. Higher temperatures as in hot working tend to make metals more plastic, but recrystallization at high temperatures causes them to retain their soft condition with its lower strength. Higher strengths and hardnesses are obtained by deforming the metal at normal temperatures. Permanent plastic deformation can occur only at stresses above the elastic limit or yield point (Figure 11.1). As the cold work continues, more force is required and the hardness increases until a certain stress is reached; at this point the metal ruptures or splits apart. This is called the **breaking point** and it is often slightly lower than the **ultimate strength.** When metals are cold worked to a certain point, the next operation requires forces greater than those previously applied to deform the metal further. Each operation brings the particular metal closer to its ultimate strength and point of rupture. The object in cold working metals, therefore, is to stop well short of failure.

The degree of deformation (amount of cold working) determines the level of toughness, strength, hardness, and remaining ductility. Different specifications for a manufactured product can thus be obtained. In the case of sheet steel, for example, it is possible to produce from one-quarter-hard steel to full-hard steel. Quarterhard steel can be bent back 180 degrees without breaking, whereas half-hard steel can be bent only 90 degrees and full-hard steel can be bent only 45 degrees on a radius of about the material thickness. If more deformation is needed, then a **process anneal** is used to restore plasticity, which at the

FIGURE 11.1
Above the yield point more stress is required as the cold work continues. The *d* on the diagram indicates where failure occurs. (White, Neely, Kibbe, Meyer, *Machine Tools and Machining Practices,* Vol. II, © 1977 John Wiley & Sons, Inc.).

same time reduces the strength of the metal and lowers the force or stress required to deform it further. Process anneal on the cold-worked steel is often carried out in a closed container of inert gas to avoid scaling problems. When this is done, it is called **bright anneal.** See Chapter 4, Heat Treatments of Metals, for a discussion of annealing processes. Some of the advantages of cold working as compared to hot working are:

1. Closer dimensional tolerances.
2. Better surface finish.
3. Better machinability.
4. Superior mechanical properties.
5. Enhanced directional properties.

Some of the disadvantages of cold working are:

1. The metal is less ductile.
2. Annealing is sometimes required.
3. The material may contain residual stresses, causing warping when machined.
4. More massive and powerful machinery is required.
5. Subsequent heating or welding will change the cold-work properties and weaken the metal.

Elastic Recovery

When a metal is placed under stress within its elastic range, it will return to its former shape when the load is removed. If the metal is made to deform or take on a per-

manent set by loading it beyond the elastic limit into its plastic range, it will be permanently deformed, but will bounce back to some extent because of its elastic properties. This characteristic of metals, called **springback** (elastic recovery) (Figure 11.2), is a design consideration for forming and bending dies and fixtures. Parts are thus overbent several degrees, ironed, or bottom beaded to counteract springback.

Ductility and Malleability

Ductility is the property of a metal to deform permanently or to exhibit plasticity without rupture while under tension. Any metal that can be drawn into a wire is ductile. The ability of a metal to deform permanently when loaded in compression is called **malleability.** Metals that can be rolled into sheets or upset cold forged are malleable. Most ductile metals are malleable, but some very malleable metals such as lead are not very ductile and cannot be easily drawn into wire. Some ductile metals are steel, aluminum, gold, silver, and nickel. A few nonferrous alloys such as brass and Monel are also quite ductile, but most alloys of steel are less ductile than plain carbon steel. Some metals and alloys, such as stainless steel, high manganese steel, titanium, zirconium, and Inconel, tend to **work harden,** that is, to quickly increase hardness as cold working progresses. However, most metals tend to work harden to some extent. Austenitic stainless steels will tend to remain ductile until a very high hardness is achieved. This is an engineering property that is valuable when higher hardness as well as toughness is a requirement for a cold-formed product. Annealing will restore a softer condition as well as restoring the ductility of these work hardening metals. Work hardening is often a troublesome difficulty in machining operations.

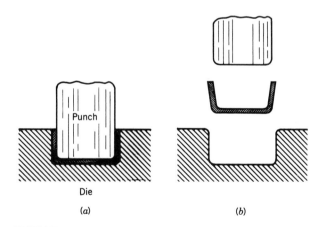

FIGURE 11.2
Cold working operation showing springback. When sheet metal is formed as in *(a)* and is removed from the die as in *(b),* it will partially revert back to its original shape.

FIGURE 11.3
Straightening rolls. The rolls
alternately bend the metal
slightly beyond its elastic limit
in two axes so that when it
emerges it is straight.

Preparing Hot-Rolled Steel for Cold Rolling

Hot-rolled steel bars and plates must be sufficiently oversize because the cold-finishing process reduces them in cross section. These are also sometimes straightened in a set of straightening rolls (Figure 11.3). The surface must be clean and free from scale. The hot-rolled bars or sheets are placed in a hydrochloric or sulfuric acid dip that removes the scale; this is called **pickling.** The acid is washed off and the steel is dipped in lime water to remove any acid residue. Sometimes the hot-rolled steel requires full annealing to make it as soft and ductile as possible prior to the cold-working operation.

COLD ROLLING IN THE STEEL MILL

The grain or crystal structure of the metal is permanently altered, flattened, and lengthened by the process of cold rolling (Figure 11.4). After a process anneal, the recrystallized metal can be further cold worked since the soft condition of the grains allows for more deformation to take place.

The coils of pickled hot-rolled steel pass through a series of rolling mills at a high rate of speed; 1 to 2 mil/min is not uncommon. The strip begins with a thickness that is a little less than $\frac{1}{8}$ in. and a length of about $\frac{3}{4}$ miles. Within 2 minutes it is squeezed to the thickness of two playing cards and it is more than 2 miles long (Figure 11.5). After annealing, the strip passes through a temper mill. There it is given the desired thickness, flatness, and surface quality, and the desired temper or hardness. Sometimes the full coil is shipped to customers; other times the coiled strip is cut into lengths for shipping flat.

Thin-gage, high-precision steel sheets were made possible by the Sendizmer mill (Figure 11.6). It was first

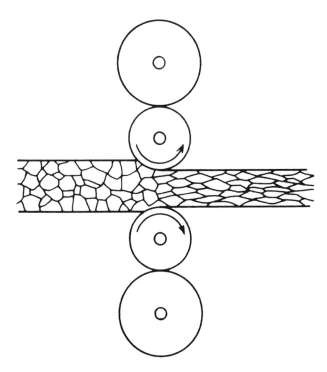

FIGURE 11.4
Cold rolling flattens and lengthens the grains in the direction of rolling (Neely, *Metallurgy,* 2d ed., © 1984 John Wiley & Sons, Inc.).

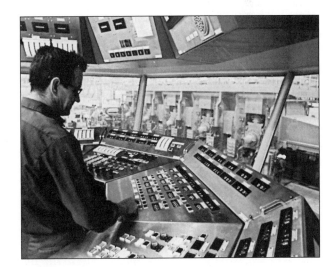

FIGURE 11.5
Cold strip mill showing the operator at his console in the foreground and the rolling mill stands in the background (American Iron & Steel Institute).

FIGURE 11.6
Sendizmer mill for rolling steel strip to close tolerances. Only the two small central rolls touch the metal; the outer cluster of rolls is for backup and bearings. These mills are used to make thin strip and foil (Courtesy of American Iron & Steel Institute).

developed to produce thin sheets for lightweight radar equipment used in aircraft. Today these thin sheets are needed for space programs. This mill also produces steel foil of .003 in. thickness or less for packaging and other uses. The work rolls on these mills are quite small in diameter, from 1 to 2½ in., and they are sometimes several feet in length. These are supported by a cluster of back-up rolls. High surface quality of steel strip can be obtained by planishing and polishing with abrasives. Stainless steel sheet and strip are sometimes given a mirror finish before being shipped to the customer.

Metals other than steel, such as aluminum, copper, and titanium, are also rolled into bars, plates, and sheet stock, using similar methods. Both hot and cold rolling processes are used for most nonferrous metals.

Surface Coatings on Sheet Steel

Since iron and steel have a tendency to rust in the presence of oxygen and moisture, many protective coatings for them have been developed. One of the first of these was the coating of iron with tin by hammering it onto steel sheets. This method can be traced back to the thirteenth century. Tin does not corrode in the presence of moisture and will readily bond to clean iron by hot dipping or by plating by the electrolytic method—the way most tin plate is produced today (Figure 11.7). The tin plating on steel may be less than .001 in. thick. Since available supplies of tin are dwindling and the demand for it is so great worldwide, steel companies began to search for substitutes for tin-coated steel. Chromium plate is

FIGURE 11.7
Steel coated with tin by the electrolytic process leaves the plating line (Courtesy of Bethlehem Steel Corporation).

sometimes used as a substitute and other coatings are now being used for some purposes. Galvanized sheet steel (Figure 11.8) is used for corrugated roofing material, culvert pipe, and many small articles such as steel pails. Manufactured steel parts for boats, street lighting fixtures, and other exposed metal objects are often galvanized. Although zinc can be deposited by the electrolytic process, by far the most prevalent plating method is the hot-dip process. In the sheet steel mill this process is continuous and often includes a final coating of paint that is dried in bake ovens as the steel moves through them. These galvanized and painted sheets are capable of being formed or bent without destroying their protective coatings. Many other coatings are also put on steel to protect it from deterioration or to enhance its appearance. These will be covered in Chapter 18, Corrosion and Protection for Materials.

BLANKING AND PRESSING

One of the most versatile forms of metal working is that of converting flat sheet and strip metals into useful articles. Sheet metal can be pierced, punched, or blanked to produce flat shapes. It can be drawn or cupped to form such items as pots and pans. It can be rolled into cones or cylinders or pressed into automobile body shapes; it also can be shaped by many other forming operations.

Although the cutting of a blank is actually a shearing operation, it is included here because it is the first step in forming a product from sheet stock. **Blanking** is the operation of cutting a flat shape out of a strip of sheet metal in which the hole material is saved for further operations. If the hole material is scrap, it is called **punching.** By this process any shape hole can be made. **Piercing** is the operation of cutting (usually small) round holes in sheet metal; if the holes are small and close together, it is known as **perforating.** Blanking with a conventional punch and die (Figure 11.9) is usually employed where large quantities are required. It usually is used only for thin sheet, but thicker material can be punched or blanked, depending on such factors as press size, blank size, and material thickness (Figure 11.10). Some blanks may be more economically cut by the circular shears (Figure 11.11) or nibbling machines (Figure 11.12). Where square or rectangular shapes are needed, blanks may be cut with sheet metal shears (Figure 11.13). Blanks may be sawed in a vertical band saw, several pieces stacked together (Figure 11.14 on p. 229). Hand-held power shears can also be used to cut blanks (Figure 11.15 on p. 229). It is obvious that using any of these methods to make blanks is much slower than using a die set and they are useful only when a few parts are to be produced.

FIGURE 11.8
Minimized spangle steel coils await shipment. These coils are produced on the plant's modernized galvanized line (Courtesy of Bethlehem Steel Corporation).

The basic components of piercing and blanking die sets, as shown in Figure 11.9, are a punch, a die, and a stripper plate. The punch holder (top plate) of the die is attached to the press ram, which moves the punch in and out of the die. The die holder of the die set (bottom plate) is attached to the bolster plate of the press. Guideposts in the die set on which bushings slide maintain precise alignment of the cutting members of the die. The stripper

FIGURE 11.9
Blanking die set in a press showing a punch making round blanks (Courtesy of Aluminum Company of America

plate removes the material strip from around blanking and piercing punches. The material strip is advanced after each punching stroke by a feeding mechanism.

An alternate method of blanking is the use of steel-rule dies (Figure 11.16). This method was first used for cutting softer materials such as cardboard, cloth, and plywood, but the process is now used for metal. A steel band is fastened on edge in a groove in the upper die plate. Neoprene rubber pads take the place of the usual stripper plate. Steel rule dies have an expected life of about 100,000 parts.

Conventional blanking dies may produce three times that many parts before resharpening is required. An open lower die can be used or a solid maple or urethane slab can be used to cut against, lessening the cost.

When mechanical presses (Figure 11.17) are fitted with progressive punching and forming dies, a continuous

FIGURE 11.10
Hardness and clearance of blanking tools may vary with the type of sheet metal used (Courtesy of Aluminum Company of America).

FIGURE 11.12
Duplicating nibbler. This machine cuts sheet metal by means of a small oscillating punch. Here it is being moved along a template to duplicate that shape (Courtesy of Heck Industries, Inc.).

FIGURE 11.11
Circular shears cutting a large disk (Courtesy of Niagara Machine & Tool Works).

FIGURE 11.13
Guillotine shear for shearing sheet and plate.

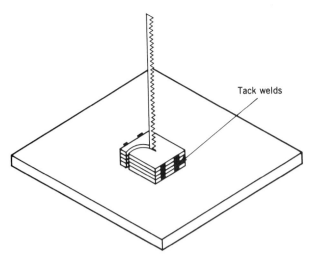

FIGURE 11.14
Band sawing a stack of blanks.

FIGURE 11.16
Steel-rule die blanking technique (Courtesy of Aluminum Company of America)

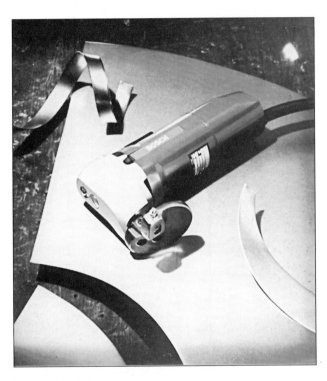

FIGURE 11.15
A portable hand-held power shears for cutting sheet metal blanks (Courtesy of Robert Bosch Power Tool Corp.).

FIGURE 11.17
Mechanical presses such as this one are extensively used for punching, blanking, and forming of sheet metal strip (Photo courtesy of The Minster Machine Company).

FIGURE 11.18
Electrical outlet boxes are punched and formed with a progressive die (Photo courtesy of The Minster Machine Company).

manufacturing of products is possible. Parts for electrical outlet boxes are formed from steel strip (Figure 11.18).

DRAWING, FORMING, AND EXTRUDING METAL

It is easy to imagine molten metal flowing by gravity into molds and steel softened by heating to a white heat being hammered or forced into shape. Also we are familiar with the hardness and toughness of cold metals. It is more difficult to understand how cold metal can be made to flow when sufficient pressure is applied. Virtually every operation performed on hot metal can also be done on cold metal, but the limit to the size of these parts is in the massiveness and power of the machinery. Since the cost of heating metal is eliminated in cold forming, it is a less expensive method, especially for the smaller piece parts, and there is no thermal damage to dies.

Combination hot, cold, and/or warm forming has the advantage of reducing equipment costs where several operations are involved and often reduces the need for secondary machining and operating personnel. Larger

parts can be hot formed and later finish formed by cold forming methods. There is a new trend toward warm forming in which metals that are difficult to cold work, such as titanium, stainless steel, and some alloy steels, can be formed without the great pressures and stresses of cold forming. For example, bearing races are now being warm formed of tough alloy steel that will subsequently be hardened and ground. Steels that contain more than .40 percent carbon are difficult to cold form but can easily be formed while warm, that is, below the transformation temperature but at or above the recrystallization temperature, about 900°F (482°C) for steel. Warm forming can substantially increase production rates for some materials.

Coining and embossing are stamping operations that form the surface of metal. Impressions of letters, figures, and patterns are formed by pressing them onto the metal. Usually knuckle-joint presses are used for coining and embossing because these machines are of rigid construc-

FIGURE 11.19
Forces applied to a blank during cold drawing operation (Courtesy of Aluminum Company of America).

FIGURE 11.20
Blanked and drawn parts showing progression of drawing operation (Courtesy of Aluminum Company of America).

tion, have rapid short strokes, and exert intense pressure during the last part of the stroke as a result of the knuckle-joint action. See Chapter 10 for a description of knuckle-joint presses.

Drawing Plate, Sheet, and Foil

This kind of drawing process is one in which a flat piece of metal is formed into a hollow shape by applying force with a punch to the center portion of the metal. The punch draws the metal into a die cavity (Figure 11.19). In this operation, the metal is stretched on the side walls and made to assume an exact thickness. Sheet metal drawing is usually performed in a vertical press.

Blanking and piercing are sometimes combined in one operation for the purpose of continuous manufacture of small parts such as ashtrays and bottle caps. This is usually done with progressive dies (Figure 11.20). Deep drawing is often done with several presses, each having a different die that contributes to the final shape (Figure 11.21). When the wall thickness must be further reduced and the parts lengthened, as in the manufacture of car-

FIGURE 11.21
Sequence of draw reductions for 6061 aluminum alloy rectangular case (Courtesy of Aluminum Company of America).

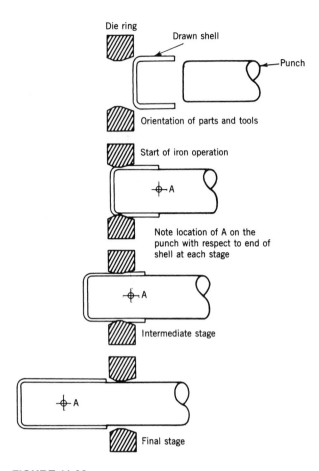

FIGURE 11.22
Ironing a drawn shell (Courtesy of Aluminum Company of America).

FIGURE 11.23
Multiple reduction with stacked ironing dies (Courtesy of Aluminum Company of America).

tridge cases, an ironing process is used (Figures 11.22 and 11.23). Sometimes an anneal is required between pressings.

Bar, Tube, and Wire Drawing

Hot-rolled bars, after pickling and cleaning to remove the scale, are reduced in cross section by pulling them through a die that is slightly smaller than they are. This *bar drawing* operation is performed on a draw bench (Figure 11.24). The drawing process hardens the metal and gives it a smooth finish suitable for machinery shafting.

Seamless tubing is also cold drawn on a draw bench; the only difference in the operation is that there is a mandrel inside the tube to thin the walls and provide an internal finish (Figure 11.25). Because of this finish operation and the higher cost of pierced seamless tubing, cold-drawn seamless tubing is considerably more expensive than butt-welded black pipe. Seamless steel tubing, after a honing operation, is widely used for hydraulic cylinder manufacture and smaller diameters are used for high-pressure pipes that carry oil or other fluids.

Wire drawing is basically the same process as bar drawing except that it involves much smaller diameters of metal and it is a continuous process (Figures 11.26 and 11.27). The wire is drawn through a series of dies, each one slightly smaller in diameter than the previous one. The wire is pulled by a rotating capstan or drum that is located between each set of dies. Coolant, acting as a lubricant and to cool the die, is flooded over the dies since this is a cold-forming operation and unwanted heat is developed in the operation. The finished wire is wound on a reel.

Cold Forming

Cold forming, also called cold heading or cold forging, is a metal upsetting process carried out on machines designed for rapid production of small parts from wire stock (Figure 11.28). Upsetting machines are frequently rated by maximum diameter cutoff capacity. For example, a ⁵⁄₁₆ in. cold header can shear alloy steel up to ⁵⁄₁₆ in. in diameter. Generally, only small parts such as screw and bolt blanks, rivets, and ball bearing blanks are produced on cold upset machines, but some machines are capable of upsetting much larger parts (Figure 11.29 on p. 234). The largest single use of upsetting machines is in forming heads on fasteners such as rivets and screws. Threads are subsequently rolled on the blanks, giving them a much greater strength and fatigue resistance than cut screws (Figure 11.30 on p. 234).

In the past, most threaded fasteners were made on automatic screw machines that turned the part to size from bar stock and then threaded it with a die, followed by a

FIGURE 11.24
Draw bench for solid-bar stock.

FIGURE 11.25
Draw bench for seamless tubing.

FIGURE 11.26
Battery of modern wire drawing machines. Rod enters at the left, is reduced in size as it passes through successive dies, and is coiled at the right (Courtesy of Bethlehem Steel Corporation).

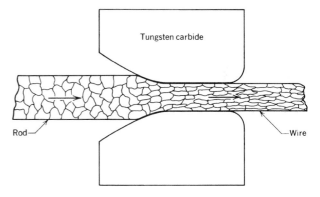

FIGURE 11.27
Enlarged cross section of wire drawing through a die (White, Neely, Kibbe, Meyer, *Machine Tools and Machining Practices,* Vol. I, © 1977 John Wiley & Sons, Inc.).

FIGURE 11.28
Wire upsetting machine (Illustration courtesy of National Machinery Co.).

FIGURE 11.29
An array of cold upset parts (Illustration courtesy of National Machinery Co.).

cutoff operation. Using automatic screw machines is necessary when high precision and complex shapes are involved, but they are much slower than cold-heading machines; a small part made in 10 sec is acceptable on an automatic machine. However, production rates on upsetting machines can be as high as 36,000/hr for small unpierced rivets, and No. 8 size screw blanks can be made at 27,000/hr. Bolts ⅜ in. in diameter can be headed, pointed, and threaded at a rate of 15,000 per hour. Another advantage of forming parts is that nothing is wasted, almost no scrap is cut away as in machining. There are also few rejects with the upsetting process.

The process and sequence of operation for a cold heading operation is shown in Figure 11.31. The wire or bar is first formed into a cone before making the head because the unsupported length tends to buckle if the heading operation is done first. Several modes of upsetting are shown in Figure 11.32; some parts are upset in the punch, some in the die, some in both punch and die, and others between the punch and die. The five basic operations performed in cold-forming machines are shown in Figure 11.33. Combinations of these give cold forming great versatility. An example of this is the spark plug shell (Figure 11.34). Figure 11.35 shows the progression of upsets and extrusions used to make the spark plug center post. This sequence of operations and the tooling involved is shown in Figure 11.36.

Nut blanks are made in four- or six-die nut formers (Figure 11.37 on p. 237). The operation is similar to upsetting and to the nut blank hot formers discussed in Chapter

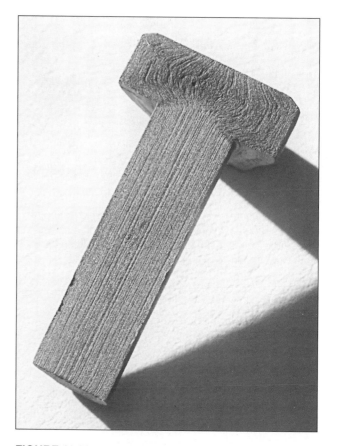

FIGURE 11.30
Grain structure of a cold-formed bolt blank (Illustration courtesy of National Machinery Co.).

FIGURE 11.31
Cone upsetting. Sequence of operations in upsetting a bolt head (Illustration courtesy of National Machinery Co.).

10. As in other cold upset operations, the advantage over hot forming is the higher production rate and lower cost. A sequence of four-die nut forming is shown in Figure 11.38 on p. 237.

Cold Forming Threads, Worms, and Gears

Although thread and gear forming is not a new idea, it is well suited to be a part of a new trend toward the continuous processing of metals. Ideally, the process would be from the raw ore to continuous casting to rolling mills to upset and roll forming to the finished product. In such a system, waste would be reduced and transferring and shipping of materials eliminated.

External thread forming by the process of cold rolling screw threads is a follow-up operation after cold heading screw blanks. However, threads may also be cut as a machining operation on cold-formed blanks. The decision whether to machine threads or roll them can be based on several factors.

1. **Plasticity** The amount of deformation required is directly related to the plasticity of the metal. Some grades of aluminum, gray cast iron and die cast metals cannot be rolled because of their insufficient elongation and reduction of area.

2. **Work hardening** Most stainless steels and high manganese steels rapidly work harden when cold worked. These metals do not readily lend themselves to cold rolling.

FIGURE 11.32
Modes of upsetting (Illustration courtesy of National Machinery Co.).

FIGURE 11.33
Five basic operations performed in cold-forming machines. Combinations and variations of these operations give the cold former a wide range of application (Illustration courtesy of National Machinery Co.).

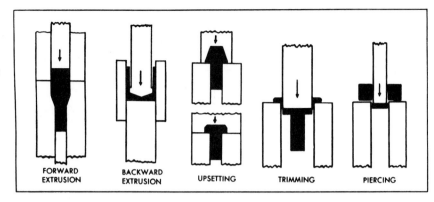

FORWARD EXTRUSION   BACKWARD EXTRUSION   UPSETTING   TRIMMING   PIERCING

FIGURE 11.34
From blank to finished spark plug shell in a six-forming die machine (Illustration courtesy of National Machinery Co.).

FIGURE 11.35
Progression of upsets and extrusions used to make spark plug center post (Illustration courtesy of National Machinery Co.).

A   AIR-LOADED PIN
     PUNCH
     PUNCH CASE
     INSERTS
B
C
D
     DIE

FIGURE 11.36
(a) Air-loaded pin punches the spark plug center post blank into die. (b) Die forces inserts into closed positions. (c) Continued advance of pin upsets metal into cavity formed by inserts. (d) As punch case withdraws, inserts open to allow them to clear largest upset diameter (Illustration courtesy of National Machinery Co.).

FIGURE 11.37
Four-die nut former (Illustration courtesy of National Machinery Co.).

FIGURE 11.38
Four-die nut forming. The top row of operation shows forming done by the punches; the bottom row shows forming done in the dies. The first blow upsets the blank and starts the hex formation. The second blow displaces metal away from the center of the blank toward the corners of the hex. The third blow finishes both faces and marks the hole for punching out. The final blow punches the slug, forming the hole. This method retains controlled metal flow for high-quality nuts (Illustration courtesy of National Machinery Co.).

3. **Workpiece design** Some parts are not adaptable for rolling, for example those with threads that are too close to a shoulder (Figure 11.39). Some tapered threads and threads on inaccurate blanks should be cut instead of rolled.

4. **Finish** Rolling is far superior to cutting if the finish is a consideration.

5. **Surface hardness** The rolling process produces a greater surface hardness (Figure 11.40) that wears well and helps to resist metal fatigue at thread roots.

6. **Higher speeds** Generally, thread rolling is a somewhat faster operation than cutting threads with dies.

7. **Tensile strength** Rolled threads are stronger than cut threads because of grain flow (Figure 11.41).

Thread rolling machines have two or three hardened, threaded rolls between which threads are formed on blanks. Several methods are used for forming threads, splines, worms, and gears: infeed forming, through-feed forming, the reciprocal method, the infeed-throughfeed-out method, and the automatic continuous method.

Internal threads are often cut using taps, but they may also be formed using thread forming taps (Figures 11.42 and 11.43). As in thread rolling, metal is displaced to form the threads, work hardening and smoothing it. Both cutting and forming methods for internal threads are very rapid processes and the choice of method can be based on the same factors as those of external roll forming of threads.

Gearing

A *worm* is a spiral or helical groove, similar to a screw thread that engages with a worm wheel or worm gear to impart motion. Worms have traditionally been machined

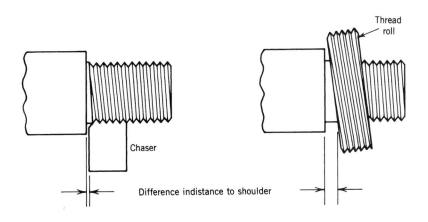

Chaser

Difference indistance to shoulder

Thread roll

FIGURE 11.39
If threads must be made close to a shoulder, cutting is better than rolling.

FIGURE 11.40
Longitudinal section of a type 304 stainless steel fastener that was cold worked by being thread rolled in the annealed condition. Cold-worked austenite (and probably some martensite) is in the area immediately below the surface of the rolled thread (50 ×) (*Metals Handbook,* Vol. 7, 8th ed., American Society for Metals, 1972, p. 134. With permission).

FIGURE 11.41
Rolled threads have a harder surface than cut threads and are generally stronger.

Rolled thread   Cut thread

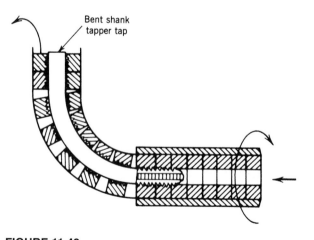

Bent shank tapper tap

FIGURE 11.42
Production tapping of nuts is usually done with bent shank taps to make it a continuous process.

Style B, Bottoming Series

Style P, Plug Series

FIGURE 11.43
Thread forming taps (Courtesy of DoAll Company).

238

on lathes or milling machines since they are highly precision parts. However, it is possible to achieve the required precision by the cold-rolling process with the added advantage of the part receiving a very smooth finish and work-hardened surface. This reduces friction and provides a longer service life. In this process, a blank is placed in a rolling machine and moved between rolls in the infeed process where hundreds of tons of pressure is applied as the rolls and blank turn and form the shape of the worm.

Helical and spur gears are formed exactly as worms are formed. Smaller gears are cold formed, but larger gears with coarser teeth are hot formed. For high precision gears, a follow-up operation is necessary. Even machine-cut gears require a finishing operation of gear shaving or grinding. This will be further discussed in Chapter 14.

Cold Extrusion

This is a process of producing long lengths of various shapes such as rod, tubing, and structural forms. Usually only soft metals with low yield strengths are extruded cold and tough metals such as steel are extruded hot.

Methods of Extrusion There are basically three methods to form extrusions: direct, indirect, and impact. In the process of extrusion, metal is forced through a die that has the desired shape at extremely high pressures, causing the metal to deform plastically and to flow through the die taking its shape. Soft metals such as lead can be extruded with low pressures of about 20 tons/in.², but higher pressures are required for aluminum, brass, and other nonferrous metals.

Direct extrusion (Figure 11.44) is more commonly used for both hot and cold extrusions. Direct or forward extrusion is widely used to produce special shapes that are not available from rolling mills. However, the cost of rolled products is considerably less than that of extruded

forms and the surface finish is better. Since the cost of extrusion dies is less than the cost of rolls, extrusions are more practical for short runs.

Indirect extrusion (Figure 11.45), also called backward extrusion, has the advantage of creating less friction, but the restricted extrusion length and extra equipment limit the method's usefulness.

Impact extrusion (Figure 11.46) is used to produce products such as collapsible tubes (for toothpaste) and soft drink cans from soft metals, usually aluminum. In the process of impact extrusion, a thick slug of metal is placed in the cavity of a die. The slug is struck by a punch having the size and shape of the inside of the part. The impact of the punch is sufficient to cause the metal in the blank to flow very quickly backward through the space between the punch and die cavity. The part adheres to the punch and is removed by a stationary stripper as the punch is withdrawn. Impact extrusion is a very rapid production method of forming cold metal.

BENDING, STRAIGHTENING, AND ROLL FORMING

Bending refers to a simple bend in one axis. When two or more bends are made simultaneously, as with a die, the process is usually called forming or drawing. Bending metals causes a plastic deformation about a linear axis, making little or no change in the surface area of the bend. The neutral axis of the bend (Figure 11.47) is not located at the center, equidistant between the outer and inner surfaces of the bend, because the yield strength of metals in compression is somewhat higher than in tension and this causes the inner compression sides of the bend to bulge, whereas the outer tension side is thinned and reduced in width. Also, as in other cold-forming operations, there is a

FIGURE 11.44
Direct or forward extrusion. The flow lines show how the metal is pushed out of the die by the use of great forces. Softer metals are usually extruded cold, whereas harder metals are brought to a forging heat before being extruded (Neely, *Metallurgy,* 2d ed., © 1984 John Wiley & Sons, Inc.).

FIGURE 11.45
Indirect or backward extrusion.

FIGURE 11.46
The method of forming thin-walled tubes by impact extrusion. A flat blank is placed in the die *(a)* and the punch is brought down rapidly with a single blow *(b)*. The material in the blank then "squirts" upward around the punch. When the punch is withdrawn, a stipper plate removes the tube (Neely, *Metallurgy,* 2d ed., © 1984 John Wiley & Sons, Inc.).

(a) *(b)*

FIGURE 11.47
Simple bend in one axis.

tendency of the metal to unbend; this varies considerably in metals. For example, lead has almost no springback, soft aluminum has very little, but 301 stainless steel has considerable springback.

Machines that make relatively sharp bends or small radii in sheet metal are called bar folders (Figure 11.48). These tools are used to bend sheet metal for architectural shapes, furnace and air conditioning pipes, and sheet metal products. When bends in heavier sheet metal are required or more complex bends are needed, press brakes are used (Figures 11.49 to 11.54 on pp. 241 and 242). These machines are designed to make short strokes to a predetermined depth, and they are powered either with hydraulic cylinders or in much the same way as a crank punch press.

Roll bending (Figures 11.55 to 11.57 on p. 243) is used to form curved shapes, cylinders, or rings. The metal used can range from thin sheet metal to massive bar or plate stock. Plates, bars, and structural shapes are formed this method.

Cold-roll forming (Figure 11.58 on p. 243) consists of bending a flat strip into a complex shape by a series of rolls, each contributing to the final shape. Channels, moldings, building trim, and rain gutters are formed by this method (Figure 11.59 on p. 243).

FIGURE 11.48
Bar folder. Such a device is used for bending light sheet metal.

Straightening or *flattening* is a process designed to remove unwanted bends from metal. Sheet or bar stock can be straightened by being passed through a series of rolls, as shown in Figure 11.3. The metal is bent back and forth to a point slightly beyond its elastic limit, thus removing previous permanent deformations. Heavy sections are often straightened in a press by applying pressure on the bend while the part is supported at two points.

Stretch forming light-gage sheet metal can be done by stretching it over a contoured form or die block. This is done on a machine that has gripping jaws that are mechanically or hydraulically powered (Figure 11.60 on p. 244).

240

FIGURE 11.49
Vertical press brake (Courtesy of Clausing Machine Tools).

FIGURE 11.50
Typical press brake tools (Courtesy of Aluminum Company of America).

Stretch-wrap forming (Figure 11.61 on p. 244) is similar to stretch forming. The blank is first stretched beyond the yield point and then wrapped around a form block. Stretch-draw forming (Figure 11.62 on p. 244) is a shallow stamping process in which bottom and top dies are brought together to form a stretched sheet. The principal advantages of stretch forming are that it greatly reduces springback, virtually eliminates wrinkles, and reduces tooling cost.

High Energy Rate Forming Process

High energy rate forming processes (HERF) involve a high rate of workpiece deformation. The most important of these techniques are chemical (explosive), electrohydraulic, electromagnetic, and high energy mechanical forming.

Explosive forming (Figure 11.63 on p. 244) is used for operations such as shaping metal parts, cladding, joining, and forming powdered metal parts.

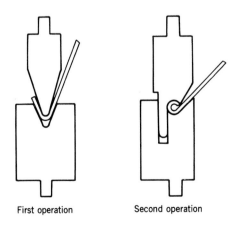

FIGURE 11.51
Press-brake forming of bead in two operations (Courtesy of Aluminum Company of America).

FIGURE 11.52
Press-brake forming of a
corrugated sheet (Courtesy of
Aluminum Company of
America).

FIGURE 11.53
Brake tool with spring-loaded pressure pad in bottom die
(Courtesy of Aluminum Company of America).

FIGURE 11.55
Roll bending is used to form cylindrical parts (Courtesy of
Aluminum Company of America).

FIGURE 11.54
Brake tools used with urethane die (Courtesy of Aluminum
Company of America).

FIGURE 11.56
Vertical roll-bending machine. Heavy bars may be rolled into rings with this machine (Courtesy of Buffalo Forge Company).

FIGURE 11.58
Roll-forming machine (Courtesy of Aluminum Company of America).

FIGURE 11.59
Typical roll-formed shapes (Courtesy of Aluminum Company of America).

FIGURE 11.57
The pinch-type roll bender is used to form shapes, such as channels, into rings (Courtesy of Buffalo Forge Company).

FIGURE 11.60
Stretch forming (Courtesy of Aluminum Company of America).

FIGURE 11.62
Stretch-draw forming (Courtesy of Aluminum Company of America).

FIGURE 11.61
Stretch-warp forming (Courtesy of Aluminum Company of America).

FIGURE 11.63
Explosive sheet and plate forming.

FIGURE 11.64
Components in electrohydraulic metal-working.

Electrohydraulic metalworking (Figure 11.64) is similar to explosive forming. Instead of a chemical explosion to produce shock waves, an electric arc along a bridgewire immersed in a liquid provides the high energy needed to form the metal.

With explosive forming there is virtually no limit to the size or thickness of plate that can be formed. This is a low-volume production process, but the original investment is also low. Springback is reduced. Large dome shapes and curved gore segments are produced by this means.

Magnetic pulse compression and expansion (Figure 11.65) is used for metals that are good electrical conductors. A magnetic field acts like a compressed gas to deform the metal. It is used to blank, bulge, compress, perforate, dimple, swage, flange, and assemble parts.

METAL SPINNING AND FLOW FORMING

Some metals such as titanium alloys do not draw well in press forming. However, they can be formed into cylindrical, tapered, or curved shapes by metal spinning (Figure 11.66). *Metal spinning* is a process in which a disk of metal is rotated and forced against a form, called a chuck (Figure 11.67), with no intended reduction of wall thickness. In contrast, in *flow forming* (Figure 11.68) the blank is intentionally reduced in thickness while being lengthened. Objects such as seamless tubes and oxygen tanks and many industrial products and consumer items can be cold formed with this

Coil around tubing and insert

Tubing is compressed into grooves of fitting

Coil inserted inside of tubing surrounded by die

Tubing is expanded into die to form beading

Coil placed under flat sheet with die on top

Flat sheet is formed to contours of die

FIGURE 11.65
Typical applications of magnetic pulse compression, expansion, and flat forming coils (Courtesy of Aluminum Company of America).

method. Shear spinning (Figure 11.69) is similar to flow forming in that the cross section of the blank is reduced. In shear spinning the blank is usually a disk instead of a tube.

Shearing Operations

Shearing is a sheet metal cutting operation that makes no chips. It consists of two straight- or square-edged hardened blades, the bottom one stationary and the top one fastened to a ram. Other forms of metal working that also shear metal are blanking, trimming, piercing, notching, and shaving. These shearing operations differ from simple shearing in that the cutting edges are curved, whereas shear blades are always straight.

FIGURE 11.66
Products formed by metal spinning (Courtesy of Aluminum Company of America).

Slitting is a shearing process in which rolls of sheet metal are cut into narrower widths. This is done by means of rotary shears in which the larger diameter slitting blades on one cylindrical roll fit into grooves on the mating roll.

FIGURE 11.67
Diagram of metal spinning. Flat circular blanks are often formed into hollow shapes such as photographic reflectors. In a spinning lathe, a tool is forced against a rotating disk, gradually forcing the metal over the chuck to conform to its shape. Chucks and follow blocks are usually made of wood for this kind of metal spinning (Neely, *Metallurgy,* 2d ed., © 1984 John Wiley & Sons, Inc.).

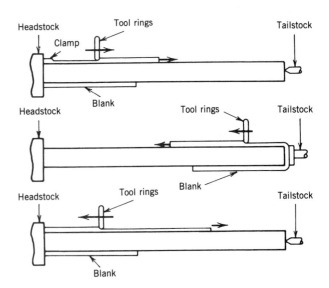

FIGURE 11.68
Flow forming diagram—spinning tubular parts (Courtesy of Aluminum Company of America).

FORMING PIPE AND TUBING

Pipe and tubing are formed with rolls in a manner similar to that of cold-roll forming. Smaller sizes of square, rectangular, and round tubing are rolled to shape and the edges are electric-resistance welded together (Figure 11.70). This process makes use of the natural resistance of the metal when an electric current is applied close to the edges to be joined to generate sufficient heat for welding. Larger-diameter pipes, such as those used for gas and oil pipelines, are made in special hydraulic presses (Figures 11.71 to 11.73). A final O-forming press makes the cylindrical shape (Figure 11.74). The edges are joined by arc welding processes (Figure 11.75) and the pipe is tested hydrostatically by capping the ends and pumping in water under pressure (Figure 11.76). For some uses the pipe is galvanized or covered with a corrosion resistant coating.

RUBBER PAD PRESSES

Certain products that do not require deep draws can be produced on rubber pad presses (Figure 11.77). Forming is done with an elastic pad that presses sheet metal against forms made of metal or of dense laminated wood (Figure 11.78). The elastic pad is usually made of polyurethane or similar material. The forming principle is shown in Figure 11.79. The major advantages of this method of sheet metal forming are that manufacturing costs for either short or long runs are low, dies are inexpensive, and large objects can be made, up to 5 by 10 ft plan area. This process is particularly useful for aircraft component manufacturers. Materials such as titanium and stainless steel can also be formed at slightly elevated temperatures by this method.

FIGURE 11.69
The principle of the shear spinning process (Courtesy of Aluminum Company of America).

FIGURE 11.70
In the electric-resistance welding process, skelp coming through the forming rolls is welded automatically (Courtesy of Bethlehem Steel Corporation).

FIGURE 11.71
Pipe forming press-crimping tool ("Presses for All Applications of Metalforming" Schirmer-Plate-Siempelkamp, Hydraulische Pressen GmbH, Krefeld, Germany).

FIGURE 11.72
Second stage in large pipe forming—U-canning transport ("Presses for All Applications of Metalforming" Schirmer-Plate-Siempelkamp, Hydraulische Pressen GmbH, Krefeld, Germany).

FIGURE 11.72
Second stage in large pipe forming—U-canning transport ("Presses for All Applications of Metalforming" Schirmer-Plate-Siempelkamp, Hydraulische Pressen GmbH, Krefeld, Germany).

FIGURE 11.73
Steps in formation of large steel pipe ("Presses for All Applications of Metalforming" Schirmer-Plate-Siempelkamp, Hydraulische Pressen GmbH, Krefeld, Germany).

FIGURE 11.74
O-forming press gives the pipe its final shape ("Presses for All Applications of Metalforming" Schirmer-Plate-Siempelkamp, Hydraulische Pressen GmbH, Krefeld, Germany).

FIGURE 11.75
The final step in making a large diameter steel pipe is welding the curved plates together by the submerged arc process (Courtesy of Bethlehem Steel Corporation).

FIGURE 11.76
Hydrostatic pipe tester
("Presses for All Applications
of Metalforming" Schirmer-
Plate-Siempelkamp,
Hydraulische Pressen GmbH,
Krefeld, Germany).

FIGURE 11.78
A large number of different components can be made
simultaneously during one press cycle with rubber pad
presses ("Rubber Pad Presses," Schirmer-Plate-
Siempelkamp, Hydraulische Pressen GmbH, Krefeld,
Germany).

FIGURE 11.77
Rubber pad press showing forming tools on the press
table ("Rubber Pad Presses," Schirmer-Plate-
Siempelkamp, Hydraulische Pressen GmbH, Krefeld,
Germany).

FIGURE 11.79
Principle of rubber pad
forming process ("Rubber
Pad Presses," Schirmer-
Plate-Siempelkamp,
Hydraulische Pressen GmbH,
Krefeld, Germany).

Guerin Forming (Rubber Pad)

1 Pad Container
2 Elastic Pad
3 Sheet
4 Tool
5 Tooling Plate
6 Press Table

249

Review Questions

1. What is the property of metals that makes possible the cold deformation of metals?

2. When considerable deformation is required in a metal, what can be done to avoid brittleness or rupture?

3. Are closer dimensional tolerances and better finish advantages of hot forming, or of cold forming?

4. In which type, cold forming or hot forming, is more massive machinery required to form a given size part?

5. What is springback?

6. What must be done to hot-rolled steel in the mill before it can be cold finished?

7. Cold-rolled steel strip is the material used for many products such as auto bodies and kitchen appliances. How is it made?

8. Aluminum foil is used every day in our homes. How is it made?

9. What is the difference between blanking and punching?

10. How can a sheet metal blank be formed into a cup shape?

11. How is the process of making wire similar to that of cold drawing bars?

12. Cold-forming upsetting machines can make parts from wire stock at a high rate of production, as much as 36,000 parts per hour. What kind of parts are usually made on these machines?

13. Larger, more complex products than those of Question 12 are made from bar stock on cold-forming machines that do combined operations in a sequence of operations. Name two products made on these machines.

14. How are threads, worms, and gears cold formed?

15. Threads cannot be rolled on gray cast iron, die cast metals, and 301 stainless steel. Why not?

16. In what ways are rolled threads superior to cut threads?

17. Would extrusion or cold rolling be more practical for producing various odd-shaped cross sections from soft metals such as aluminum in short runs? Why?

18. When press brakes or bar folders are used, what is the process called: forming, shaping, bending, or drawing?

19. What is the difference between metal spinning and flow forming?

20. Large steel pipes are electric arc welded after being formed. How are small round, square, or rectangular tubes joined at the seams after being formed?

21. If thin aluminum panels, 4 by 8 ft, needed a shallow, ribbed form to give them rigidity and 300 were required, which would be more economical to use, standard drawing dies, or a rubber pad press?

Case Problems

Case 1: Cold Working versus Hot Forming

Automotive wrist pins were once made only on automatic screw machines, an operation that turned 40 percent of the starting stock into scrap. Wrist pins are usually less than 1 in. in diameter, about 3 or 4 in. long, and hollow. These have been formed of steel for the past 30 years. They must have a good finish and accurate dimensions. Which method, hot forming or cold forming, do you think the manufacturers chose for making this part?

Case 2: Stainless Steel Nuts

A manufacturer of fasteners uses a four-die nut former to produce blanks. They received an order for 80,000 stainless steel Type 305 hex nuts. A trial run was made with the stainless steel bar stock and the piercer punch began to break down after only a few hundred nuts were produced. What change in the operation can be made so the manufacture of these nuts can be made profitable with no breakdowns?

Case 3 (Part 1): Punching Pressure

The rule for finding the approximate pressure for punching circular holes in sheet steel is:

$$P = D \times t \times 80$$

where

$$D = \text{diameter of punched hole}$$
$$t = \text{thickness of sheet steel}$$
$$P = \text{pressure (force in tons)}$$

You are given a die set having four punches: two are 1¼-in. in diameter and two are ¾-in. in diameter. You are asked to determine the press tonnage required to punch holes in ½ in. sheet steel. What size press will be required for this operation?

Case 3 (Part 2): Punching Pressure

When a punched part is not circular, the same formula as in Part 1 may be used but instead of the diameter of the hole, one-third the length of the perimeter or outline of the hole is used. You are asked to determine the pressure required to punch a rectangular hole in ¼-in. steel plate that is 3½ in. long and 2½ in. wide. What size press will be needed?

Case 4: Force Required to Shear Plate

Squaring shears have an upper and lower blade or knife (Figure 11.80). The lower one is in a level position and is stationary, whereas the upper one moves up and down and has rake to allow for a gradual cutting action. Most plate shears have ½ in./ft rake and scrap shears have up to 2 in./ft rake because distortion of the cut piece is not important and larger rakes require less force. One formula to determine the force required to shear mild steel plate is:

$$\text{maximum } KP = \frac{174,000 \times T^{1.6}}{R^{0.79}}$$

where

$$KP = \text{knife pressure, lb}$$

$$R = \text{rake, in./ft}$$

$$T = \text{plzate thickness, in.}$$

Using this formula, calculate the pressure required to shear 1¼ in. steel plate with ½ in./ft rake, and with 2 in./ft rake. Also determine KP for ½ in. plate with ½ in./ft rake.

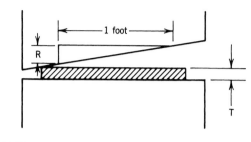

FIGURE 11.80
See Case 4.

Powder Metallurgy

Powder metallurgy (P/M) is one of the four major methods of shaping metals (machining, hot and cold plastic deformation, casting, and P/M). The P/M process is essentially the compression of finely divided metal powder into a briquette of the desired shape that is then heated but not melted to form a metallurgical bond between the particles.

Although the P/M manufacturing method dates back to the nineteenth century, it was not until past decades that this field has gained wide acceptance and use. Technological advances in P/M have been growing very rapidly also in the past decades. Products that are difficult if not impossible to produce by other means are being manufactured with P/M at high production rates at very competitive cost.

Parts manufactured by the P/M process have found a widespread use in a variety of applications. Over half of the P/M products manufactured are used in the transportation industry (automobiles and trucks). Parts used in farm and garden equipment and household appliances are also made by P/M. Many new applications will be found in the future for this unique method of forming metals.

Objectives

This chapter will enable you to:

1. Explain the methods by which metal powders are produced.

2. Understand and explain the processes involved in simple die compaction.

3. Describe the metallurgical principles involved in bonding of powders during the sintering process.

4. State the advantages of several advanced powder metal processes.

5. Make a decision when to use P/M processes for manufacturing a part.

HOW P/M PARTS ARE MADE

The basic traditional process of making P/M parts consists of two basic steps—compacting (molding) and sintering (Figure 12.1). In addition to these two basic manufacturing steps involved in the P/M process, secondary operations are commonly performed to impart final desired properties to the P/M product (e.g., coining, sizing, repressing).

In the first step **(compacting),** loose powder (or a blend of different powders) is placed in a die and it is then compacted between punches. This operation is commonly performed at room temperature. The compacted part (Figure 12.2), called a **briquette** or green compact, is now a solid shape. However, green compact can easily be broken or chipped by mishandling and requires careful handling. In the second step **(sintering)** the briquette is heated in an appropriate atmosphere to a temperature high enough to cause the powder particles to bond together by a solid state diffusion (Figure 12.3) and to homogenize any alloy constituents in the powder. Melting does not normally occur. The P/M part is now ready for use unless other finishing operations are needed.

Secondary operations may include sizing, machining, heat treating, tumble finishing, plating, or impregnating with oil, plastics, or liquids. Secondary operations can significantly increase the cost of the finished part. Therefore, designers should limit the use of secondary operations

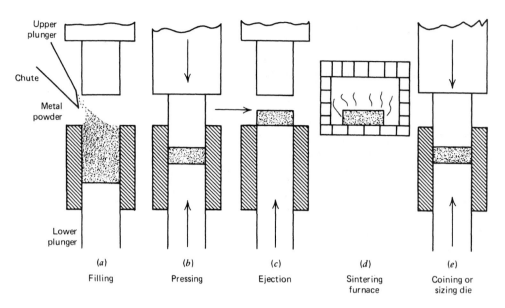

FIGURE 12.1
Sequence of operations for powder metallurgy. The depth of die cavity and the length of plunger stroke is determined according to the density required. *(a)* A measured amount of metal powder is placed in the die cavity. *(b)* Pressure is applied. *(c)* The briquette is ejected from the die cavity. *(d)* The parts are sintered at a specified temperature for a given length of time. The parts are now ready for use. *(e)* If more precision is needed, they can be sized in a coining die (Neely, *Metallurgy,* 2d ed., © 1984 John Wiley & Sons, Inc.).

and, if possible, complete the product in the first two basic steps. However, the sintering process tends to deform and shrink the shaped briquette slightly. Thus, some parts (e.g., precision gears) must have a finishing operation in which they obtain desired tolerances.

METAL POWDERS

A number of different metals and their alloys are used in P/M (e.g., iron, alloy steel, stainless steel, copper, tin, lead). The three most important methods of producing metal powders are: (1) atomization, (2) chemical methods, and (3) electrolytic processes.

Atomization is a process in which a stream of molten metal is transformed into a spray of droplets that solidify into powder. Molten metal spray can be produced in several ways. The most common method is to use a stream of high velocity gas to atomize the molten metal. This method has several variations. In one method, the gas stream is expanded through a venturi tube, which siphons the molten metal from the crucible located below the tube. The gas breaks the stream of molten metal into small droplets that then solidify as they are carried by the gas stream. In another variation, the crucible with bottom gate

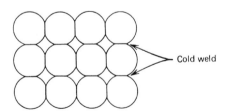

FIGURE 12.2
The compressed particles in the briquette are cold welded together at this stage with very weak bonding (Neely, *Metallurgy,* 2d ed., © 1984 John Wiley & Sons, Inc.).

FIGURE 12.3
After sintering, the particles in the briquette become fused together (Neely, *Metallurgy,* 2d ed., © 1984 John Wiley & Sons, Inc.).

is located above the gas tubes. The metal flows under the influence of gravity and passes through the gas stream, which breaks the molten metal stream. The solidified metal droplets are then collected in a collection chamber. In addition to gas, water and synthetic oils can also be used in the atomization process.

Several chemical methods can be used to make metal powders, such as reduction and precipitation. Chemical reduction is a process in which metal powders are formed by chemical reaction between metal oxides and reducing agents (e.g., hydrogen or carbon monoxide). Hydrogen or carbon monoxide reacts with oxygen in metal oxide, resulting in pure metal.

The electrolytic process begins in the electrolytic cell where the source of desired metal is the anode. As the anode is dissolved, the desired metal is deposited on the cathode. After this step is complete, the metal deposit is removed from the cathode and is washed and dried.

In each process, the powders may be ground further to a desired fineness, usually in a ball mill (Figure 12.4). Metal powders are screened and larger particles are returned for further crushing or grinding. The powders are classified according to particle size and shape in addition to other considerations such as chemical composition, impurity, density, and metallurgical condition of the grains. Particle diameters range from about .002 in. to less than .0001 in. Test sieves are used to determine particle size. This method of testing has been standardized throughout the industry.

Various powders are often blended by tumbling or mixing (Figure 12.5). Lubricants (e.g., graphite) are added to improve flowability of material during feeding and pressing cycles. Deflocculants are also added to inhibit agglomeration of powders and improve flowability during feeding.

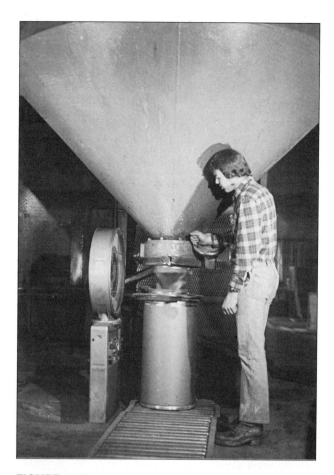

FIGURE 12.5
Blending metal powders (Pennsylvania Pressed Metals Inc.).

POWDER COMPACTION

Compacting or pressing gives powder products their shape. Metal powder forming can be separated into two groups, conventional and alternative pressing and sintering techniques. The method most commonly used today is the conventional approach, which consists of the pressing operation first, followed by sintering. The alternative techniques can be classified into (1) alternative compaction methods, (2) combined compaction and sintering, and (3) alternative sintering methods.

In the conventional compacting process, the powder is pressed unidirectionally in a single- or a double-acting press. Unlike liquids, which flow in all directions under pressure, powders tend to flow mainly in the direction of the applied pressure (Figures 12.6 and 12.7). Die compaction can be done either hot or cold. Engineering properties such as tensile and compressive strength depend to a great extent on the density of compacted material. Hot pressing, in which the powder is pressed in the die at a

FIGURE 12.4
Action of the ball mill is shown as a continuous grinding as the drum rotates (Neely, *Metallurgy,* 2d ed., © 1984 John Wiley & Sons, Inc.).

FIGURE 12.6
Custom-blended powder is precision metered into a die cavity and compacted under high pressure (Pennsylvania Pressed Metals Inc.).

high temperature, produces a density approaching that of rolled metal.

Mechanical presses (Figure 12.8) are favored when the required load is not too high. Eccentric (crank) presses rarely exceed 30 tons (60 tons would be considered quite large). Toggle-type presses may reach a 500-ton force. The advantage of mechanical presses is that they are high-production machines. Hydraulic presses (Figure 12.9) can exert forces in excess of 5,000 tons. They have the advantage of a long stroke and easily adjustable stroke length. However, these machines provide low productivity compared to mechanical presses.

Compaction of powders with various presses has the advantage of speed, simplicity, economy, and reproducibility. Such compaction produces a strong, dimensionally accurate, and relatively inexpensive product. However, it does have limitations. The aspect ratio (length to diameter) must be relatively small. Parts with a large aspect ratio will have uneven densities, being denser nearest the punches. Hence, these parts may have nonuniform and uncertain properties, and should not be made by die compaction. Grooves or undercuts or parts with thin sections cannot be made by simple die compaction. Thus, not every part is a good candidate for powder metallurgy. However, some of these limitations are overcome by alternative forming techniques such as split die techniques to provide undercuts, isostatic pressing, and densification methods.

Advanced Processes

Since conventional presses can compact powder along only one axis, such presses cannot make some shapes, such as hollow hemispheres, long parts, and internal

FIGURE 12.7
Green compacts of gears are produced and stacked ready to be taken to sintering furnaces (Pennsylvania Pressed Metals Inc.).

FIGURE 12.8
Automatic presses have controlled molding speeds up to 50,000 parts per hour (Pennsylvania Pressed Metals Inc.).

FIGURE 12.9
Large hydraulic press used for briquetting metal powders
or sponge for experimental purposes (Bureau of Mines).

threads. However, one method allows pressure to be applied from all directions: isostatic pressing. In **cold isostatic pressing (CIP),** the powder is loaded into molds made of rubber or other elastomeric material and subjected to high pressures at room temperature (Figure 12.10). Pressure is transmitted to the flexible container by water or oil. The compacted parts are removed and sintered, followed by secondary operations if needed. With **hot isostatic pressing (HIP),** an inert gas such as argon or helium is used in a pressure chamber to provide the squeeze. This gas is reclaimed between each batch of pressings. Hot isostatic pressing provides more density and achieves a finer microstructure than the cold process. Powders are often preformed to an oversize shape prior to placing them in the isostatic chamber. Heat is applied to the preform by induction for a short time while the gas pressure compacts the preform. Temperatures may be as high as 1600 to 2000°F (871 to 1093°C) with pressures in excess of 15,000 psi. Isostatic pressing is useful only for certain special applications. CIP is a comparatively slow process, and HIP is even slower. Parts made by either CIP or HIP are not limited by the shape constraints of rigid tooling.

Powder Forging

Fully dense P/M parts equaling or surpassing mechanical properties of wrought products are being produced in commercial quantities by powder forging (P/F). The green compact or preform is made in a conventional press and then sintered. These operations are then followed by a

FIGURE 12.10
Isostatic pressing. *(a)* Prepared powder is placed inside a flexible container or mold. A vacuum is drawn in the mold and it is then sealed. *(b)* The powder and mold are then placed in a pressure chamber into which water or oil is pumped under pressures of 15,000 psi or more. *(c)* The green compact is removed from the pressure chamber and the container is stripped off.

restrike (forge) that brings the part to the final density. Mechanical properties may sometimes exceed those of wrought metals because a more uniform composition is achieved in P/M processes. Fatigue and impact strengths are particularly high in powder forgings compared to conventional P/M parts. P/M bearing races have been shown to outlast wrought steel races by a factor of 5 to 1.

Metal Powder Injection Molding

A P/M technology that borrows a plastic injection molding process shows great promise for production of small precision parts. In fact some variations of this process can use plastic molding machinery. In order to inject powders into molds, the particle size must be much finer than that used for conventional P/M processes. This "dust" is combined with a thermoplastic binder. The molding step is performed at injecting pressures of about 900 psi and about 325°F (163°C). The result is a green compact that is sintered in the conventional fashion after the thermoplastic binder has been removed in an oven at about 400°F (204°C). Thin walls, high densities, unsymmetrical shapes, and accurate dimensions are possible with this method. Compared to die pressing, injection molding is not a high-production method.

Metal Powder-to-Strip Technology

Direct rolling of metal strip from a powder slurry (powder-to-strip process) is a process in which thin strip is directly produced without numerous hot or cold rolling operations. In this process, an appropriate powder mix is blended with water and a cellulose binder to form a fine slurry. The slurry is deposited on a moving band as a continuous film (Figure 12.11). After drying, the moving strip is compacted between rolls and then sintered, first to remove the binder and then to bind the particles. It is rolled a second time and resintered to remove porosity. As in all of these advanced P/M processes, metals or alloys not possible to form in any other way can be produced with powdered metals. Bimetal alloys can be produced in a strip, and high-strength titanium strip is being produced for the aircraft industry.

Powder Extrusion

Metal powders can be hot extruded with or without presintering. Metal powders are placed inside a can that is then evacuated and sealed. The unit is then heated and extruded. Metal billets and tubing from powder are made by this process.

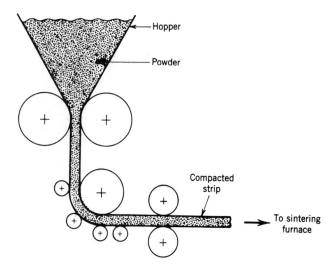

FIGURE 12.11
Powder rolling can produce a compacted strip of difficult-to-work, refractory, or reactive metals.

SINTERING

In *solid phase sintering* the green compact part must be heated from 60 to 80 percent of the melting point of the constituent with lowest melting point. This usually requires from 30 min to 2 hr in a sintering furnace to produce metallurgical bonds. The important changes that take place during the process of solid phase sintering are:

1. **Diffusion** This takes place on or near the surface of the particles as the temperature rises. For example, any carbon present in the voids between the particles will diffuse (penetrate) into the metal particles.

2. **Densification** Particle contact areas increase considerably. Voids decrease in size, therefore lowering porosity. As a result, there is an overall decrease in the size of the part during the sintering process. Therefore, a green compact must be made larger to allow for this shrinkage.

3. **Recrystallization and Grain Growth** Because sintering is usually carried out well above the recrystallization temperature of metals, grain growth can occur within and in between particles. Methods of inhibiting excessive grain growth are often used since large grains tend to weaken metals.

Alternately, *liquid phase sintering* is carried out above the melting point of one of the constituents. When one of the blended metal powders has a melting point below the sintering temperature, a liquid phase of that metal fills the voids between the particles that do not melt.

Infiltration is a process in which the pores or voids of a sintered or unsintered compact are filled with a metal or alloy of a lower melting point. For example, a steel-copper compact is heated at a temperature lower than the melting point of the steel compact and higher than the melting temperature of copper. The molten copper is drawn into the pores of the compact and fills the voids. This process increases densities and tensile strengths considerably.

High density and low porosity are not always desirable. Porous filters or prelubricated bearings are produced by loose sintering or by combining the powder with a combustible or volatile substance, which is later removed by sintering after the green compact is made. Very large parts are sometimes formed with very low pressures or none at all, and then loose sintered (called pressureless sintering) followed by a cold forging operation.

Sintering furnaces on production lines are usually of the continuous type (Figure 12.12). Furnace atmospheres consist usually of a hydrocarbon gas. However, with certain metals or alloys other gases may be used. Some manufacturers use a nitrogen gas atmosphere for both ferrous and nonferrous metals.

SECONDARY OPERATIONS

For many products the slight dimensional changes that take place during sintering are acceptable. However, where close dimensional tolerances must be maintained the product must be finished after sintering (Figure 12.13). Common secondary operations are densification, sizing, impregnation, infiltration, heat treatment, and finishing. Some of these secondary operations are described in the following paragraphs.

Sizing is similar to forging; the part is forced into a sizing die of the correct shape by a sizing punch. This operation is performed to improve dimensional accuracy.

Small electric motors and many other mechanisms have sintered, porous, prelubricated bearings that last the lifetime of the motor. These bearings are made by impregnation of the sintered P/M part with oil, which is stored in the pores of the metal. A thin film of oil covers the shaft and, when the temperature rises from rotation of the shaft and friction, more lubricant is supplied. When the shaft stops and the temperature drops, the oil is reabsorbed by capillary action, so that none is lost by dripping or leakage.

P/M articles can be finished by plating. However, because of the porosity of sintered products, precautions should be made to avoid trapping the plating electrolyte in the pores, which could cause corrosion and ultimate failure of the part. Sintered parts that are infiltrated with lower melting point alloys are less likely to have this problem. These articles can be painted or coated using the same processes used for rolled or cast metals.

Machining and precision grinding are common secondary operations. Drilling and threading holes and reaming operations are often necessary for plain die compacted products because they cannot be done in the die. Grinding is commonly performed on cylindrical or flat surfaces because many P/M parts are relatively hard.

FIGURE 12.12
Molded briquettes are sintered in controlled atmosphere furnaces to bond metal powder particles at temperatures up to 2200°F (1204°C) (Pennsylvania Pressed Metals Inc.).

FIGURE 12.13
Precision powder parts such as gears must be finished by resizing to increase their accuracy and density (Graebener Press Systems, Inc., Providence, RI).

Depending on the content, P/M parts can be heat treated by the same methods as parts formed by other processes. Steel parts can be hardened by quenching and then tempered. Annealing is sometimes necessary before machining can be performed.

P/M PRODUCTS AND THEIR USES

A wide array of small parts made by the P/M process can be seen in Figure 12.14. Manufacturers often choose P/M over other manufacturing methods because of the following characteristics of P/M parts:

1. Superior engineered microstructures and properties with precise control
2. Consistent properties and quality
3. Controlled porosity for filters and self-lubrication
4. Very low scrap loss
5. Wide variety of shape designs
6. Unlimited choice of alloys and composites
7. Low-cost, high-volume production
8. Good surface finishes
9. Close dimensional tolerances
10. Little or no machining required.

A single part can be made having different properties in different areas of the part. A part can be made in one piece by the P/M process whereas several pieces would be needed with other methods. Internal and external gear teeth that end at a shoulder can easily be formed by P/M methods.

Surprisingly, P/M steels are almost as strong as machining steels, and powder forging processes bring the tensile strength up further. The versatility of the P/M process allows parts to be made lighter than with other manufacturing processes. P/M processes allow parts to be made of very hard metals, such as tungsten carbide cutting tools for machine tools. The P/M process can also be used to make friction materials (in the form of bimetal powder materials that are bonded to a steel base) and aluminum-based antifriction materials containing graphite, iron, and copper. Copper-nickel powders are often formed as a layer on steel strip and then sintered. The sintered strip is impregnated with babbitt metal and formed into insert bearings for automobile and aircraft engines.

Some disadvantages of P/M are found in the conventional cold die compacting and sintering process. Since P/M products are somewhat porous and present a larger internal surface to any corrosive atmosphere, they have lower corrosion resistance than solid metals. P/M products also tend to have reduced plastic properties (ductility and impact strength) than conventionally produced metals.

FACTORS FOR DESIGN OF P/M PRODUCTS

When parts are made by the conventional process, several elements of design should be observed. Thin sections and feather edges should be avoided (Figure 12.15). Generous fillet radii should always be provided in a die (Figure 12.16), and internal holes should have rounded corners (Figure 12.17). External corners should be chamfered (Figure 12.18), and narrow deep slots should be avoided (Figure 12.19). Splines or keyseats should have rounded roots.

FIGURE 12.14
These are some of the thousands of intricate shapes and designs that are produced by the powder metallurgy process (Pennsylvania Pressed Metals Inc.).

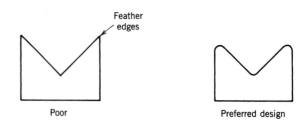

FIGURE 12.15
Metal powders do not fill dies with sharp feather edges well. A preferred design is one with rounded corners.

FIGURE 12.16
Sharp corners on shoulders should be avoided. A fillet radius should be used.

FIGURE 12.20
Undercuts as in *(a)* are difficult to make, requiring special expensive dies. The design for an undercut at *(b)* serves the same purpose and is easily done with simple pressing dies.

FIGURE 12.17
Avoid sharp corners. Internal shapes should have rounded corners.

Using secondary machining processes is common practice for making precision P/M parts. Holes, tapers and drafts, countersinks, threads, knurls, and undercuts usually must be machined after the parts are sintered. However, holes in the direction of pressing are readily produced in P/M parts. Round holes are easiest to produce but shaped holes, keys, splines, hexagonals, squares, and any blind holes can also be made.

Holes that are not in the direction of pressing generally have to be machined later. Some redesign often can eliminate secondary machining [Figures 12.20*(a)* and 12.20*(b)*]. The purpose of the undercut must be kept in mind. In this case, it is simply a relief for a sleeve whose internal edge should not contact any radius on the flange. A preferred arrangement that will still do the job and be even stronger is shown in Figure 20*(b)*. Vertical (straight) knurls and splines can be made in the direction of pressing but diamond or angled knurls cannot be pressed because they would interfere with part ejection.

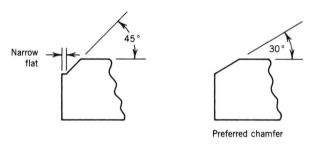

FIGURE 12.18
Both of these chamfers are used, but the 30-degree angle is the better form.

P/M is the ideal way to make cams and gears. Clusters of gears or gears and cams together are easily produced by die compaction. Gears with blind corners, both internal and external, are very difficult to make by other processes, such as machining or forming, but are produced easily by P/M processes. Helical gears with angles up to 45 degrees can theoretically be made, but gears with angles up to 30 degrees are manufactured in quantity. Hubs for gears and sprockets should be located as far as possible from the root diameters of the gear; that is, the hub should be as small as possible while retaining sufficient strength.

Probably the greatest advantage in adopting the P/M process over other methods of manufacture is that it allows for redesign that makes use of the great versatility of the P/M process. Many small mechanisms such as those found in pneumatic drills, electronic printers and sequencers, door locks, firearms, and sewing machines

FIGURE 12.19
Deep slots are not easily made by the P/M process. Shallow grooves with rounded bottoms are better.

have a number of small parts that can often be combined into one piece by redesigning for powder metallurgy. Of course a great savings in production can be realized and the mechanism itself can be simplified as well. P/M is not suitable for every metal product, but its use should always be investigated when making a decision for designing a part to be manufactured.

Review Questions

1. In simple die compacting, what are the three steps in producing a precision finished part?

2. Name the three most important methods of producing metal powders.

3. A heat-resistant alloy part that has a melting point higher than that of steel is to be used to make 20,000 small symmetrically shaped parts. Both P/M and die casting processes have high production rates and making the dies would cost about the same for each. Which method would you choose? Why?

4. Why would a bushing or sleeve of 1 in. diameter and 4 in. in length have uncertain mechanical properties and probably fail in use if it were produced by simple die compaction?

5. Name two methods of increasing the density of P/M products.

6. Simple die compaction presses metal powders in only one axis, making undercuts, threads, and cross holes impossible. By what two methods of compaction can these operations be done?

7. Name three processes of metalworking other than die compacting of powders that have been adopted for P/M.

8. In liquid-phase sintering, does all of the metal powder reach the melting point? Explain.

9. How does infiltration increase density?

10. What method is used to make metal filtration devices and prelubricated porous bearings?

11. Is it possible to harden and temper a P/M part to make a wear-resistant mechanical part?

12. A manufacturer must tool up to produce 20,000 metal handwheels with a knurl on the outer edge of the wheel. The parts will be produced with P/M simple die compaction. The type of knurl is not specified. What kind of knurl, diamond or straight, should be used? Why?

Case Problem

Case 1: Improving a Good Product

A drive hub and flange has been produced successfully by the P/M process from iron powder. However, under some shock load conditions, the hub failed. It was found to have considerable porosity that contributed to its somewhat low impact strength. How can this product be made more dense with increased impact strength without changing the process from plain die compaction and without changing present die size?

CHAPTER 13

Principles of Machining Processes

Objectives

This chapter will enable you to:

1. Understand the principles of material removal and the basic mechanisms of machining.

2. Explain the basic machining motions and the purpose of speeds and feeds.

3. Understand the effect of heat and pressure upon tool wear and their influence upon cutting tool design.

4. Describe the properties of commonly used cutting tool materials including the economic justification for the use of expensive cutting tool materials.

5. State the importance of rake, clearance, cutting edge, and lead angle for cutting tools.

6. Show why cutting fluids are used.

Machining is essentially the process of removing unwanted material from wrought (rolled) stock, forgings, or castings to produce a desired shape, surface finish, and dimension. It is one of the four major types of manufacturing processes used to create product components. Machining is done by shaving away the material in small pieces, called chips, using very hard cutting tools and powerful, rigid machine tools. The cutting tool may be held stationary and moved across a rotating workpiece as on a lathe, or a rigidly held workpiece may move into a rotating cutting tool as on a milling machine.

Historically, machining processes probably began with Besson in 1569 when he invented the screw-cutting lathe and later when a practical version was built by Henry Maudslay in 1800. Horizontal milling machines first appeared in 1820 when Eli Whitney used them for the manufacture of firearms, and the vertical milling machine first appeared in 1860. Through continuous improvement these primitive machine tools have developed into the variety of manufacturing machinery that is used today. Many modern machine tools are controlled by a computer (numerical control) and can perform complex machining operations without the guidance or constant attention of a machinist. Some specialized machines are designed primarily for manufacturing at high production rates. The more versatile machines are designed to make one-of-a-kind prototype parts for research and development (R&D), tool and die making, or to repair existing machinery. These versatile machines are usually found in tool rooms or local machine shops and are operated by general machinists or tool and die makers. The less highly skilled workers who operate production machinery are called machine operators.

Few manufacturing technologies can achieve the precision of the various machining processes. The common .001 in. precision and .0001 in. precision that is possible for modern machining processes are far superior to the precision of common casting, molding, and forming operations. Large heavy sections such as those used for dies are machined to precise dimensions, a virtual impossibility for many forming processes that handle only thin sheet materials. The basic machining processes are shown in Table 13.1

Machining processes remove material in the form of chips that are disposed or recycled. Machining is more costly than casting, molding, and forming processes, which are generally quicker and waste less material, but machining is often justified when precision is needed. For example, gears for an alarm clock may be entirely acceptable if stamped out of sheet metal or produced by plastic injection molding, which would be far less expensive than machining, but gears for an automobile transmission must be hard, strong, and precisely made so they will function

TABLE 13.1
Basic machining processes

Operation	Diagram	Characteristics	Type of Machine
Turning		Work rotates, tool moves for feed	Lathe and vertical boring mill
Milling (horizontal)		Cutter rotates and cuts on periphery. Work feeds into cutter and can be moved in three axes.	Horizontal milling machine
Face milling		Cutter rotates to cut on its end and periphery of vertical workpiece.	Horizontal mill, profile mill, and machining center
Vertical (end) milling		Cutter rotates to cut on its end and periphery, work moves on three axes for feed or position. Spindle also moves up or down.	Vertical milling machine, die sinker, machining center
Shaping		Work is held stationary and tool reciprocates. Work can move in two axes. Toolhead can be moved up or down.	Horizontal and vertical shapers
Planing		Work reciprocates while tool is stationary. Tool can be moved up, down, or crosswise. Worktable cannot be moved up or down.	Planer

TABLE 13.1

Operation	Diagram	Characteristics	Type of Machine
Horizontal sawing (cutoff)		Work is held stationary while the saw either cuts in one direction as in bandsawing or it reciprocates while being fed downward into the work.	Horizontal bandsaw, reciprocating cutoff saw
Vertical bandsawing (contour sawing)		Endless band moves downward, cutting a kerf in the workpiece, which can be fed into the saw on one plane at any direction.	Vertical bandsaw
Broaching		Workpiece is held stationary while a multitooth cutter is moved across the surface. Each tooth in the cutter cuts progressively deeper than the previous one.	Vertical broaching machine, horizontal broaching machine
Horizontal spindle surface grinding		The rotating grinding wheel can be moved up or down to feed into the workpiece. The table, which is made to reciprocate, holds the work and can also be moved crosswise.	Surface grinders, specialized industrial grinding machines
Vertical spindle surface grinding		The rotating grinding wheel can be moved up or down to feed into the workpiece. The circular table rotates.	Blanchard-type surface grinders

(continued)

TABLE 13.1 *Continued*
Basic machining processes

Operation	Diagram	Characteristics	Type of Machine
Cylindrical grinding		The rotating grinding wheel contacts a turning workpiece that can reciprocate from end to end. The wheelhead can be moved into the work or away from it.	Cylindrical grinders, specialized industrial grinding machines
Centerless grinding		Work is supported by a workrest between a large grinding wheel and a smaller feed wheel.	Centerless grinder
Drilling and reaming		Drill or reamer rotates while work is stationary.	Drill presses, vertical milling machine
Drilling and reaming		Work turns while drill or reamer is stationary.	Engine lathes, turret lathes, automatic screw machines
Boring		Work rotates, tool moves for feed on internal surfaces. (On some horizontal and vertical boring machines, the tool rotates and the work does not.)	Engine lathes, horizontal and vertical turret lathes, and vertical boring mills

quietly and efficiently when subjected to variable forces and heavy shock loads. These gears must be made by machining processes and then heat treated and precision ground.

The performance of the cutting tool used to remove workpiece material determines the efficiency and cost of a machining operation. The geometry of the cutting edge controls the shearing action as a chip is torn away from the part. The cutting tool material determines how fast the operation may progress, and since time is money in manufacturing activities this is an important factor in the cost of the operation.

MOTION AND PARAMETERS: SPEED, FEED, AND DEPTH OF CUT

Machining operations require two basic simultaneous motions; one motion creates cutting speed, and the other is the feed motion. Cutting speed is the rate at which the workpiece moves past the tool or the rate at which the cutting edge of the tool moves past the workpiece. Regardless of whether the tool moves or the workpiece moves the relative cutting speed between the two is basically unchanged. English units for cutting speed are feet per minute (fpm), which is often called surface feet per minute (sfpm or sfm). Metric units are meters per minute (m/min). Table 13.2 suggests cutting speeds for machining several metals using high-speed steel cutting tools. Higher

FIGURE 13.1
This micrograph shows the surface of the specimen that was turned at 100 sfm. The surface is irregular and torn, and the grains are distorted to a depth of approximately .005 to .006 in. (250 ×).

cutting speed shortens the time required to complete the machining cut but can greatly shorten the useful life of the cutting tool. Cutting speeds that are too low tend to tear instead of cut, produce rough finishes, and distort the grain structure at the surface of the workpiece, all of which can cause early failure of a machined part (Figures 13.1 and 13.2). Speeds should be as high as can be maintained without causing the tool to wear out too quickly. Recommended cutting speeds for machining operations

TABLE 13.2

Cutting fluid and speed table for machining various metals with high-speed tools

Material	Cutting Fluid	Speed (ft/min)
Aluminum and alloys	Soluble oil, kerosene, light oil	200 to 300
Brass	Dry, soluble oil, synthetic coolant solution, light mineral oil	150 to 300
Bronze	Dry, soluble oil, mineral oil, synthetic coolant solution	200 to 250
Cast iron, soft	Dry, air, synthetic coolant solution	100 to 150
Cast iron, medium	Dry, air, synthetic coolant solution	70 to 120
Cast iron, hard	Dry, air, synthetic coolant solution	30 to 100
Magnesium and alloys	60-sec mineral oil	300 to 600
Steel, alloy	Soluble oil, synthetic coolant solution	30 to 90
Steel, low carbon	Soluble oil, synthetic coolant solution	90 to 120

FIGURE 13.2
At 400 sfm this micrograph reveals that the surface is fairly smooth and the grains are only slightly distorted to a depth of .001 in. (250 ×).

TABLE 13.3
Lathe feeds, inches per revolution

Material	Roughing Feed	Finishing Feed
Cast iron	.010 to .020	.003 to .006
Low carbon steel	.010 to .020	.003 to .005
High carbon steel, annealed	.008 to .020	.003 to .005
Alloy steel, normalized	.005 to .020	.003 to .005
Aluminum alloys	.015 to .030	.005 to .010
Bronze and brass	.010 to .020	.003 to .010

can be found in commonly available tables. These suggested speeds vary based upon the workpiece material, cutting tool material, and type of machining operation.

Most machine tools have rotating spindles, so the rate of rotation (the spindle speed) for the machining operation must be established. This rotation of the machine spindle generates the cutting speed necessary for the cutting tool to efficiently shear material from the workpiece. The spindle speed is calculated based upon the cutting speed to be used in the operation, and units are revolutions per minute (rpm). (See the Appendix at the end of this chapter for spindle speed calculations.)

The feed motion is the advancement of the cutting tool along the workpiece (for lathes) or the advancement of the workpiece past the tool (for milling machines). The rate at which the tool or workpiece moves is the feed rate. The basic measure of the feed rate is the distance advanced per revolution of the machine spindle. Units for this feed per revolution are inches per revolution (ipr) or millimeters per revolution (mm/rev). (See the Appendix at the end of this chapter for feed rate calculations.) Table 13.3 suggests feed rates for machining several different metals. Faster feeds shorten the time required to complete the machining cut but create rougher workpiece surfaces. Feed rates for roughing (rapid material removal) should be as heavy as the tool material and the machine tool can withstand without failure, and finishing feeds should be fine enough to produce the desired smooth surface finish on the workpiece.

The rate at which the workpiece advances (measured in inches or millimeters) per minute is an alternate way to specify the feed rate. The units are inches per minute (ipm) or millimeters per minute (mm/min). The feed per minute is calculated based upon the spindle speed and the

feed per revolution. (See the Appendix at the end of this chapter for feed rate calculations.)

The tool must be engaged in the workpiece to remove material. The amount of engagement, called the depth of cut, is equal to the thickness of the layer of material removed from the workpiece. The depth of cut also determines the width of each chip removed from the part as it is machined.

MACHINING: SHEARING CHIPS FROM THE WORKPIECE

Pressure, Heat, and Friction in Machining

Machining is similar to whittling wood with a sharp knife. The cutting tool exerts enough force on the workpiece to smoothly cut away material in the form of chips. This action results in the continuous shear failure of the work material at the cutting edge of the tool (Figures 13.3 and 13.4). Imagine how much pressure is required to shear chips from a steel workpiece and make it appear to be as easy as cutting butter. This tremendously high pressure causes the tool and workpiece to flex or deflect away from each other. This movement affects precision, surface finish, and tool life, so deflection must be opposed by rigidity. The machining system is only as strong as its weakest link, so the workpiece, workholding tooling, machine tool, toolholder, and cutting tool must all be extremely stiff to oppose deflection.

The high forces in machining create a considerable amount of heat near the cutting edge. Most of this heat is generated within the shearing process, and some heat is created by friction between the tool and the workpiece. Most of the heat is carried off in the chips, but the remainder stays in the tool and workpiece creating a large amount of thermal stress and softening the tool. Cutting fluid (coolant) is often used to bathe the tool and workpiece to remove much of the heat and minimize damage to the tool and part.

FIGURE 13.3
Micrograph of chip formation showing shear plane. Point of negative rake tool producing a discontinuous chip (100 ×) (Neely, *Metallurgy,* 2d ed., © 1984 John Wiley & Sons, Inc.).

FIGURE 13.4
Positive rake tool showing less distortion ahead of the tool than produced by a negative rake tool (100 ×) (Neely, *Metallurgy,* 2d ed., © 1984 John Wiley & Sons, Inc.).

Workpiece Materials and Their Effects

The workpiece material plays an important part in chip formation. Greater cutting forces are required to shear harder and higher strength materials, causing more deflection between the cutting tool and workpiece. More heat is also generated, which creates higher operating temperatures in the tool and part. The increased friction and tool temperature encountered when machining hard or abrasive materials can accelerate tool wear, shortening the amount of time that the tool will survive while cutting (tool life). Softer, lower strength materials shear more easily and may be machined faster while causing less tool wear, but very soft materials such as soft pure aluminum or hot rolled (HR) low carbon steel are somewhat "gummy" when cut on machines and tend to cause **built-up edge (BUE)** on the cutting tool. Built-up edge (Figure 13.5) is a common machining problem in which workpiece material becomes welded to the cutting edge of the tool, changing the geometry of the cutting edge. BUE causes chips to be torn away rather than cleanly cut, resulting in rough part surfaces, and it may damage the tool. Correct cutting speed and tool geometry helps to eliminate built-up edge and produce a clean cutting action with improved surface finish.

Fragmented chips called discontinuous chips are produced when brittle materials are machined. Brittle gray cast iron chips break immediately when sheared from the part and have the appearance of a powder. When ductile materials are machined the chip is formed as a coil or helix, a continuous chip. Because the tool is in constant contact with the chip added friction causes operating temperatures to be higher when continuous chips are produced.

When ductile materials are machined, heavier (larger) feed rates produce chips that curl tightly and break readily, and lighter (smaller) feeds tend to produce long stringy chips. Chips that break after one curl, forming a 9-shape, are the most desirable form of continuous chips for ease of handling and for safety's sake.

The **machinability** of a workpiece material is the ease with which it can be machined. Machinability ratings compare various materials in terms of cutting ease or difficulty resulting from the forces required and heat generated in the shearing of a material. The relative machining difficulty of various materials may be compared by measuring the power required and cutting tool life for each workpiece material. Commonly machined materials have been machinability rated using a scale based upon low carbon soft steel (AISI B-1112), which was arbitrarily given a rating of 100 percent on the machinability scale (Table 13.4). Metals that are more easily machined have a higher rating than 100 percent, and materials that are more difficult to machine are rated lower.

CUTTING TOOL MATERIALS

Commonly machined materials include plastics, aluminum, many varieties of steels including heat-treated tool steels that are as hard as a knife blade, ceramics, and many others. Cutting tools must be capable of retaining their hardness at high temperatures (hot hardness). Better hot hardness permits tools to operate at higher cutting speeds, thereby improving productivity. A variety of cutting tool materials is needed. Some must be very hard for long tool life and to machine hard workpiece materials. Others must be very shock resistant (tough) to withstand interrupted cuts (intermittent cutting action). An example of an interrupted cut would be turning a bar which has a hexagonal cross section. The cutting tool would, on the

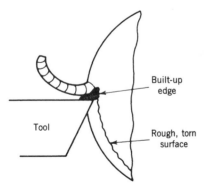

FIGURE 13.5
Tool with built-up edge.

TABLE 13.4
Machinability ratings and approximate hardness of annealed steels

AISI Classification	Machinability Rating, %	Approximate Hardness—HB (Brinell)
B-1113	135	200
B-1112	100	205
C-1118	80	160
C-1020	65	150
C-1040	60	200
A-8620	50	220
A-3140	55	200
A-5120	50	200
C-4140	50	200
Cast iron	40 to 80	160 to 220
302 stainless	25	190

(*Source:* John E. Neely, *Practical Metallurgy and Materials of Industry,* 2d ed., Wiley, New York, © 1984.)

first cut, contact only the corners of the rotating bar, creating six impacts each time the bar rotates. The mechanical shock of this intermittent cutting action tends to fracture brittle cutting tools.

Cutting speed has a pronounced effect upon the useful life of a cutting tool. Higher speeds increase both friction and the shear rate, generating added heat that makes cutting tools wear faster or break down. This increases tool costs and tool changing delays, but higher speed also decreases the time required to perform a machining operation, and that improves productivity. The chosen cutting speed strikes the balance between tool wear and productivity. Any cutting tool materials that can operate at higher cutting speeds with no increase in the rate of tool wear enable manufacturers to increase productivity without paying a penalty in tooling costs and delays. Millions of dollars are spent each year to develop new cutting tool materials that can produce such rewards.

High-Speed Steel Cutting Tools

One hundred years ago hardened plain carbon tool steel was used for metal cutting tools. Cutting speeds were very slow because carbon steel would permanently lose its hardness if it got very hot. Today, plain carbon tool steel is used only for low-speed, low-temperature applications such as hand tools, including files, knives, and chisels. A type of tool steel developed around 1900 was called high-speed steel (HSS) because it could get hot, cool down, and retain its hardness. High-speed steel is less expensive than most cutting tool materials, and steel tools can be machined to complex shapes in their soft state, heat treated for hardness, and then ground to a sharp cutting edge. Because of its versatility and low cost, HSS is today the most commonly used cutting tool material in machining applications. High-speed steel drills, milling cutters, and lathe tools are widely used. After the cutting edge dulls HSS tools are sharpened using a grinder, greatly increasing the useful life of the tool.

Carbide Tools

Most cutting tools used for production machining are made of tungsten carbide. Compared to high-speed steel tools, carbide cutting tools have much better hot hardness, so they can machine at higher temperatures without softening and destroying the cutting edge. Cutting speeds are three to four times faster for carbides than for HSS tools. In the making of a carbide tool, tungsten carbide particles are mixed with a cobalt powder, compressed into a briquette of the required tool shape, and then sintered in a furnace, causing the cobalt to bind the tungsten carbide particles into a very hard, strong, solid material. Carbide is made in grades of varying hardness and

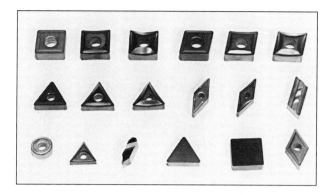

FIGURE 13.6
Carbide inserts (Photo courtesy of Kennametal Inc.).

toughness, and titanium carbide and tantalum carbide are sometimes added to the mixture to provide greater hardness for wear resistance. Virtually all carbide tools used today in manufacturing operations are throw-away inserts that have several indexable cutting edges (Figure 13.6). Carbide inserts are clamped into different toolholders for external and internal machining (Figures 13.7 and 13.8). Milling cutters, turning tools, boring bars, and many other types of toolholders equipped with carbide inserts have long life and the inserts can be quickly rotated (indexed) to expose a fresh cutting edge or replaced when they wear out.

Ceramic Cutting Tools

Ceramic tools are often used to machine hard workpiece materials and have better hot hardness than carbide. These operate at much higher speeds than carbide inserts but require a very rigid setup (machine, cutting tool, work-

FIGURE 13.7
Carbide toolholders for turning machines (Photo courtesy of Kennametal Inc.).

FIGURE 13.8
Boring bar holder with interchangeable heads for internal turning operations (Photo courtesy of Kennametal Inc.)

piece, work-holding system) and will not endure the impact of heavy interrupted cuts. Ceramic cutting tools are manufactured in insert shapes similar to carbide inserts.

Coated Carbide and Cermet Cutting Tools

Tungsten carbide inserts coated with titanium nitride (TiN), ceramic (aluminum oxide), and/or titanium carbide combine the qualities of carbide and the wear resistance of the coating material to permit increased cutting speeds and added productivity. Cermet, a mixture of carbide and ceramic that is sintered into inserts, competes closely with the productivity of coated carbide tools.

Diamond and CBN Cutting Tools

Diamond cutting tools can produce exceedingly smooth surface finishes and hold very close tolerances. Since diamond is the hardest material known to man it retains a sharp, stable cutting edge, but it is prohibitively expensive for many applications. Cubic boron nitride (CBN), a man-made material, is second in hardness to diamond. CBN is used to "hard turn" steel workpieces that are too hard to be machined by less exotic cutting tool materials, but it is also very expensive.

TOOL GEOMETRY

Tool geometry (shape) varies considerably depending upon the machining application. The shape and angles of the cutting tool must be accurate in order to have sufficient strength at the cutting edge, to ensure good chip formation and flow, and to provide sufficient relief (clearance) so the tool will cut into the work material and not rub. An exam-

ple of a right-hand cutting tool is illustrated in Figure 13.9. There are many shapes of cutting tools, including left-hand, threading, and form tools, but all of them must have correct angles for relief and rake. Chip-breakers are sometimes needed on tools to control chip formation (Figures 13.10 and 13.11).

At what angle would you hold a knife blade when peeling an apple? If you hold the knife blade perpendicular to the surface of the apple (a neutral or zero rake angle) the knife would scrape the skin off the apple, requiring a large amount of force and cutting very poorly. If you hold the knife blade at an angle greater than 90 degrees (a negative rake angle), which would seem to be backwards, the blade would cut even less effectively, requiring a very large force to push the skin off, possibly crushing the apple. The most effective knife blade angle, of course, is the acute angle that one would normally use when peeling an apple (a positive rake angle). A positive rake angle will shear material smoothly and efficiently while requiring less force. This rake angle relationship also holds for machining operations although the angles become more complex. Positive rake angles are always preferred, but the poor tensile strength of many cutting tool materials makes the use of negative or neutral rake angles necessary in many cases.

All cutting tools must have relief angles to prevent the flank of the tool, which is located next to and beneath the cutting edge, from rubbing the workpiece. If the flank of the tool rubs the workpiece it prevents the cutting edge from reaching the workpiece. Grinding positive rake tools in the flank area provides the needed relief but weakens the cutting edge. The operating orientation of negative and neutral rake tools provides a natural relief angle. It is not necessary to grind the flank of these tools, so negative and neutral rake tools are stronger than positive rake tools. This extra strength can be extremely important, especially in heavy machining operations where large forces are required to remove large amounts of material or in interrupted cuts where impact resistance is needed.

Chip Control

Breaking chips is very important. Chips that do not break readily and form strings longer than the ideal 9-shaped chip create handling and safety problems. Chips are as sharp as knives or saws and can cause deep cuts. Chips should not be handled with bare hands, and long chips are especially difficult to handle and dispose, creating a workplace hazard. Many automated manufacturing machines carry chips away for disposal on a conveyor. Short, broken chips are a necessity for this automatic machinery because stringy, wiry chips would entangle machine parts and eventually jam the conveyor. Chips tend to fly from the work at high speed, sometimes as far as 20 to 30 feet

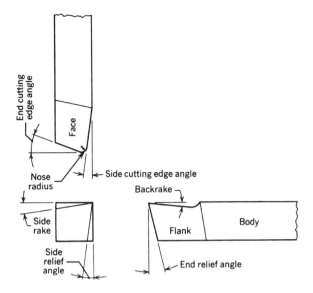

Important Elements of Cutting Tool Geometry

1. The tool shank is the area where the tool is held and clamped by the toolholder.
2. Back rake, a component of the rake angle, is very important to the smooth chip flow that is needed to have a uniform cut and a smooth surface finish.
3. The side rake, the other component of the rake angle, directs the chip flow away from the cutting edge. Large back and side rakes are used for machining soft workpiece materials to minimize built-up edge.
4. The end relief angle prevents the front flank of the tool from rubbing on the workpiece.
5. The side relief angle prevents the side flank of the tool from rubbing on the workpiece and allows the tool to feed into the work material.
6. The side cutting edge angle (SCEA) may vary considerably depending upon the application. For roughing it should be approximately 15 to 30 degrees as shown in Figure 13.9, but tools used for cutting square shoulders or for light machining could have reversed angles from 5 to 30 degrees depending on the application. The side cutting edge angle may be provided by angling the toolholder (creating a lead angle), by grinding the desired angle on the tool, or by a combination of both. When a large nose radius and a small depth of cut are employed in finishing operations the SCEA has no effect on the cut. The side cutting edge angle shown in Figure 13.9 directs the cutting forces into a stronger section of the tool. It also helps to direct the chip flow away from the workpiece (avoiding damage to the workpiece surface) and reduces the thickness of the chip (decreasing the forces required to cut the chip).
7. The nose radius of cutting tools is established to help provide the required surface finish. A larger tool nose radius provides a smoother surface.

FIGURE 13.9
The parts and angles of a tool. Angles given are only examples. The listing to the right describes elements or tool geometry. (White, Neely, Kibbe, Meyer, *Machine Tools and Machining Practices,* Vol. I, © 1977 John Wiley & Sons, Inc.).

FIGURE 13.10
Stringy chips.

FIGURE 13.11
Chips that are 9-shaped are best.

FIGURE 13.12
High-speed steel chip breakers.

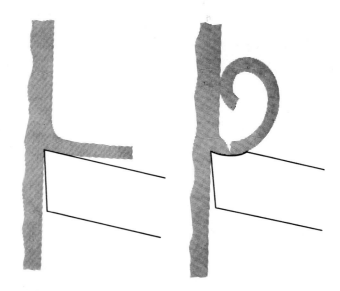

FIGURE 13.14
How chips are broken. Chip flow with plain tool and with a chip breaker (White, Neely, Kibbe, Meyer, *Machine Tools and Machining Practices*, Vol. I, © 1977 John Wiley & Sons, Inc.).

away. They can easily injure the eye if safety glasses are not worn, and hot chips can penetrate clothing or soft-soled shoes.

Many cutting tools have special chip breaking geometry called chipbreakers (Figures 13.12 and 13.13) that break the chips by sharply curling them (Figure 13.14). The shape of chipbreakers is vital to chip control. Chip breaking geometry is most needed when machining soft workpiece materials or when finish machining with a light (small) feed rate.

CUTTING FLUIDS

Cutting fluids are usually liquid, but air is sometimes used to avoid contamination for certain materials. In general, cutting fluids serve to dissipate heat, to lubricate between the tool and workpiece for lower friction, and to carry away

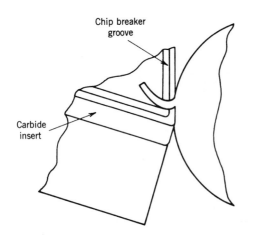

FIGURE 13.13
Carbide chip breakers.

chips from the cutting area. There are essentially two types of cutting fluids, cutting oils and coolants. Some cutting operations, such as using a tap to cut threads, create a considerable amount of friction and need a cutting oil for good lubricating qualities. Cutting oils can be either animal fats or petroleum-based. Higher speed cutting operations require relatively less lubrication and more of the heat removal qualities that are found in a coolant. Coolants are usually water-based soluble oils that have excellent heat transfer qualities. Petroleum oils or waxes are treated so that they emulsify when mixed with water to produce a milky-white solution. Other chemicals such as wetting agents, rust inhibitors, antibacterial agents, and polarizing agents are added to improve the coolant. Many synthetic coolants have been developed for use as emulsions or cutting oils for specific and general uses.

Some machine tools used in tool rooms or for low-volume production are operated without coolant, but all production machines are fitted with coolant pumps and tanks. In some production machining operations, however, coolant is not used. For example, in an intermittent cutting setup with a carbide tool where there is alternate heating and cooling of the carbide, coolants would add to the thermal shock, which can crack the carbide insert, so no cutting fluid is used. Also, due to the risk of thermal shock and the superior hot hardness of ceramics, cutting fluids are not normally required with ceramic tools. This practice of "dry machining" is becoming more commonplace due to environmental concerns involving the disposal of spent cutting fluid.

APPENDIX: CALCULATION OF SPEEDS AND FEEDS FOR MACHINING OPERATIONS

Let: V = cutting speed

N = spindle speed

f_r = feed per revolution

f_m = feed per minute

Recommended cutting speeds may be obtained from commonly available tables or by utilizing machinability ratings. Recommended feeds per revolution may also be obtained from charts. These recommended speeds and feeds represent only a beginning point, so they must be tempered considering the conditions of each machining operation and should be optimized in production.

Speed Calculation

The spindle speed may be calculated given the cutting speed and the diameter of the rotating tool or workpiece.

$$\text{English units: } N = \frac{3.8V}{D}$$

where: N is spindle speed in rpm

V is cutting speed in feet per minute

D is diameter in inches

$$\text{Metric units: } N = \frac{320V}{D}$$

where: N is spindle speed in rpm

V is cutting speed in meters per minute

D is diameter in millimeters

Feed Rate Calculation

The feed per minute may be calculated given the feed per revolution and the spindle speed.

$$\text{English units: } f_m = f_r N$$

where: f_m is inches per minute feed rate

f_r is inches per revolution feed rate

$$\text{Metric units: } f_m = f_r N$$

where: f_m is millimeters per minute feed rate

f_r is millimeters per revolution feed rate

Review Questions

1. In terms of interchangeability of manufactured parts, what characteristic of machining processes is virtually unique and most important?

2. Define the process of machining metal.

3. Describe the process by which metal is removed by a cutting tool in the machining process. Is the metal removed in chips by splitting the metal ahead of the tool, or by plastic deformation?

4. The most common cutting tools used in machining operations of all kinds are high-speed steel, ceramic inserts, diamond tools, and varieties of carbide tools. Of these, which type of tool is by far the most commonly used in production machining operations?

5. Stringy, wiry chip formations are hazardous and cannot be handled or disposed of easily. How can this kind of chip be avoided? What is a more desirable chip form?

6. Holes made by twist drills are usually rough and often oversized. How can a drilled hole be made to have a more accurate dimension with a smoother finish?

7. How can a built-up edge cause a reamer or other cutting tool to produce a rough finish and inaccurate dimensions?

8. Cutting speeds (sfm) for manufacturing should be as high as possible without burning the tool or workpiece. If the speed for a lathe operation has been determined to be 300 rpm for high-speed tools, how fast should the machine turn for carbide tools?

9. Cutting fluids come in two types. Name each and describe its special qualities.

10. When are lathe operations sometimes performed without cutting fluid?

Case Problem

Case 1: Chip Control Problem

High-speed steel cutting tools were used for production turning of stainless steel fittings on an automatic screw machine. The setup person originally sharpened the cutting tools by offhand grinding on a pedestal grinder. A problem developed in which wiry chips entangled the mechanism causing lost machining time in shut-downs for cleaning out the entangled chips. What would you recommend to correct this problem?

CHAPTER 14

Machine Tool Operations

This chapter will enable you to:

1. Decide when to use toolroom machinery and when to use production machinery for making a machined part.

2. Decide which type of drilling or sawing machine is needed for certain drilling or sawing operations.

3. Explain the special uses of vertical and horizontal spindle turning, milling, and boring machines.

4. State how machine threads are cut and how they are designated.

5. Describe the processes of making many kinds of internal and external gears and splines.

6. State the principles and uses of abrasive machining.

7. Determine how grinding machines and processes are used in machine shops and in manufacturing.

The following factors are obviously considered when choosing the best manufacturing process to perform a needed machining operation: finished part configuration, tolerances, surface finishes, cost, and available equipment. Less obvious considerations that must not be overlooked are production volume and the skill level of the work force. Low-volume and one-of-a-kind toolroom machining are very labor intensive and utilize basic, relatively simple manually operated machine tools such as drill presses, engine lathes, and milling machines. Medium-volume and high-volume (mass production) machining are called production machining. Production machining is repetitive, and frequent repetition of machining operations justifies more sophisticated machines and tools that produce parts more quickly and save labor. Production machining is a good application for CNC (computer numerical control) machine tools and automatic machine tools.

Low-volume or one-of-a-kind operations require the constant attention of a skilled machinist. General machinists require much more training than production machine operators because of the many skills needed to produce complete, precision machined parts from mechanical drawings. An even more highly skilled trade is that of a tool and die maker who is a specialized machinist, trained in toolmaking. Usually the most highly regarded and highest paid of all machinists in the trade, toolroom machinists are always in demand. Besides making new tooling these machinists often repair existing tools and dies, repair machine parts for local industries, and make prototypes of new mechanisms (Figure 14.1). When these new prototypes have been developed sufficiently in the toolroom machine shop, a manufacturing plant can then put the machining process into production using numerically controlled machines, automatic machines, or flexible manufacturing systems (FMS) to produce the parts in the desired volumes.

Many large manufacturing plants have their own machine shops for toolmaking, machinery repair, prototype work, and low-volume machining. Small privately owned job shops can be found in virtually every city. They do one-of-a-kind work and produce short runs of products as suppliers to larger companies that assemble and market the end product.

BASIC MACHINE TOOLS

Basic machine tools such as metal cutting saws, drill presses, lathes, and milling machines are used in conventional machining. They are not normally used for produc-

FIGURE 14.1
Robot positioner. Robot arms and components such as this one must first be made by conventional machining processes before being put into production (Barrington Automation Ltd.).

FIGURE 14.2
Vertical band saw (Courtesy of DoALL Company).

tion purposes unless attachments such as jigs and fixtures are used with them to guide tools and hold workpieces.

Metal Sawing Operations

The vertical band saw (Figure 14.2) removes unwanted metal without turning all of it into chips. For some operations the saw cut can be made and no subsequent machining is required. In other cases the blueprint specifies precision machined surfaces, so finishing operations are required after the part is sawed. A band saw blade has a carbon steel body, and high-speed steel teeth are welded into most blades to create cutting edges. Blades are cut to length and welded on a resistance but welder (Figures 14.3 and 14.4), sometimes called a "flash" welder, to form endless bands. Internal features can be cut by inserting the blade through a drilled hole in the workpiece and then welding the blade together. Contours, small or large radii, and many other shapes can be cut with vertical band saws.

Cutoff Saws

Metal bar stock for manufacturing is usually shipped in 12- to 40-foot lengths and range from ¼-in. diameter round bars to those 12 in. in diameter and over. Universal

milled flat bar stock is rolled in sizes up to 14 in. wide, and there is a great variety of round, square, and rectangular bar stock and a wide variety of round, square, and rectangular tubing, plus many other shapes. All of these must be cut to the length needed for specific part manufacturing. The most common type of cutoff saw used for this purpose is the horizontal band saw (Figure 14.5). Automatic band saws, often controlled by microprocessors (Figure 14.6), cut off long bars to programmed lengths, feeding the bar after each cut. Most of these saws use cutting fluids to prolong saw blade life. These machines are mainly used for production.

The reciprocating cutoff saw uses a straight blade and is often called a power hacksaw since its action resembles the operation of a hand hacksaw with its back-and-forth motion. It has a slower cutting action than the band saw, but it is still used in many shops because it can cut harder material than the band saw with less blade breakdown. It is useful for cutting off tough, hard metals such as titanium, zirconium, stainless steel, and tool steels.

FIGURE 14.3
Flush-mounted blade and grinder enable the operator to weld saw blades for external and internal sawing (Courtesy of DoALL Company).

FIGURE 14.5
Horizontal band saw (Courtesy of DoALL Company).

FIGURE 14.4
Ends of band being welded in a resistance "flash" butt welder.

FIGURE 14.6
Automatic band cut-off saw (Courtesy of DoALL Company).

Drilling Machines and Equipment

Drilling machines are designed primarily for making holes in metal and other materials; however, other operations such as counterboring, spotfacing, reaming, and tapping are often performed on them. Since hole drilling operations range from a few thousandths of an inch diameter (size of a hair) to over a foot in diameter, it follows that several kinds of drilling machines are needed. Some drilling machines are suited to low-volume work and others are production machine tools. The three basic types of drill presses are sensitive, upright heavy duty, and radial arm drills. Other variations used are multiple spindle drill presses, turret-type drills, and deep hole drilling machines.

The sensitive drill press (Figure 14.7) is so named because the operator "senses" or "feels" the cutting action of the drill while holding the handle that feeds the drill into the work. These machines have no power feed and are usually limited to holes of up to ½ in. diameter. Sensitive and upright heavy duty drill presses are sized by the largest diameter of a circular piece in which a centered hole can be drilled. Thus if the distance from the column to the center of the spindle is 10 in., it is a 20-in. drill press. Ordinary sensitive drill presses used in machine shops can drill holes no smaller than about 1/16 in. diameter due to the higher spindle speed (rpm) required to maintain the cutting speed for smaller drill sizes. For example, with a cutting speed of 100 sfm, a 10-in. diameter drill would need to rotate at 40 rpm, which would be realistic for this diameter drill. If the drill were only .010 in. in diameter the spindle speed would have to be 40,000 rpm. However, it has been found that these high rotational speeds are not realistic. Even at 10,000 rpm, which would be conservative for a .010-in. drill, fine, powdery chips often develop which can quickly dull the drill. Holes smaller than approximately 1/16 in. must be drilled on a microdrill press, and most microdrilling takes place at speeds between 5,000 and 12,000 rpm.

Heavy duty, upright drilling machines (Figure 14.8) have power feeds and reversing capability. On some machines tapping accessories are provided that feed the tap with the correct thread pitch and then rapidly reverse to remove the tap from the workpiece. A coolant pump is often utilized with these heavy duty drill presses to provide an ample supply of pressurized cutting fluid. A wide range of hole sizes can be drilled on these machines, often from ¼ to 3½ in. in diameter, the largest twist drill size. Larger holes are normally made on radial arm drilling machines with flat spade drills.

Radial arm drill presses (Figure 14.9) are made in a variety of sizes. They are sized by the diameter of the column and the length of the arm as measured from the center of the spindle to the outer edge of the column.

Small workpieces can be shifted around by the operator until the desired hole location is directly under the twist drill, as is done on sensitive drill presses. However, when the workpiece is a large casting that weighs many tons it is not convenient to reposition the part for each drilled hole. The drill head of the radial arm drill press, which contains the spindle and the drilling tool, is easily moved along a rail on the arm to locate the tool over the workpiece. When it is located at the drilling position, the radial arm can be clamped to the column and the drill head can be clamped to the arm. Like the upright machine, the radial arm drill press has a power feed mechanism and a hand-feed lever. It also has a coolant

FIGURE 14.7
Sensitive drill presses are used for light duty drilling (Clausing Machine Tools).

FIGURE 14.8
Upright heavy duty drill press (Buffalo Forge Company).

FIGURE 14.9
Radial arm drill press (Clausing Machine Tools).

tank, usually located in the base, with a pump to filter and recirculate cutting fluid.

Gang drills (Figure 14.10, see page 282) are used where a series of drilling and reaming operations must be performed in a sequence. Each drill head has its own special tooling and a fixture for holding the workpiece. Time-consuming tool changing and setups are eliminated by using gang drills. Drilling machines with multiple spindles (Figure 14.11, see page 282) allow many drilled holes to be made in one pass. Drill jigs that guide the drills to precise locations are often used with multiple spindle drills. Modern CNC machining centers (Figure 14.12, see page 283) are often used for precision drilling and reaming operations due to their capability to position the tool precisely under computer control.

Cutting Tools: Drills, Taps, and Reamers

High-speed steel twist drills (Figure 14.13, see page 283) are the most common tools used for general purpose drilling in metals. The twisted flutes provide for chip clearance and removal. The helix of the flute also provides positive rake (Figure 14.14, see page 283), which is needed for most work materials. Straight flute drills (Figure 14.15, see page 283) provide zero rake and are used for plastics, brass, and bronze, which have the tendency to "grab" when drills have positive rake. Zero rake can also be provided on a standard twist drill by grinding a flat on its cutting edge (Figure 14.16, see page 284). Spade drills (Figure 14.17, see page 284) are made in sizes from very tiny to 12 in. in diameter and are widely used for drilling holes 1 in. and larger. A spade drill has the advantage a very rigid shank enabling heavier cuts to be taken, and the main body is often provided with holes to pump coolant through to the blade. Blades are easily sharpened and replaced. Spade drills are frequently equipped with a stepped blade to pro-

FIGURE 14.10
Gang drills (Buffalo Forge Company).

FIGURE 14.11
Multiple spindle drill (Buffalo Forge Company).

duce a bevel or chamfer on the edge of the hole to facilitate tapping and assembly.

Gun drills (Figure 14.18, see page 284) are capable of making very precise holes many feet deep. Frequently used for drilling holes in automotive engine blocks among other deep hole applications, they are used with special machinery that guides the drill and pumps coolant through it at high pressure.

A center drill is used for making a 60 degree conical hole in the end of a shaft for turning in a lathe. This hole accepts a center mounted in the tailstock of the lathe to provide support for the end of the long, slender work-

piece. Center drills are also used for spotting, providing a starting hole for larger drills.

Most drilling operations lack precision. Control of hole diameter, location, roundness, straightness, and surface finish is difficult. Reamers (Figure 14.19, see page 284) are used to enlarge an existing hole and give it a more accurate diameter with an acceptable finish. However, if the drilled hole is off center, rather than relocating it the reamer will follow the eccentric hole. Boring operations, unlike reaming, do not have a tendency to follow the existing hole and can restore an off-center hole to true position. Reamers are used on drill presses, lathes, and vertical milling machines.

Special tools called counterbores are used to enlarge existing holes (Figure 14.20, see page 285). Countersinks provide a bevel to receive certain bolt heads, and spotfacing is done to provide a machined seat for bolt heads or nuts.

Taps are hardened, threaded cutting tools with flutes that collect chips. Taps are brittle, and breakage is a serious problem in manufacturing processes. Standard hand taps

FIGURE 14.12
Machining centers such as this one perform many different cutting operations automatically (Cincinnati Milacron, Inc.).

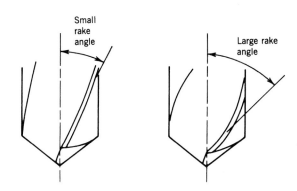

FIGURE 14.14
Helix of twist drill provides low or high rake angle.

(Figure 14.21, see page 285) include taper, plug, and bottoming taps. Threaded internal holes can be made by hand-twisting the tap into a properly sized drilled hole by means of a tap handle. Machine tapping is a higher speed operation in which plug taps are most often used. In machine tapping, spiral point taps (Figure 14.22, see page 285) are far more efficient and are less likely to break, but they are effective only for the tapping of through holes. Spiral point taps push the chips forward through the hole, causing jamming at the bottom of blind (not through) holes.

Tapping a blind hole presents problems due to incomplete threads at the bottom of the hole and because chips collect at the bottom of the hole. Bottoming taps are designed with only one or two tapered teeth to cut complete threads near the bottom of blind holes. Spiral flute

FIGURE 14.13
A variety of high-speed steel twist drills.

FIGURE 14.15
Straight flute drill.

FIGURE 14.16
A flat ground on the cutting edge of the drill is often
necessary when drilling brass, bronze, and plastics.

taps (Figure 14.23, see page 285) are also helpful when
tapping blind holes because they direct the chips upward
along the flutes and out of the hole.

Threading dies for making external threads are used
in a diestock, a handle that holds the die, for hand threading. Dies for high production threading on machine tools
are covered later in this chapter.

TURNING MACHINES AND EQUIPMENT

Turning machines rotate a workpiece against a tool that
gradually moves along the work, producing an accurate
cylindrical or tapered (conical) surface. Tracing attachments make possible almost unlimited curved shapes on the

FIGURE 14.18
Gun drill.

turned cylinder. Computer numerical control-operated turning machines (Figure 14.24, see page 285) can produce a
great variety of cylindrical, curved, and tapered shapes
(Figure 14.25, see page 286) under computer control.
These NC machines are fully automatic and require an operator only to load and unload the workpiece. Turning
machines include the engine lathe with its horizontal spindle
(Figure 14.26 see page 286) and the vertical spindle

FIGURE 14.17
Taper shank spade drill (Courtesy of DoALL Company).

FIGURE 14.19
A variety of reamers.

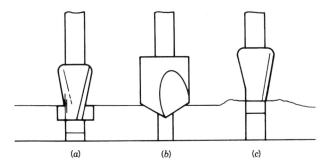

FIGURE 14.20
(a) Counterbore. *(b)* Countersink. *(c)* Spotfacing.

FIGURE 14.22
Spiral point tap (Courtesy of DoALL Company).

machines, the vertical boring mill (Figure 14.27, see page 286), and vertical turret lathe (Figure 14.28, see page 286).

The Engine Lathe

The principal function of the lathe is to remove unwanted material to form cylindrical and conical shapes. This is done in a turning operation that utilizes a single-point tool (as contrasted with a multiple-point tool such as a milling cutter). However, other tool shapes are used to make grooves, to cut off round stock, and to make various kinds of threads by chasing (taking a series of cuts in the same groove until a full thread is made).

The engine lathe got its name from the steam engine that first powered it, an improvement over the earlier foot- or hand-powered lathes. It has been the most important tool in machine shops and manufacturing plants for over 150 years. Today these valuable machines are still very much in use in tool and die shops, job machine shops, and machine repair shops, although they have been replaced by CNC lathes and automatic screw machines for produc-

FIGURE 14.23
Tapping with a spiral flute tap (Courtesy of DoALL Company).

FIGURE 14.21
Standard set of three taps. Top to bottom: taper, plug, and bottoming.

FIGURE 14.24
NC lathe (Cincinnati Milacron, Inc.).

FIGURE 14.25
Curved and tapered turning operation on a shaft in an NC lathe (Cincinnati Milacron, Inc.).

FIGURE 14.27
Vertical boring mill.

tion purposes. Lathes range in size from tiny jeweler's lathes to huge machines that turn massive forgings (Figures 14.29 and 14.30, see page 287).

Threads

Although cutting threads is one of the important functions of the lathe (Figure 14.31, see page 287), most threads are cut on production machines. There are a number of thread forms that have been standardized, and some of the more commonly used ones can be seen in Figures 14.32 through 14.43 (see pages 287–290). Screw thread sizes are expressed in terms of major (outside) and minor

(inside) diameter, pitch diameter, number of threads per inch and pitch (distance between threads) (Figure 14.44, see page 291).

Even though the American National Form as a thread standard with its common thread series National Coarse (NC) and National Fine (NF) is obsolete, it is still in use in many machine shops. Even some tap drill charts are still being produced based on this popular system. The American standard for Unified threads is the official thread standard at the present time with its basic Unified coarse thread (UNC) and Unified fine thread (UNF). American and Unified standards will interchange for the most part; fits and tolerances make up the principal difference.

FIGURE 14.26
A 15-inch swing engine lathe. Swing (largest diameter that can be turned) is one measurement of lathe size. (Clausing Machine Tools).

FIGURE 14.28
Vertical turret lathe (Ductr Mfg.).

FIGURE 14.29
Jeweler's lathes are used to make very tiny precision parts. Many attachments for toolholding can be used with these machines (Photograph courtesy of Louis Levin & Son, Inc.).

Screw thread manufacture in the United States is either in the Unified or Metric series. Both have similar thread forms but different pitch, pitch diameter, and major diameter. For metric thread forms the pitches and pitch diameters may vary by country and often are not interchangeable. Besides the British ISO, there is the Système International (SI), the Löwenherz metric thread used in Germany, SAE spark plug threads, the British standard for spark plugs, and numerous differences between manufacturers worldwide.

FIGURE 14.30
A large heavy duty engine lathe (Clausing Machine Tools).

FIGURE 14.31
Single-point threading on a lathe.

In the Unified and the ISO metric thread systems there are coarse and fine pitches, and the pitch varies with the diameter. The angle of the thread teeth is 60 degrees in both standards. Threads are designated in the Unified standard beginning with the nominal size first, followed by the number of threads per inch, the thread series, and class of thread. If the letters LH are added, the thread is left-hand. An example of this is ¾-10 UNC-2A. The symbols 1A, 2A, and 3A indicate an external thread tolerance and are right-hand; 1A is a loose fit and 3A is a tight fit. Three other tolerances are used for internal threads, 1B, 2B, and 3B. These can be adjusted to get a desired fit, for example a 2A bolt can be matched with a 3B nut. Metric threads are likewise designated with the major diameter

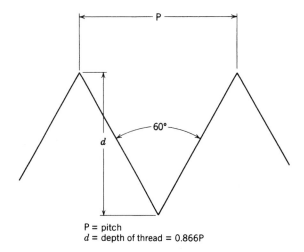

P = pitch
d = depth of thread = 0.866P

FIGURE 14.32
The 60-degree sharp V-thread form (White, Neely, Kibbe, Meyer, *Machine Tools and Machining Practices,* Vol. I, © 1977 John Wiley & Sons, Inc.).

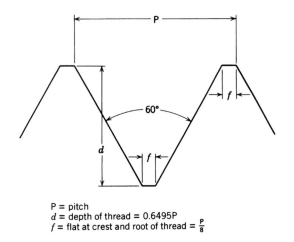

P = pitch
d = depth of thread = 0.6495P
f = flat at crest and root of thread = $\frac{P}{8}$

FIGURE 14.33
The American National form of thread (White, Neely, Kibbe, Meyer, *Machine Tools and Machining Practices,* Vol. I, © 1977 John Wiley & Sons, Inc.).

first, followed by the pitch, and then the tolerance grade. An example is M20 × 2 6H/5g6g.

Internal threads that are tapped must be predrilled (tap drilled) to a certain diameter to produce a correct percentage of thread for ease of tapping in a given material. For example, a 75 percent thread is correct for soft machine steel, but a 60 percent thread (larger tap drill)

may be better for 301 stainless steel. The tap drill size dictates the *actual minor diameter* and can be determined with the following formula:

$$\text{percentage of thread} = \frac{\text{major diameter} - \text{actual minor diameter}}{2 \times \text{basic thread height}} \times 100$$

or,

$$\text{actual minor diameter} = \text{major diameter} - 2 \times \text{basic thread height} \times \text{percentage of thread}$$

The height of the tooth (basic thread height) for Unified threads equals the pitch × .541.

Tapered threads such as those used to join pipes can be made on an engine lathe by using a taper attachment. The symbol for American standard taper pipe thread is NPT.

Workholding Devices on Lathes

It is essential that lathe workpieces are held securely to the machine spindle to avoid the possibility that the rapidly spinning part could be thrown from the machine spindle, causing damage or injury. It is also important that the desired center of the workpiece is aligned to the center of the machine spindle. Poor workpiece centering causes inaccurate location of machined diameters and cre-

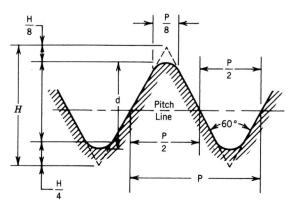

External unified threads.

P = pitch = $\frac{1}{n}$
d = depth of thread = 0.61343 × P
  = 17H ÷ 24
Flat at crest = 0.125 × P

Rounded root optional or resulting from tool wear.

Rounded or flat crest.

Unified threads.

Internal unified threads.

P = pitch = $\frac{1}{n}$
d = depth of thread = 0.54127 × P
  = 5H ÷ 8
Flat at crest = 0.125 × P
Flat at root = 0.125 × P

FIGURE 14.34
Unified thread forms.

FIGURE 14.35
American National standard taper pipe thread form (White, Neely, Kibbe, Meyer, *Machine Tools and Machining Practices,* Vol. I, © 1977 John Wiley & Sons, Inc.).

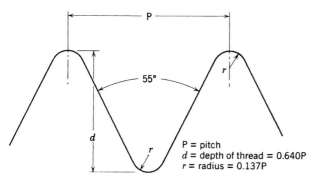

P = pitch
d = depth of thread = 0.640P
r = radius = 0.137P

FIGURE 14.36
Whitworth thread form (White, Neely, Kibbe, Meyer, *Machine Tools and Machining Practices,* Vol. I, © 1977 John Wiley & Sons, Inc.).

P = pitch
d = depth of thread = 0.703P
f = flat = 0.125P
r = radius = 0.063P

FIGURE 14.37
The SI metric thread form (White, Neely, Kibbe, Meyer, *Machine Tools and Machining Practices,* Vol. I, © 1977 John Wiley & Sons, Inc.).

P = pitch = $\frac{1}{n}$

d = depth of thread = $\frac{P}{2}$ or $\frac{.5000}{n}$

Width of flat = $\frac{P}{2}$ or $\frac{.5000}{n}$

FIGURE 14.38
The square thread form (White, Neely, Kibbe, Meyer, *Machine Tools and Machining Practices,* Vol. I, © 1977 John Wiley & Sons, Inc.).

P = pitch = $\frac{1}{n}$

$d = \frac{P}{2} + .010$ in.

$f = .456P$ (basic)

FIGURE 14.39
Modified square thread form (White, Neely, Kibbe, Meyer, *Machine Tools and Machining Practices,* Vol. I, © 1977 John Wiley & Sons, Inc.).

P = pitch = $\frac{1}{n}$

d = depth = $\frac{P}{2} = \frac{.500}{n} + .010$ in.

Crest = $\frac{.3707}{n}$

Root = $\frac{.3707}{n} = .0052$ in.

FIGURE 14.40
Acme thread form (White, Neely, Kibbe, Meyer, *Machine Tools and Machining Practices,* Vol. I, © 1977 John Wiley & Sons, Inc.).

ates an imbalance condition that interferes with smooth operation of the lathe.

Tooling used to locate and hold a workpiece on a lathe include independent (usually four-jaw) chucks (Figure 14.45, see page 291), universal (two-, three-, and six-jaw) chucks (Figure 14.46, see page 291), and collets (Figure 14.47, see page 292). Independent chucks have jaws that must each be adjusted separately to grip and locate the workpiece Square, rectangular, and odd-shaped parts are usually held in four-jaw independent chucks. Most universal chucks have three jaws that all move together or apart the same amount when a chuck wrench is turned. Round and hexagonal parts or bar stock can be quickly gripped and centered with these chucks, so they are often called self-centering chucks.

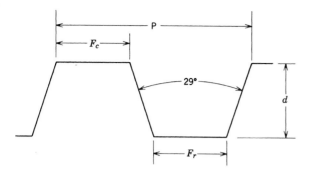

P = Pitch = $\frac{1}{n}$
d = Depth = .3P
F_c = Basic flat at crest = .4224P
F_r = Basic flat root = .4224P − .0052

FIGURE 14.41
American Standard Stub Acme thread form (White, Neely, Kibbe, Meyer, *Machine Tools and Machining Practices,* Vol. I, © 1977 John Wiley & Sons, Inc.).

P = Pitch = $\frac{1}{n}$
d = Depth = .433P
f = Flat = .25P

FIGURE 14.42
Stub Acme thread form (White, Neely, Kibbe, Meyer, *Machine Tools and Machining Practices,* Vol. I, © 1977 John Wiley & Sons, Inc.).

P = Pitch = $\frac{1}{n}$
d = Depth = .662P
f = Flat = .145P

FIGURE 14.43
Buttress thread form (White, Neely, Kibbe, Meyer, *Machine Tools and Machining Practices,* Vol. I, © 1977 John Wiley & Sons, Inc.).

Six-jaw universal chucks are able to grip round materials more securely than three-jaw chucks. Three-jaw chucks for engine lathes often have two interchangeable sets of jaws, one set for internally gripping hollow parts such as rings and one set for holding workpieces externally. However, most chucks have removable jaws (called top jaws), each attached to a master jaw. These can be quickly removed and reversed to change from external to internal gripping (Figures 14.48 and 14.49, see page 292). Soft jaws which are not hardened and can be machined while in place on the lathe are used to hold odd-shaped parts.

Power chucks (Figure 14.50, see page 293) are used on turret lathes, CNC lathes, and other production machines. The power chuck is commonly a three- or six-jaw self-centering chuck that is opened and closed by means of an air or hydraulic cylinder, which is fastened to and rotates with the machine spindle.

Most self-centering chucks, when new, will center and hold round parts within less than .001 inch of runout (eccentricity). However, when they become worn, runout may exceed .010 in. Some chucks have adjustment screws to take up wear; these can be adjusted to hold runout within .0005 in. (Figure 14.51, see page 293).

A standard collet is a spring steel sleeve with slits along its length that permit it to spring open a small amount and accept the workpiece. To close the collet, an external taper on the collet is pulled or pushed into a mating taper, squeezing the collet securely closed around the workpiece. Collets are capable of consistently holding round precision workpieces to less than .001 inch of runout. Steel collets expand very little, so they cannot hold parts that vary from the nominal size of the collet by more than a few thousandths of an inch. A large number of these collets is needed to provide the capability to hold a variety of workpiece sizes. Rubber flex collets can hold a ⅛-in. range of part sizes, reducing the number of collets needed to hold a variety of part sizes. Chucks and collets are fastened to the spindle in several methods—flange nose, camlock, long taper key drive, and threaded nose (Figure 14.52, see page 293).

When long, slender workpieces are held between centers on a lathe, a conical center is inserted into a matching depression on each end of the part. One center is held in the lathe spindle, and the other center is held in the lathe tailstock, providing support for the outer end of the workpiece. A tool called a lathe dog is required to assure that the workpiece rotates with the spindle. The dog clamps onto the part and seats against the spindle that drives the part.

Mandrels (Figure 14.53, see page 294) are a form of workholding device for parts that have a hole or bore in their centers. The mandrel holds the work concentric to the bore, so any machining done is also concentric. Workpieces are held by friction, either by means of a

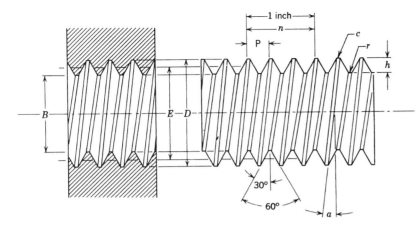

FIGURE 14.44
Thread nomenclature and dimensions (White, Neely, Kibbe, Meyer, *Machine Tools and Machining Practices,* Vol. I, © 1977 John Wiley & Sons, Inc.).

D	Major Diameter	*a*	Helix (or Lead) Angle
E	Pitch Diameter	*c*	Crest of Thread
B	Minor Diameter	*r*	Root of Thread
n	Number of threads per inch (TPI)	*h*	Basic Thread Height (or depth)
P	Pitch		

slight taper (.006 in./ft) on a mandrel that is pressed into the bore, or by an expanding mandrel. The mandrel is commonly held between centers. Stub arbors, as they are sometimes called, are a form of mandrel that is held in a chuck (Figure 14.54, see page 294). The term *arbor* correctly refers to a rotating member that holds a cutting tool such as the arbor of a milling machine, and *mandrel* correctly refers to a rotating member that holds a workpiece.

However, these terms are often interchanged in common shop terminology and in product literature.

Turning Operations

In *turning* cuts, the carriage guides the cutting tool as it feeds parallel to the centerline of rotation of the workpiece to create cylindrical shapes (Figures 14.55 and 14.56, see

FIGURE 14.45
Four-jaw independent chuck (Buck Chuck Company).

FIGURE 14.46
Three- and six-jaw chucks. These standard chucks require an extra set of jaws for gripping large diameters (Buck Chuck Company).

FIGURE 14.47
Collets are held in the spindle nose with an adaptor and a draw bar.

page 294) or at an angle to create conical shapes (tapers). When a cut is taken across the end of a piece by moving the cross-slide instead of the carriage, it is called *facing*. If a bore is required in a solid workpiece, a hole is drilled by holding the drill in the tailstock and feeding it in with the handwheel (Figure 14.57, see page 294). The hole can then be reamed (Figure 14.58, see page 295) or bored with a single-point tool (a boring bar) that is fastened to the carriage (Figure 14.59, see page 295). Counterbores and internal and external grooves and threads can be cut on the lathe.

The diameter of a reamed hole is limited to the exact size of the reamer, and if the drilled hole is eccentric the

Hardened reversible top jaws   Soft blank top jaws

FIGURE 14.49
Three-jaw chuck showing hardened reversible top jaws and soft blank top jaws (Buck Chuck Company).

reamed hole will also be eccentric as the reamer follows the existing hole. Boring has the capability to remove more material from one side of a hole than the other to make the hole concentric to the centerline of rotation of the lathe spindle. Hole diameters can be varied by merely repositioning the boring bar; however, boring precision holes requires much more time, care, and skill than reaming. For the most part, skilled machinists use engine lathes and semiskilled machine operators use turret lathes and other production machines.

Turret Lathes

Turret lathes (Figure 14.60, see page 295) are similar in many ways to engine lathes. The difference is that they are arranged for repetitive, rapid production and often perform several operations in quick succession. The workpiece is held in a chuck, which is often air operated. A turret, usually six sided, contains different tools such as drills, taps, and turning and knurling tools. The cross-slide

FIGURE 14.48
Three- and six-jaw chucks with top jaws (Buck Chuck Company).

FIGURE 14.50
Power chucks are often used on production machinery (Buck Chuck Company).

usually holds a parting or cutoff tool to sever the finished part from the bar stock.

Turret lathes require an operator at all times. The operator simply moves levers to perform machining operations in the proper sequence to make the part and checks finished dimensions. Tooling is all preset by a set-up machinist.

External threading is done on turret lathes by means of die heads. There are two basic types, radial (Figure 14.61,

Am. St. flange nose

Camlock
Camlock stud length is adjustable to suit spindle.

Long taper key drive

Threaded mounting plates

FIGURE 14.52
Four basic types of spindle nose arrangements for mounting chucks and other workholding devices (Buck Chuck Company).

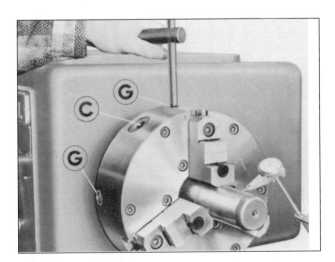

FIGURE 14.51
Adjust-Tru® chuck. Chuck is adjusted at *G* to eliminate runout using a dial indicator. Jaws are tightened at *C* (Buck Chuck Company).

see page 296) and tangent (Figure 14.62, see page 296). Both types make the full depth thread in one pass and can be quickly opened to remove the die from the part by means of a handle. Some types automatically open when the thread is completed. The cutters (thread chasers) on the radial type are arranged to cut on their side on the radius

FIGURE 14.53
Left, tapered mandrel (.006 in./ft) holds part to be turned by friction in the bore. Right, gang mandrel holds several parts for machining on a straight threaded screw.

FIGURE 14.54
Stub mandrel. Part to be machined is gripped in the bore by expanding the mandrel with a tapered screw.

FIGURE 14.56
Short workpieces can be turned while supported in the chuck only.

(Figure 14.63, see page 296). Tangent chasers cut on their ends on the tangent (Figure 14.64, see page 297). Tangent die chasers are easier to sharpen and last longer than radial chasers. Internal threads are also quickly produced in turret lathes by means of collapsing taps (Figure 14.65, see page 297). Ordinary taps, after making the thread, must be screwed out of the work by means of reversing the spindle rotation. Collapsing taps (Figure 14.66, see page 297) simply collapse to reduce their diameter when the thread is

completed and can be removed quickly. These quick-opening die heads and collapsing taps can be adapted for use in engine lathes, drill presses, and other machine tools.

Automatic Screw Machines

Automatic screw machines (Figure 14.67, see page 298) are similar to turret lathes in that they have preset, index-

FIGURE 14.55
Turning operation. Long pieces should be supported between the chuck and the center.

FIGURE 14.57
Drilling in the lathe. The drill is supported in the tailstock and is forced into the metal by means of a handwheel.

FIGURE 14.58
A drilled hole can be finished to size by reaming.

able tooling and follow a set sequence of operations to produce a part. Most screw machines are cam-driven and completely automatic, producing small parts at a rapid rate without needing constant attention by an operator. A stock feed mechanism feeds metal bars into the machine as needed. Stock feeders are kept full by an operator who is in charge of a number of machines. In addition to small threaded fasteners, screw machines can produce a large variety of small cylindrical and tubular parts.

Vertical Turning Machines

Vertical turret lathes (see Figure 14.28) have the advantage of supporting large heavy parts on a horizontal chuck surface that rotates on a vertical spindle. These machines

FIGURE 14.59
Accurate holes may be made by boring with single-point tools.

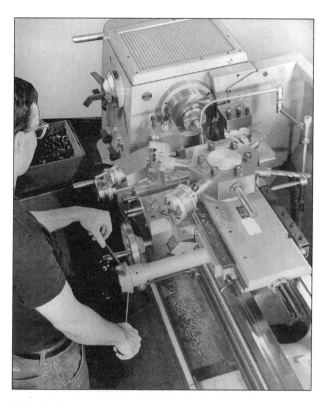

FIGURE 14.60
Turret lathes are semiautomatic production machines (Clausing Machine Tools).

are capable of taking heavy cuts in tough materials with a high stock removal rate.

Vertical boring mills (see Figure 14.27) are similar machines that are built on a much larger scale, some having rotating tables over 50 ft in diameter. Massive steel castings weighing many tons can be placed on the table and clamped in place. These are then bored, counterbored, faced, or turned. Many of these machines are equipped with indexable turrets similar to turret lathes, so various tools can be brought into use quickly.

Horizontal Boring Mills

Horizontal boring mills (Figure 14.68, see page 298) make use of a single-point tool for machining precision bores, especially in large or awkward parts. Like the lathe, they have a horizontal spindle to which boring bars may be attached, or they have long, line-type bars with an outboard bearing support. The workpiece is mounted on a sliding table that can be moved in two horizontal axes, and the spindle and outboard bearing may be raised or lowered. These useful machines are being replaced rapidly by NC machining centers.

FIGURE 14.61
Radial die head. (TRW Cutting Tools Division).

Cutting Tools: Milling Cutters

A milling machine makes use of a rotating cutter with two or more teeth to remove material from a workpiece. Milling cutters must have correct relief and rake angles (Figure 14.69, see page 298). Cutters used on horizontal milling machines for slab milling and slotting are called *plain milling cutters* (Figure 14.70, see page 299). Plain milling cutters are sharpened on their outer edge, and cutting clearance or relief is provided by the sharpening or grinding operation. *Form-relieved cutters* have a custom form such as a radius or gear shape and must be sharpened on the tooth face instead of on the periphery of the cutter. Form-relieved cutters are used on horizontal spindle machines for cutting gears, producing radii, and making other special shapes (Figure 14.71, see page 299).

FIGURE 14.62
Tangent die head (TRW Cutting Tools Division).

Large vertical spindle milling machines use *face mills* for machining flat surfaces and utilize smaller diameter two flute or multiple flute end mills (Figure 14.72, see page 299). *End mills* cut on the sides (periphery) and on the end (face), so vertical plunge cuts can be made as in drilling. This can be followed by horizontal milling cuts using the same tool; thus a blind keyseat (Figure 14.73, see page 299) can be machined into a shaft or an internal pocket can be milled on a part with this tool.

MILLING MACHINES AND EQUIPMENT

There are basically two types of milling machines, those with horizontal spindles (Figure 14.74, see page 300) and those with vertical spindles (Figure 14.75, see page 300). There are many combinations of these two types, and some machines incorporate both orientations. Sizes range from small bench-top models to huge planer-type mills that can machine castings weighing many tons. Large profiling mills (Figure 14.76, see page 300) are exten-

FIGURE 14.63
Principle of radial thread cutting.

FIGURE 14.64
Tangential principle of thread cutting.

Lead angle

sively used in the aircraft industry for machining aluminum alloy spars and other structural parts.

Conventional horizontal spindle milling machines are used in machine shops and tool rooms. Many types of milling cutters are used on horizontal milling machines: plain milling cutters for slab milling, side milling or staggered-tooth cutters for deep slots, saws or slitting cutters for grooves, and face mills for milling flat surfaces. These cutters are either made of high-speed steel or they have inserted carbide cutting edges. Horizontal mills are sometimes used in machine shops for making splines and gears (Figure 14.77, see page 301).

Vertical spindle milling machines are very useful machines in tool rooms and machine shops, and they utilize tooling ranging from end mill cutters to face mills and fly cutters. These versatile machines can be used for drilling, reaming, and boring operations as well as for milling operations. Vertical mills are sometimes operated by numerical control for drilling hole patterns or for milling operations.

SHAPERS AND PLANERS

Shapers (Figure 14.78, see page 301) are designed to cut metal by forcing a single point tool across a workpiece in a reciprocating action. These machines are primarily used for low-volume machining and toolroom work. Shaping is also used in gear manufacture, where it is very useful for producing certain types of gears. Planers (Figure 14.79, see page 301) are virtually obsolete as machine tools, but they are still useful for machining some extremely large machine parts. On the shaper, the ram moves back and forth, holding the tool that cuts the stationary workpiece.

FIGURE 14.65
Collapsing taps make possible the removal of the tap without stopping or reversing the spindle rotation (TRW Cutting Tools Division).

FIGURE 14.66
Here a collapsing tap is mounted on the turret of a turret lathe so it can be quickly swung into position for making an internal thread (TRW Cutting Tools Division).

FIGURE 14.67
Automatic screw machine.

In contrast, on the planer, the workpiece is damped to the table and moves back and forth past a stationary tool. Some large planers have been retrofitted with a milling head and the table speed is altered to convert them into planer mills, which are much faster and more versatile machine tools.

BROACHING

Broaching is the precision cutting of a material by passing over, or through it, a series of progressively stepped teeth. Flat surfaces, contours, internal splines, and external splines are some of the shapes produced by broaching. Broaching is usually a one-stroke operation in which roughing, semifinish, and finish cuts are all done in a single machine stroke. However, when excessive stock prevents a one-stroke application, multiple strokes can be used, either using several different internal broaches or by adjusting the position of an external broach.

Broaching is ideally suited for high-production machining (Figures 14.80 to 14.82, see pages 302–303). For example, forged diesel engine connecting rods require a machining operation to make them ready for use (Figures 14.83 and 14.84, see page 303). They have 12 surfaces that are flat, round, and serrated. These connecting rods are finish machined by broaching at the rate of 112 per hour. Another high-production broaching operation on a vertical machine is seen in Figures 14.85 and 14.86 (see page 304). Where quite long broaches are needed, horizontal broaching machines (Figures 14.87 and 14.88, see page 305) are used. External broaching of large flat surfaces is an economical means of machining when a large number of piece parts are involved.

GEARS

One of the most efficient methods of transferring rotary motion from one shaft to another is that of gearing. Large amounts of torque with a definite ratio can be transmitted

FIGURE 14.68
Horizontal boring mill (Ductr Mfg.).

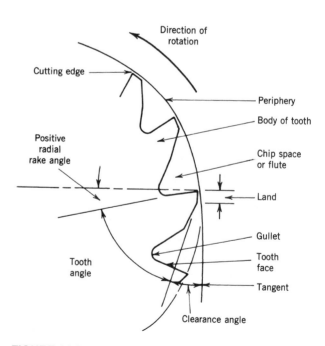

FIGURE 14.69
Milling cutter angles.

FIGURE 14.70
Plain milling cutters.

FIGURE 14.72
A variety of end mills.

in a small space by using gears. When they are properly made and kept lubricated, they are quiet, dependable, and long lasting.

Two friction disks or crossed belts can be used as examples to explain the operation of gears (Figure 14.89, see page 305). The involute curve has been proven by long experience to be the most satisfactory profile for spur and helical gear teeth. This gear tooth form transmits motion smoothly and uniformly. The involute can be described as a curve traced by a point on a crossed belt as it moves from one pulley to another without slippage (Figure 14.90, see page 305). If the crossed belts are superimposed on the friction disks (Figure 14.91, see page 306), it can be seen that the outer diameters of the disks

represent the pitch circles of the gears; the pulleys are the base circles, and where the belt crosses the line of centers is the pitch or tangent point of the friction disks. Figure 14.92 (see page 306) shows the terms used to identify parts of spur gears.

Many types of gears are made for various purposes. *Spur gears* are used to transmit motion between parallel shafts where they turn in opposite directions. *Helical gears* are used for the same purpose, but they are quieter running gears. *Bevel gears* connect intersecting shafts on

FIGURE 14.71
Form-relieved gear and radius cutters.

FIGURE 14.73
Blind keyseat.

FIGURE 14.74
Horizontal milling machine (Cincinnati Milacron, Inc.).

FIGURE 14.75
Vertical milling machine (Courtesy of DoALL Company).

FIGURE 14.76
Profiling mill. Aircraft plants
use these NC machines for
milling parts (Lockheed Co.).

FIGURE 14.77
Horizontal milling machine, gear-cutting operation.

FIGURE 14.78
Shaper.

the same plane but at an angle. *Miter gears* are bevel gears that connect shafts at a 90 degree angle. *Worm gears* connect shafts at 90 degree angles but not on the same plane. *Hypoid* gears such as those used in automotive differentials are similar to worm and bevel gears. They are quiet running like worm gears and have some of the efficient characteristics of bevel gears. *Crossed helical gears* also connect two shafts that are at an angle but not on the same plane. Internal gears can be helical or spur types. *Herringbone gears* consist of two sets of helical teeth with opposite angles on the same gear. Their main advantage is that they eliminate the end thrust characteristic of helical gears. *Gear racks* are designed to convert rotary motion to linear motion using a rack and pinion.

Gear Cutting Operations

Several methods are used to manufacture gears. Where high production is not a requirement, external gears are cut one tooth at a time on conventional horizontal milling machines with form cutters. A single gear cutter mounted on an arbor of a milling machine removes material on the periphery of a gear blank to form the gear teeth. This is usually done in two cuts, first a heavy roughing cut at each tooth location all around the blank and then a light finishing cut. These precise tooth divisions are positioned by means of a dividing (indexing) head. This is a relatively slow gear-making process that is useful when a special gear is needed for a prototype or a repair operation.

High production rates for external gears can be realized in the process of hobbing. Hobbing (Figures 14.93

FIGURE 14.79
Planer shop floor. Planers are useful for machining extremely large heavy parts, such as castings for presses (Illustration courtesy of National Machinery Company).

FIGURE 14.80
Twin-ram with in-line transfer progressive broaching machine that cuts blind external splines in two different parts. One part is loaded while the other is broached (Apex Broach & Machine Co.).

and 14.94, see page 306) is a generating process in which a rotating cutter is slowly fed into a rotating workpiece (called a gear blank). Extremely large gears can be generated by the hobbing process (Figures 14.95 and 14.96, see page 307). Racks (Figure 14.97, see page 307) can be produced by hobbing, and the hobbing process is also used for making small gears, splines, and roller chain sprockets. Often, when the blanks are thin as with plate sprockets, many blanks are stacked together on a mandrel and cut in one operation. Large herringbone gears (Figure 14.98, see page 308) and large bevel gears (Figure 14.99, see page 308) can also be produced on special machines.

Where internal gears, helical teeth, and special gears are required at higher production rates, gear shaping is used. The principle of gear shaping is based on the fact that the teeth on two meshing gears rotating together will fol-

FIGURE 14.81
Part being loaded under the front ram while rear ram is stroking down (Apex Broach & Machine Co.).

FIGURE 14.82
Progressive tooling for two different parts (Apex Broach &
Machine Co.).

FIGURE 14.83
Broaching machine making connecting rods and caps
(Apex Broach & Machine Co.).

Before
broaching.

After first
operation.

After second
operation.

FIGURE 14.84
Connecting rods and caps
(Apex Broach & Machine
Co.).

FIGURE 14.85
Blind spline broaching machine. Inset shows blank and internal broached part. These automatic transmission converter-clutch hubs are splined at the rate of 240 hubs per hour, 12 times faster than conventional gear shaping methods (Apex Broach & Machine Co.).

low an involute curve. If one gear is hardened and provided with suitable clearances, it can be used as a generating cutter for a mating gear blank (Figure 14.100, see page 308). The cutter is made to reciprocate like the tool in a vertical shaper, while the tool and blank are slowly rotating. The cutter is fed into the blank until the gear teeth are machined to the correct size and shape (Figures 14.101 and 14.102,

FIGURE 14.86
Closeup of turntable on broaching machine. Two parts are broached at the same time under a single ram. Automated pick-and-place arms unload the machine in just 10 sec (Apex Broach & Machine Co.).

see pages 308–309). These machines are often computer controlled (CNC). Both internal and external gears can be produced in these machines (Figures 14.103 to 14.106, see pages 309–310). Very large and very small gears as well as helical and special shapes can be produced on gear shapers. Internal splines and some external gears that are easily made on gear shapers cannot be made by conventional gear cutting processes such as hobbing.

Hypoid gears are generated on special machines (Figures 14.107 and 14.108, see page 310). Like other automotive gears, they must be finished by other processes after they are machined (Figure 14.109, see page 310).

Finishing operations include gear shaving, flame hardening, and gear grinding. A rotating cutter, shaped like a gear, is made to mesh and turn with a machine cut gear before it is heat treated. This eliminates most indexing errors and roughness. A flame or induction hardening process (Figure 14.110, see page 311) is often performed on the teeth of a gear or splines of a shaft, followed by a finish grind to restore accuracy. This assures strong, shock- and wear-resistant gears that mesh together accurately and run quietly.

ABRASIVE MACHINING

Grinding processes remove metal by means of abrasive grains. Abrasives can be silicon carbide for cast iron and nonferrous metals, aluminum oxide for ferrous metals,

FIGURE 14.87
Horizontal broaching machine (Apex Broach & Machine Co.).

FIGURE 14.88
Surface-to-air missile casings. Internal grooves are made with the horizontal broach in Figure 14.87 (Apex Broach & Machine Co.).

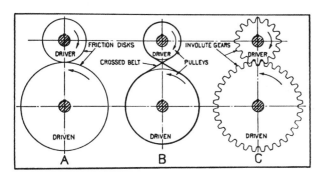

FIGURE 14.89
Three means for transmitting rotary motion from one shaft to another (Courtesy Fellows Corporation).

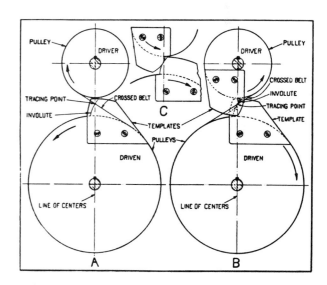

FIGURE 14.90
Diagram illustrating that the involute is that curve traced by a point on a crossed belt as it moves from one pulley to another (Courtesy Fellows Corporation).

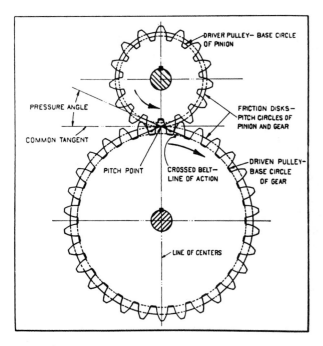

FIGURE 14.91
Diagram showing gears, pulleys, and crossed belt superimposed on friction disks, illustrating basic gear elements (Courtesy Fellows Corporation).

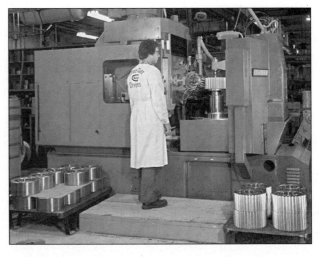

FIGURE 14.93
Hobbing machine producing gears (Illinois Gear, a Household International Company).

and boron nitride and diamond for grinding hard materials such as tungsten carbide. Since abrasives are harder than hardened metals, the grinding process is ideally suited to the removal of unwanted material on hardened workpieces. Also, since there are many small cutting edges traveling at a high surface speed, extremely good finishes can be obtained by abrasive machining. Accurate dimensions can be maintained on a production basis with these precision machines. Precision grinding operations often

follow a conventional machining process or heat treatment process to provide the final dimension and finish to the workpiece.

Many variables exist in the production of uniform precision parts by grinding. The complexity in this area of machining is greater than that of single-point tooling for metal cutting. Grinding wheels tend to break down and become smaller as they are used. This must be compensated for by adjusting downfeed or infeed. Grinding wheels tend to be self-sharpening. When the grains dull,

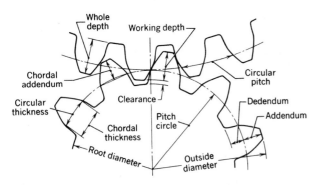

FIGURE 14.92
Spur gear terms (White, Neely, Kibbe, Meyer, *Machine Tools and Machining Practices,* Vol. II, © 1977 John Wiley & Sons, Inc.).

FIGURE 14.94
CNC gear hobbing machine (The Gleason Machine Division).

FIGURE 14.95
Extremely large gears can be produced by the hobbing process (Illinois Gear, a Household International Company).

they break away, exposing new sharp grains. Wheels must be trued and dressed occasionally to maintain their shape and remove any grains that fail to self-sharpen. This is usually done with a diamond dresser on production machines. Feed rate, wheel speed, sparkout time, and wheel dress can be set and adjusted on the grinding machine. Production factors of grinding machines include surface finish, surface integrity, cycle time, problems of burning the workpiece, and cost per piece. Furthermore, a large variety of grinding wheels is available, from soft to hard and from open to dense bond, and various grit sizes are used. Choosing the correct wheel for the job requires extensive study of abrasive wheels and their uses.

Grinding Machines

Grinding machines range from shop pedestal grinders for sharpening small tools to large specialty machines such as

FIGURE 14.96
Hobbing a large helical gear for a hot strip mill (Illinois Gear, a Household International Company).

FIGURE 14.97
Cutting large heavy-duty racks for steel mill equipment (Illinois Gear, a Household International Company).

FIGURE 14.98
Herringbone gears are made on special machines (Illinois Gear, a Household International Company).

FIGURE 14.100
Illustration showing how the gear shaper employs the "molding-generating" process for cutting gears (Courtesy Fellows Corporation).

crankshaft grinders. Vertical spindle machines are useful for grinding surfaces on many small parts at one time. Horizontal spindle machines such as small surface grinders (Figure 14.111, see page 311) are useful to produce flat surfaces and other forms. Workpieces are usually held on a magnetic base or table if they are ferromagnetic. The table passes under the rotating abrasive wheel in a reciprocating motion while the table is fed crosswise. Cylindrical grinders (Figure 14.112, see page 311) are designed to grind finish surfaces on rotating parts that are cylindrical in shape. Many

of these machines are equipped to do internal grinding of parts that are held in a chuck. Special production machines are available for internal grinding in large volumes.

Production Grinding Processes

Three methods of grinding diameters on cylindrical parts are used in manufacture: infeed grinding, plunge feed grinding, and centerless grinding. *Infeed grinding,* in

FIGURE 14.99
Large bevel gears are also produced on special machines (Illinois Gear, a Household International Company).

FIGURE 14.101
CNC gear shaper (Courtesy Fellows Corporation).

1 Spindle positioning
2 Infeed
3 Rotary feed
4 Stroking speed

FIGURE 14.102
Schematic of CNC gear shaper (Courtesy Fellows Corporation).

FIGURE 14.104
External gear shaping. Here a crowned gear (higher in the center) is being generated on a gear shaper (Courtesy Fellows Corporation).

which the grinding wheel moves along the work axis (Figure 14.113, see page 311), is usually used where long cylinders or shafts must be ground to a uniform diameter. This method is most often used in small machine shops.

Plunge grinding (Figures 14.114 and 14.115, see page 312) is most often used in production grinding where large numbers of parts are manufactured to exacting specifications and where short lengths are involved. In high-volume production grinding, automatic workpiece loading on the fly (without stopping shaft rotation) greatly reduces cycle

FIGURE 14.103
Internal gear shaping (Courtesy Fellows Corporation).

FIGURE 14.105
Quite large external gears may be produced by the gear-shaping process (Courtesy Fellows Corporation).

FIGURE 14.106
A large internal ring gear being generated by the gear-shaping process (Courtesy Fellows Corporation).

FIGURE 14.107
Spiral bevel and hypoid gear generator (Illinois Gear, a Household International Company).

FIGURE 14.108
Close-up of a hypoid gear being cut from a solid blank on a gear generator (The Gleason Machine Division).

FIGURE 14.109
Lapping of intermediate size spiral bevel and hypoid gears (Illinois Gear, a Household International Company).

time. Wheels are dressed with diamond nibs or wheels at predetermined intervals under computer control, and automatic compensation is used to correct for wheel imbalance. In-process gaging (Figure 14.116, see page 312) allows workpiece diameters to be checked as the grinding process proceeds. With NC programmable machine controls (Figure 14.117, see page 312) dimensions can be controlled and measured to .00004 in. or .001 mm.

Cylindrical grinders with swiveling grinding heads are called angle grinders (Figure 14.118, see page 313).

FIGURE 14.110
Flame hardening gear teeth with pyrometer-controlled machine (Illinois Gear, a Household International Company).

FIGURE 14.111
Surface grinders are used to produce flat surfaces (Clausing Machine Tools).

FIGURE 14.112
Cylindrical grinder.

FIGURE 14.113
Infeed grinding in which the workpiece traverses across the wheel to a preset stop (Rodco Precision, Honing & EDM).

On these machines both diameters and shoulders can be finished in one plunge grinding operation. Grooves in wheels or rings can be plunge ground by fastening them on an arbor (Figure 14.119, see page 313). Many dimensions on one shaft can be finish ground in one operation with multiwheel production grinders (Figures 14.120 and 14.121, see page 313).

Centerless grinding is done by means of a large grinding wheel, a regulating wheel (Figures 14.122 and 14.123, see page 314), and a work rest to support the rotating workpiece. Centerless grinding is a high-production method of obtaining precision rod, shafting, and small parts. Commercial ground and polished (G and P) shafting is produced on centerless grinders. Centerless internal grinding is accomplished with three rolls supporting and rotating the workpiece while the inside surface is ground (Figure 14.124, see page 314).

FIGURE 14.114
Straight plunge grinding on a shaft. In this operation the wheel is fed straight in (Photo Courtesy Landis Tool).

HONING, LAPPING, AND SUPERFINISHING

When smooth internal finishes are required, such as in tubes for industrial hydraulic cylinders, a preferred method of finishing is *honing* (Figure 14.125, see page 314). The honing head follows the existing hole or work surface and rotates while it oscillates in the bore. If the hole needs to be realigned, a boring operation must first be performed before honing. Multiple spindle honing machines are used in manufacturing. An example is honing engine blocks in which all cylinder bores in a block are honed at one time.

Lapping is an abrasive finishing process in which fine abrasives are charged (embedded) into a soft material called a lap. Lapping can be done by hand (Figure 14.126, see page 314) on a cast iron surface or it can be done on a special machine (Figures 14.127 to 14.129, see page 315).

FIGURE 14.115
Plunge grinding on a crankshaft (Photo Courtesy Landis Tool).

FIGURE 14.116
In-process gaging. Dimensions can be checked while the machine is in operation with this setup (Photo Courtesy Landis Tool).

Superfinishing (Figure 14.130, see page 315) is similar to honing. It can be done on a flat or cylindrical surface. In superfinishing, a lubricant is used to maintain separation of a honing stone and the workpiece. Tiny projections causing roughness on the workpiece are cut off by the honing stone because they extend through the lubricant. By maintaining a controlled pressure, a desired degree of smoothness or finish is obtained and the controlled-viscosity lubricant prevents further cutting action.

FIGURE 14.117
Small crankshaft grinder. Grinding process can be programmed on this machine for a precision automatic operation (Photo Courtesy Landis Tool).

FIGURE 14.118
Angular plunge grinding on a crankshaft. Two diameters, fillets, faces, radii, and a tapered diameter are made in one plunge grind (Photo Courtesy Landis Tool).

FIGURE 14.119
Grooves can be made by the plunge grinding process. Grinding the diameter of a large bearing race mounted on an arbor (Photo Courtesy Landis Tool).

FIGURE 14.121
Programmable machine control for the multispindle grinder allows the operator to control wheel feed, in-process gaging, wheel balance, and monitoring dimensions (Photo Courtesy Landis Tool).

FIGURE 14.120
Multiwheel grinding machine increases productivity (Photo Courtesy Landis Tool).

FIGURE 14.122
Centerless grinding. Through feed grinding of bearing rings (Lidkopings Mekaniska Verkstads AB, Sweden).

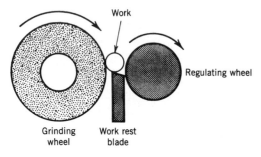

FIGURE 14.123
Principle of centerless grinding.

FIGURE 14.124
Centerless internal grinding.

FIGURE 14.125
Honing machine (Rodco Precision, Honing & EDM).

FIGURE 14.126
Lapping by hand. Grooves hold the abrasive. A circular motion is used to ensure flatness.

FIGURE 14.127
Lapping machine (Courtesy of DoALL Company).

FIGURE 14.129
These typical lapping workholders are custom designed for each individual flat or cylindrical workpiece application (Courtesy of DoALL Company).

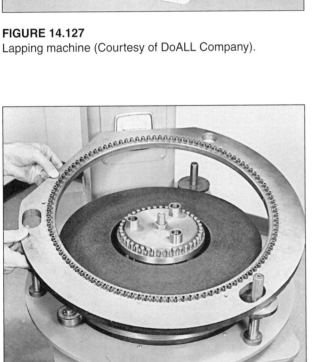

FIGURE 14.128
The planetary lapping attachment provides for complete coverage of laps during operation (Courtesy of DoALL Company).

Abrasive honing stick

Workpiece

FIGURE 14.130
Superfinishing.

Review Questions

1. Saw blades (bands) for vertical and horizontal band saws are made in long strips that are coiled for packaging and shipment. How are these strips made into endless bands?

2. What is the greatest advantage of vertical band sawing over other machining processes?

3. Two drilled holes are required on a part, one .010 in. diameter and one .100 in. diameter. Which drilling machine or machines should you use to drill these holes?

4. Using the following formula, calculate how fast the drill press should rotate a ½-in. diameter drill if the cutting speed is 90 sfm:

$$N = \frac{3.8V}{D}$$

5. A massive steel casting weighing 3 tons requires that several 3-in. diameter holes be drilled into it. Which type of drilling machine would be best suited for this job?

6. An NC lathe produces accurate shafts having several diameters without the need for an operator to be in constant attendance. Would you call this an engine lathe or a production machine? Explain.

7. Threads are often designated omitting the tolerance symbols. Explain the meaning of the following: ⅜ – 16 UNC, and M16 × 2.

8. Internal and external threads can be made with hand taps, dies, or on an engine lathe with a single-point tool. How are threads cut at rapid production rates on turning machines?

9. How can a round or hexagonal bar be securely and accurately gripped for turning operations in a lathe? How can a square or rectangular piece be gripped for turning?

10. On which type of turning machines are air-operated chucks commonly used?

11. What principal advantage does a turret lathe have over an engine lathe?

12. A blind keyseat is a short groove that begins and ends anywhere along a shaft but not on the very ends. Which machine, a vertical or a horizontal spindle milling machine, would be most appropriate for cutting a blind keyseat?

13. Name two methods by which internal splines can be cut.

14. Hobbing is used almost exclusively for what machining operation?

15. How can hardened workpieces that cannot be machined with ordinary cutting tools be finish machined to a precise dimension with a very smooth finish?

16. How are steel workpieces usually clamped to the reciprocating table on a small surface grinder?

17. What is the difference between plunge grinding and plain feed grinding on cylindrical grinders?

18. A precision part must be held to a tolerance of ± .0001 (± .0025 mm). Would you attempt to finish such a part on a turning machine? Explain your answer.

19. Fifty thousand hardened capscrews are made a few thousandths of an inch oversize and must be ground to a diameter of ± .0001 in. on a cylindrical surface. Select which grinding process and machine should be used to get the job done most economically: (a) center-type cylindrical grinder, (b) multiwheel grinder, or (c) centerless grinder.

20. Where is honing most likely to be performed, on flat surfaces, outside surfaces of cylinders, or in bores? What is the principal purpose of honing?

Case Problems

Case 1: The Manufacture of a Transmission Gear and Shaft

A manufacturer is asked to design and build a small tractor with a four-speed manual transmission. You are asked to decide upon the means of manufacture, on a cost effective production basis, of a transmission shaft (Figure 14.131). One end of the shaft has a spline and the other end has a spur gear. The gear end is hollow and has a blind end internal spline. The gear, the internal and external splines, and the two bearing surfaces must be hardened. The gear must also be a precision one with close tolerances. The

FIGURE 14.131
Transmission shaft. See Case 1.

blank will be forged from a high chromium carbon steel that can be hardened. What processes would you use to produce and finish this geared transmission shaft?

Case 2: The Robotic Wrist Component

A prototype (one only) of a robotic wrist component similar to the one shown in Figure 14.132 is required before it is put into production. It must be tested and perhaps revised and retested before it is ready to be manufactured in quantity. The prototype is mostly made of aluminum alloy. Where and how should it be made? Should the parts be:

a. Cast in a foundry?

b. Machined from aluminum alloy bar stock on production machines?

c. Machined from aluminum alloy bar stock by broaching the holes and surfaces?

d. Machined from aluminum alloy bar stock in a tool room (machine shop) one piece at a time?

Explain why you chose your answer.

Case 3: RPM for Drilling Operations

A production drilling machine needs to be adjusted to the correct rpm for several drill sizes, ¼-, ½-, and 1½-inch diameter. The drilling operations are all in steel, which requires a cutting speed of 90 sfm. Using the formula

$$N = \frac{3.8V}{D}$$

where

      N is spindle speed in rpm

      V is cutting speed in sfm (feet per minute)

      D is diameter of the drill in inches

list the correct rpm settings for each drill size.

FIGURE 14.132
Robotic wrist component. See Case 2 (Bird-Johnson Company).

CHAPTER 15

Nontraditional Manufacturing Processes

Objectives

This chapter will enable you to:

1. List common nontraditional machining processes.
2. Describe in general terms how these processes work.
3. Describe applications of the processes.

The previous chapters described machining processes in which workpiece material is removed in the form of chips, by direct contact of a cutting tool. These processes are often called traditional machining processes and account for the majority of materials processing.

The design and manufacturing engineer, in the continuing search for improvements in designs, both stimulates development of alternate materials processing methods and takes advantage of new processes that evolve naturally throughout manufacturing industries.

Machining processes making use of electricity, laser light, electrochemical methods, and extremely high temperature gas plasmas have developed in more recent years. These processes have become well established and because of their unique capabilities have opened many new avenues of design to the engineer. The processes described in this chapter are often referred to as nontraditional machining processes, only because they do not involve the chip-shaving process of conventional machining. They are nonetheless well established as standard manufacturing processes.

Nontraditional machining processes include the following:

1. Electrodischarge machining (EDM)
2. Electrochemical machining (ECM) and electrochemical deburring (ECDB)
3. Electrolytic grinding (ELG)
4. Laser machining
5. Ultrasonic machining
6. Water jet machining
7. Electron beam machining (EBM)
8. Plasma beam machining

ELECTRODISCHARGE MACHINING

Electrodischarge machining, commonly known as **EDM,** removes workpiece material by an electrical spark erosion process. This process is accomplished by establishing a large potential (voltage) difference between the workpiece to be machined and an electrode. A large burst (spark) of electrons travels from the electrode to the workpiece. When the electrons impinge on the workpiece, some of the workpiece material is eroded away. Thus the machining is accomplished by electrical spark erosion.

The EDM process (Figure 15.1) takes place in a dielectric (nonconducting) oil bath. The dielectric bath concentrates the spark and also flushes away the spark-eroded workpiece material. A typical EDM system consists of a power supply, dielectric reservoir, electrode, and

FIGURE 15.1
Basic EDM system components.

FIGURE 15.2
CNC wirecut EDM machining center (Japax Inc.).

workpiece. The EDM machine tool has many of the same features as its conventional counterpart. These include worktable positioning mechanisms and measurement devices. Many EDM machine tools (Figure 15.2) are equipped with computer numerical control systems. Thus the versatility of CNC for workpiece positioning and tool control functions can be used effectively in the process.

EDM Electrodes

In the EDM process, the shape of the electrode controls the shape of the machined feature on the workpiece. EDM electrodes may be made from metal or carbon (graphite) and shaped to the desired geometry by molding or machining. Erosion of the electrode takes place as well during the EDM process. In time the electrode will become unusable. To circumvent this problem, a roughing electrode may be used to generally shape the workpiece and then a finish electrode may be applied to complete the process and establish final dimensions and geometry.

Wirecut EDM

Wirecut EDM uses a slender wire as an electrode and is extremely useful for making narrow slots and detailed internal features in the workpiece (Figure 15.3) and for detailed tool and die making. Wirecut EDM is similar in some ways to band sawing in conventional machining. Whereas the saw teeth do the cutting in band sawing, in wirecut EDM a spark occurs between the wire electrode and the workpiece. Wire feed is continuous during the process and the wire may not be reused.

Different metals are used for wire electrodes. Brass is one popular material. EDM wire can be of very small diameter, on the order of .005 in. Sapphire guides maintain the alignment of the wire as it passes through the work. Wire tension is maintained by weighted pulleys or other mechanical tension schemes.

Wirecut EDM using such small diameter wires permits extremely narrow slots to be machined in the workpiece. The kerf is only slightly wider than the wire diameter. This process is highly suited to cutting intricate slot features in tools and dies.

Advantages and Applications of EDM

EDM processes can accomplish machining that would be impossible by the traditional methods. Any odd electrode

FIGURE 15.3
Blanking die for watch parts made with wirecut EDM (Japax Inc.).

FIGURE 15.4
Mold machined with ram-type
EDM machine (Japax Inc.).

shape will be reproduced in the workpiece. Fine details can be made, such as those needed in tool, die, and mold work (Figures 15.4 and 15.5). When EDM is coupled to numerical control, an excellent machining system for the tool and die maker is available.

Since machining is accomplished by spark erosion, the electrode does not actually touch the workpiece and is therefore not dulled by the hard workpiece materials that would dull conventional cutters. Thus, the EDM process can spark erode very hard metals, and this process has found wide application in removing broken taps without damaging the workpiece.

Disadvantages of EDM

The EDM process is much slower, from a metal removal standpoint, as compared to conventional machining processes involving direct cutter contact. There is also a possibility of overcutting and local area heat treating of the parts being machined.

However, EDM can accomplish many machining tasks that could never be done by conventional machin-

FIGURE 15.5
Intricate internal geometry machined with wirecut EDM (Japax Inc.).

ing. Thus, the EDM process has become well established as a very versatile manufacturing process in modern industrial applications.

ELECTROCHEMICAL MACHINING (ECM) AND ELECTROCHEMICAL DEBURRING (ECDB)

Electrochemical machining (ECM) is essentially a reverse metal-plating process (Figure 15.6). The process takes place in a conducting fluid or electrolyte that is pumped under pressure between electrode and workpiece. As workpiece material is deplated, it is flushed away by the flow of electrolyte. Workpiece material is removed from the electrolyte by a filtration system.

Advantages and Applications of ECM

As in EDM, ECM can accomplish machining of intricate shapes in hard-to-machine material. The process is also burr free, and does not subject the workpiece to distortion and stress as do conventional machining processes. This makes it useful for work on thin or fragile workpieces. ECM is also used for part deburring in the process called ECDB or electrochemical deburring. This process is very useful for deburring internal workpiece features that are inaccessible to traditional mechanical deburring processing.

ELECTROLYTIC GRINDING (ELG)

In the process of **electrolytic grinding (ELG)** an abrasive wheel much like a standard grinding wheel is used. The abrasive wheel bond is metal, thus making it a conducting medium. The abrasive grains in the grinding

FIGURE 15.6
Basic ECM system components.

wheel are nonconducting and aid in removing oxides from the workpiece while helping to maintain the gap between wheel and work. ELG, like ECM, is a deplating process and workpiece material is carried away by the circulating electrolyte.

The ELG System

The basic ELG system (Figure 15.7) consists of the appropriate power supply, the electrode (metal-bonded grinding wheel), workholding equipment, and the electrolyte supply and filtration system. Workpiece material is deplated and goes into the electrolyte solution.

Advantages and Applications

Since this process is primarily electrochemical and not mechanical as is conventional grinding, the abrasive

wheel in ELG wears little in the process. ELG is burr free and will not distort or overheat the workpiece. The process is therefore very useful for small precision parts and thin or fragile workpieces.

LASERS AND LASER MACHINING

Lasers

Laser is an acronym for *l*ight *a*mplification by *s*timulated *e*mission of *r*adiation. By electrically stimulating the atoms of certain materials, such as various crystals and certain gasses, the electrons of these atoms can be temporarily displaced to higher electron shell energy level positions within the atomic structure. When the electrons fall back to their originally stable levels, photons of light energy are released (Figure 15.8). This light

FIGURE 15.7
Basic ELG grinding system.

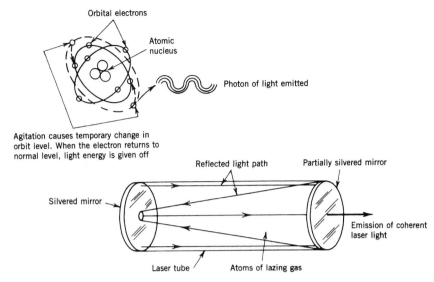

FIGURE 15.8
Photon emissions and laser generation.

energy can be enhanced and focused into a coherent beam and then used in many manufacturing, medical, measurement, and other useful applications. The applications of lasers in manufacturing are widespread. Laser energy finds many applications other than use as a cutting tool.

Laser Machining

In laser machining, the coherent laser light becomes the cutting tool. When the laser is integrated with CNC machine tool positioning and CRT process data display (Figures 15.9 and 15.10), an extremely versatile system with many capabilities is available.

Applications of laser machining include cutting plate (Figure 15.11) for shapes such as saw blades (Figure 15.12), slotting stainless tubing (Figure 15.13), and marking systems (Figure 15.14 on p. 325). Lasers are also effectively applied in welding processes. These are discussed in Chapter 16.

ULTRASONIC MACHINING

Ultrasonic machining is similar to an abrasive process like sand blasting. High-frequency sound (Figure 15.15 on p. 325) is used as the motive force with which to propel abrasive particles against the workpiece. Advantages of this process include the ability to machine very hard mate-

FIGURE 15.9
CNC laser machining center (Photo courtesy of Coherent, Inc., Palo Alto, California).

FIGURE 15.10
CRT on CNC laser machining center showing process
status during production (Laserdyne).

FIGURE 15.11
Laser plate cutting (Laserdyne).

FIGURE 15.12
CNC laser cutting of circular
saw blades (Coherent).

rial with little distortion. Good surface finishes may be
obtained and part features of many different shapes can be
machined.

WATER JET MACHINING

The cutting tool in the **water jet** machining process is an
extremely high pressure water jet. Pressures of several
thousand pounds per square inch are used to pump water
through a small nozzle, thus directing a narrow stream at
the workpiece. This process is extremely effective in cutting fabric, foam rubber, and similar low-strength materials. The efficiency of the process is enhanced by adding
an abrasive material to the water. Thus both pressure and
mechanical abrasion join forces to accomplish cutting and
machining tasks. The process may be used for hard metallic materials. However, there are limitations as to the
thickness of the materials being machined.

ELECTRON BEAM MACHINING

Electron beam machining (EBM) is in some ways
related to EDM and to electron beam welding. In EDM,

FIGURE 15.13
Stainless steel tube slotting (Laserdyne).

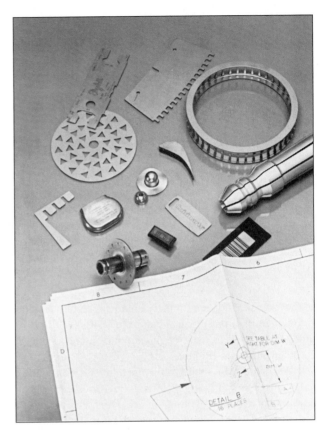

FIGURE 15.14
Laser marking and other laser machined products (Laser Fare).

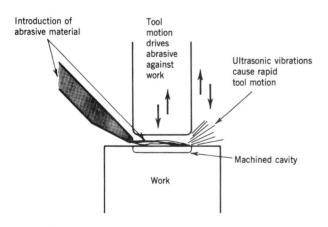

FIGURE 15.15
Ultrasonic machining.

however, a burst of electrons (spark) impinges on the workpiece, whereas electrons in a continuous beam are used in the EBM process. The workpiece material is heated and vaporized by the intense electron beam. This process, like electron beam welding, must occur in a vacuum chamber, and appropriate shielding must be employed to protect personnel from X-ray radiation.

PLASMA TECHNOLOGY

A great deal of interest in and development of **plasma** technology has occurred in recent years. Plasma is created by passing a gas through an electric arc. The gas is ion-

Cool gas flow

Electrode

Orifice

High-temperature Plasma gas
directed on workpiece is concentrated
into a narrow beam

FIGURE 15.16
Plasma arc.

ized by the arc and an extremely high temperature results. Temperatures in plasma arcs can exceed 40,000°F, many times hotter than any arc or flame temperature produced by other processes (Figure 15.16).

Plasma arcs of such high temperatures have a multitude of uses, not only in machining, but for other applications such as ore smelting and waste metal recovery operations. There is ongoing research on the applications of high-temperature plasmas for incinerating industrial wastes.

The process is useful for cutting, welding, and machining nonferrous metals and stainless steel. Plasma torch machining is extremely fast and will produce smooth surface finishes to close dimensions.

Review Questions

1. Describe in general terms how the following processes work: EDM, ECM, ELG, laser machining, ultrasonic machining, hydrojet machining, and plasma technology.

2. Which processes might be used for deburring applications?

3. Which process would be used for detail die slotting?

4. What temperatures can be reached in a plasma system?

5. Which process is essentially deplating?

6. What is the primary function of the grinding wheel in ELG?

7. What provides the energy in ultrasonic machining and what does the cutting?

8. Which process is accomplished in a dielectric fluid?

9. Which process uses a conducting fluid?

10. Describe in general terms how a laser functions.

Case Problems

Case 1: Punch and Die Tooling

A punch and die set is required as shown in Figure 15.17. Which process discussed in this chapter would be most applicable to the manufacture of this tooling?

Case 2: Internal Deburring

A complex part with several intersecting drilled holes must be as burr free as possible. Since many of the internal features are inaccessible for mechanical deburring, which process discussed in this chapter might be used to accomplish the task?

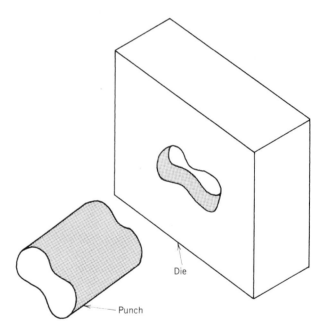

Die

Punch

FIGURE 15.17
Punch and die set. See Case 1.

Joining Processes

Objectives

This chapter will enable you to:

1. Describe common methods of joining materials and cite their advantages and disadvantages.
2. Describe common welding processes and their applications.

Manufacturing processes commonly require some type of assembly involving several different pieces made of the same or different materials. In manufacturing, it is almost always necessary to join the materials at some point in the process. Plastics and wood material are glued, fabrics are stitched, and metal components are welded, bolted, or riveted. The purpose of this chapter is to survey the major methods of joining materials in manufacturing. In general, joining processes can be divided into the following categories:

1. Mechanical fasteners
2. Adhesive bonding
3. Welding
4. Brazing and soldering

MECHANICAL FASTENERS

Many separate mechanical fastening systems and components join materials. These include threaded fasteners, nails, staples, rivets, stitching, tying, snaps, pins, retaining rings, pressing, crimping, and specialty fastening systems.

Threaded Fasteners

Threaded fasteners include screws, bolts, and nuts. These fasteners are very common in all mechanical assemblies. All threaded fasteners make use of the inclined-plane principle from mechanics to put pressure on the parts that are to be assembled. Threaded fasteners will withstand large loads, are available in a wide variety of sizes and materials, and can be designed to meet many specialty fastening applications.

Bolts and nuts are found in almost every mechanical assembly. The bolt is designed to work in conjunction with a nut, with the material to be joined compressed between the nut and bolt head (Figure 16.1). Bolts with threads all the way to the head, called cap bolts or cap screws, are designed to screw into tapped holes, capturing the parts to be joined between bolt head and base material.

Machine screws are similar to bolts, but are usually smaller (Figure 16.2). They have many of the same fastening applications as bolts and nuts. Sheet metal screws are made with coarse threads and are designed for joining thin sheet materials. All threaded fasteners have features that permit them to be secured with various tools such as wrenches and screwdrivers.

The two major advantages of using threaded fasteners for joining are that they can be disassembled and they come in a wide variety of types and sizes. These advan-

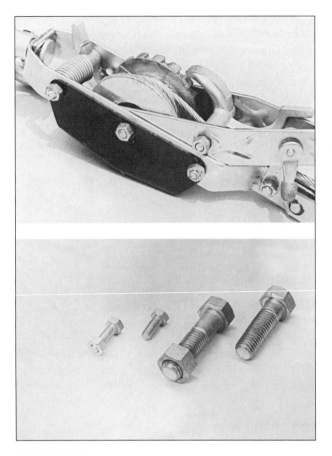

FIGURE 16.1
Common bolt and nut assembly.

tages make the system extremely versatile and likely to remain in wide use.

Disadvantages include labor intensive assembly, requirements to stock a large number of types and sizes, and the possible loss of the hardware during assembly, disassembly, or service. Easily damaged threads on fasteners can make assembly or disassembly difficult or impossible. Any threaded fastener will require a predrilled or punched hole in the workpiece before the materials can be assembled or joined.

Nails and Staples

The foremost application of fastening by nails is in wood building construction. Threaded fasteners may also be used in some wood construction applications. However, the additional time required to drill holes for bolts and to make up assemblies usually limits the use of threaded fasteners to large wood beam constructions that are designed to support large loads.

A typical nail consists of a stiff wire with a head on one end for driving a point on the other end to penetrate the material to be joined. Nails come in a variety of sizes, styles, and materials including steel, brass, and aluminum (Figure 16.3). As the nail is driven into wood it pushes aside the grain of the material. The frictional forces exerted by the wood recompressing holds the fastener in place. Nailing at an angle to the forces trying to separate the material can also add some additional mechanical holding capacity. More holding capacity can be obtained by using fast helix nails. These have some of the characteristics of threaded fasteners and are used in crate and pallet manufacturing.

Nails are extremely well suited for static assemblies such as houses. They do require considerable energy for driving, which is often done by hand. However, nailing by

FIGURE 16.2
Machine screw assembly.

FIGURE 16.3
Common types of nails.

machine and by explosive charges can expedite this joining process.

Nailing is not suitable for dynamic loads in an assembly. The fastener will pull out if the structure is sufficiently distorted. A material assembled with nails must also be compressive by nature so that the fastener will not fall out.

Stapling is related to nailing in that a staple is basically a two-pointed nail. Stapling is a versatile mechanical fastening system, highly adaptable to high production of furniture, paper boxes, and installation of many materials such as insulation and ceiling tiles. Stapling is best used where loads on the structure are moderate. It is not suitable for heavy loads because of the usually short penetration of the fastener into the material being joined. Because stapling can be most unsatisfactory in this respect, this system should be selected only after a careful engineering evaluation of the loads the assembly will encounter.

Threaded fasteners, including sheet metal screws, wood screws, and many types of specialty hollow wall fasteners, are available to the building industry (Figure 16.4).

Rivets

Rivets are a very common and extremely versatile mechanical joining system. They are used to join materials ranging from heavy steel structurals to light gage sheet metal, plastics, and composites. Rivets are widely used in the aircraft industry, where it is desirable to have as little projection of the fastener head as possible above the surface of the materials being joined. This greatly reduces air friction on skin and cowl surfaces subjected to airflow during flight (Figure 16.5). However, since supersonic aircraft require almost totally smooth skins, riveting is giving way to adhesive bonding in this application.

Riveted structures are used extensively in internal aircraft structure (Figure 16.6, see page 330). Extremely strong but very light ribbed structures can be fabricated by riveting (Figures 16.7, see page 330 and 16.8, see page 331). Machine riveting is widely used in aircraft production. Since rivets are retained on the back side as are many threaded fasteners, they are well suited to dynamic loads and will not come loose through distortion of the assembly.

Much of the riveting is accomplished with power machinery. The equipment used may be portable or may be bench type. The equipment illustrated in Figure 16.9 (see page 331) swages rivets using a kneading action from the center of the rivet outward. Multipoint riveting machines can swage several rivets at one time (Figure 16.10, see page 332). Many riveting tools and rivet styles are available to meet a broad range of riveting applications (Figure 16.11, see page 333).

Advantages of riveting are strength with a minimum-sized fastener and the ability to join material by insertion and attachment of the fastener from only one side of the

FIGURE 16.4
Wood and sheet metal screws and special hollow wall fasteners.

FIGURE 16.5
Riveting flush with surface is used extensively in aircraft cowl and skin applications in order to reduce airflow turbulence (Lockheed Co.).

FIGURE 16.6
Riveting internal aircraft structures makes them very strong but also very light.

assembly. An example of this is the popular pop rivet used in many common fastening applications. Disadvantages include labor intensive installations, requirements for predrilled or punched holes in the material, and difficulty in disassembly. Most rivets must be drilled out in order to take an assembly apart. Reassembly requires replacement of the fastener with new hardware.

Stitching, Tying, and Snaps

Stitching and tying are found primarily in the fabric industry. Machine lock stitch is an essential process in the making of clothing. The modern sewing machine has perfected this method of material joining. Other applications include sack sewing with an easily unraveling chain stitch. Tying with rope, string, or cord has many temporary applications. Mechanical snaps are frequently used for temporary joining of fabrics and canvas.

Pins and Retaining Rings, Pressing, and Crimping

Assembly of mechanical components often makes use of pins and retaining rings. These fasteners make disassembly possible, but precision machining of the parts is often required. Retaining rings may be used to hold mechanical parts together (Figure 16.12, see page 334). Pins are often found where parts must be joined and disassembled frequently (Figure 16.13, see page 334).

Many parts in mechanical assemblies can be joined by pressing (press fitting). In this method, the mating part dimensions overlap slightly and the parts are forced together under mechanical pressure. The method has many

FIGURE 16.7
Strong but light-weight internal riveted rib structures in aircraft (Lockheed Co.).

FIGURE 16.8
Riveting major aircraft components (Lockheed Co.).

FIGURE 16.9
Bench rivet machine forms rivets with a center-outward (radial) action causing little distortion of the parts being joined (Bracker Corporation/Leland J. Balber Associates).

applications, especially in assemblies where bearings and bushings are to be retained without the need for additional mechanical hardware (Figure 16.14, see page 334).

Also related to mechanical pressing are shrink and expansion fitting, where parts in an assembly are joined by heating or shrinking one or the other, fitting the parts, and then returning them to normal temperature. This process makes use of the normal expansion and contraction of material as temperature is varied.

Crimping is another method of material joining, where mechanical force is applied to compress or expand one piece of material against another piece to join the parts. Generally, crimping requires a groove or other feature to provide a place where the material of the mating part can grip (Figure 16.15, see page 335).

Seaming is related to crimping. This joining process is widely used in the sheet metal industry for closing duct work and other sheet metal fabrications. Common sheet metal seams include those illustrated in Figure 16.16 (see page 335). The industry makes use of a number of hand- and power-operated machinery to roll, form, and crimp the various types of joint seams.

Specialty Fastening Systems

Specialty fastening systems have found wide applications. One example, plastic barbs, consists of a pad of plastic barbs designed to grip into mating materials (Figure 16.17, see page 336). This fastening system has many uses where temporary joining of materials desirable.

FIGURE 16.10
Multipoint radial riveting machine for die cast metals and plastics (Bracker Corporation/Leland J. Balber Associates).

Plastic barbs are not suitable for joining materials that will be subjected to heavy pull-apart loads. However, this system is extremely useful for temporary joining of fabrics and other light-duty holding applications.

ADHESIVE BONDING

It is naturally desirable to join material without the need of labor intensive preparation such as hole drilling or punching and the insertion of mechanical hardware. Processes involving adhesives have developed rapidly in recent years and are now widely used. Research and development into adhesive joining of metals plastics, and composite materials is under continual development. Adhesives are divided into two major groups: (1) natural adhesives, and (2) plastic resin adhesives. Advantages gained by adhesive bonding include:

1. Dissimilar materials can be joined that cannot be joined by other processes.
2. Mechanical preparation such as hole drilling is eliminated.
3. Adhesive bonding joins materials with no gaps as there would be between bolts or rivets in a mechanical assembly. Overall strength of the joint can therefore be enhanced.
4. Many adhesives are not subject to deterioration from service environments and therefore can be used in a wide range of temperature and corrosive environments.
5. Joined materials are not subjected to mechanical or heat distortion as they would be if joined with mechanical or welding processes. Hence, precision parts can be joined without risk of damage or distortion.

Disadvantages of adhesive bonding include:

1. Specific joint preparation may be required.
2. Adhesive material may be expensive and require application in controlled environments.
3. Curing time may be long and may require a special environment.
4. Mechanical strength of the adhesive-bonded joint may depend greatly on how the joint is loaded. Shear loading is best followed by tension. Where loading of the joint tends to peel materials apart, adhesive bonding may prove unsatisfactory (Figure 16.18, see page 336).

Natural Adhesives

Natural adhesives include: (1) vegetable and animal glues and casein, (2) sodium silicate, and (3) natural gums.

Vegetable glues are used in the paper industry and have applications in woodworking. Examples include wallpaper paste made from starch and the gum arabic used on envelopes and postage stamps. Animal glues made from bone and hide are adequate for some applications, but they tend to deteriorate in moist environments. Casein, made from milk, is a water-based adhesive.

Sodium silicate shows good temperature characteristics and it can be used as a binder to make some types of composite material.

Natural gums have long been used as adhesive materials. These include natural rubbers, sealing waxes, and asphalt.

Plastic Resin Adhesives

Synthetic resins (plastics) used for adhesives have undergone much development in recent years and have found an important place in material joining processes. In many

Tool styles

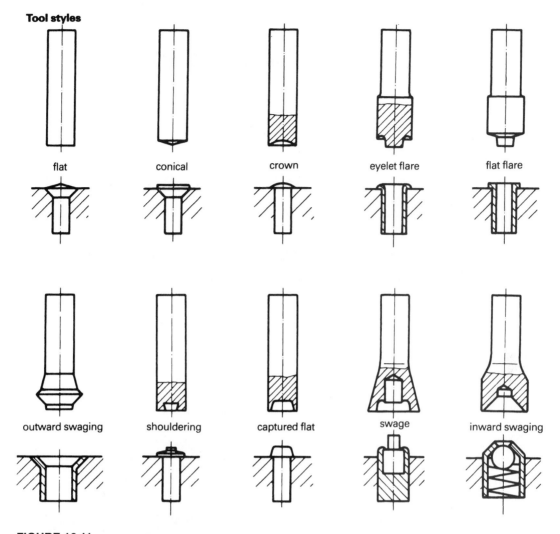

flat	conical	crown	eyelet flare	flat flare

outward swaging	shouldering	captured flat	swage	inward swaging

FIGURE 16.11
Rivet types and applications (Bracker Corporation/Leland J. Balber Associates).

cases plastic resin adhesives have proved to be superior to mechanical joining systems.

Resins used for adhesives come from both major plastic classifications: thermosetting and thermoplastics. The thermosetting plastics undergo fundamental chemical change while hardening and cannot be reused or recycled. Thermoplastics, on the other hand, can be heated, resoftened, and reused.

Thermoplastic resins used for adhesives include cellulose nitrate, acetate, acrylic, cyanoacrylate, and vinyl. Thermosetting resins used for adhesives include polyesters, epoxies, phenolics, ureas, and silicones.

Adhesive Bonding in Aircraft Production

Adhesive bonding in aircraft assembly has undergone much development in recent years. The bonding of aircraft skin panels has numerous advantages over joining them with screws or rivets, such as:

1. Bonding resists corrosion (Figure 16.19, see page 337).
2. Adhesive bonds are as strong as or stronger than riveted joints.
3. Joint loading is more uniform with adhesive bonding (Figure 16.20, see page 337).
4. Joint fatigue and crack development are reduced using bonded joints.
5. Bonded joints are more effective in sound and vibration dampening in aircraft structures.
6. Bonded joints eliminate the need for many fasteners and thus may result in a substantial weight saving, essential to modern fuel-efficient aircraft.

The requirements of aircraft skin panel bonding are stringent. The adhesive used must have high resistance to

FIGURE 16.12
Mechanical assembly using
retaining rings.

External retaining ring   Internal retaining ring

peel apart and shear loads. The bond joint must be able to
resist penetration to corrosive fluids and must also be able
to withstand high temperatures in its service environment.

Another requirement of bonded joints is consistency
of the bond throughout the joint area. In addition, the
adhesive material must contain a corrosion inhibitor and
must not crack or crumble when drilled for attachment to
any mechanically fastened components. The most versatile adhesives that demonstrate these properties necessary
for aircraft bonding requirements come from the epoxy
group.

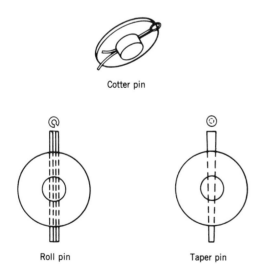

Cotter pin

Roll pin   Taper pin

FIGURE 16.13
Mechanical assembly using pins.

FIGURE 16.14
Mechanical assembly using press fitting.

FIGURE 16.15
Mechanical assembly with crimping.

Aircraft Bonding Technology

Skin panels are first sprayed with a protective coating. Then they are formed by pressing to the contour shape of the aircraft fuselage. The protective coating previously applied is not affected by the forming process. The formed skin panels are then cleaned (Figure 16.21, see page 338). Cleaning, which is essential to a proper bond, is done with an alkaline solution and water rinse. This is followed by an acid rinse used to slightly etch the surface of the skin panel. Acid etching provides an effective bonding surface. The panel then goes to a temperature- and humidity-controlled adhesive bonding room where the epoxy-based adhesive primer is applied. Air spraying is used, with the spray air having been carefully cleaned of all oil and moisture. A full and uniform coating, required for proper bonding, is applied at this stage in the process. Primed panels are then cured and stored in a clean room environment. Bonding occurs in a clean-room environment where the atmosphere is maintained at a positive pressure. The epoxy-based adhesive is applied in a layer .01 in. thick. Parts are joined and held in position by nylon tape (Figure 16.22, see page 338). The joined assemblies are then placed on bonding fixtures in plastic bags and a vacuum is drawn, creating a downward pressure on the parts. The next process phase, bonding and curing, is accomplished in a large-capacity

FIGURE 16.16
Joining sheet metal by various seams.

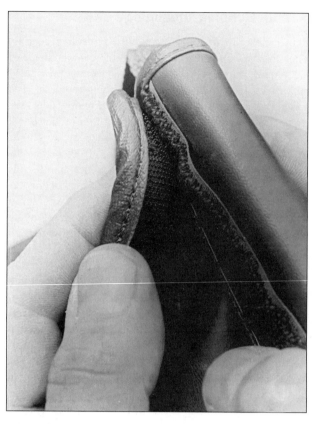

FIGURE 16.17
Plastic barbs, a unique material-joining method.

autoclave (Figures 16.23 and 16.24, see page 338). The autoclave is closed, heated, and purged with an inert gas under pressure. After a suitable period, pressure and temperature are reduced and the bonded assembly is removed from the autoclave.

After completion of the bonding process, assemblies are cleaned and prepared for inspection. Samples of the bonded assemblies undergo peel and shear testing. Bonded joints undergo ultrasonic inspections (Figure 16.25, see page 339) to determine joint integrity. The final phase is the painting and assembly of the skin panel on the aircraft (Figure 16.26, see page 339). Large aircraft skin panels are precision components and can be damaged unless handled carefully. It is interesting to note the methods by which this is accomplished. An array of vacuum cups is used to pick up and handle the skin panel. The method is very secure, and it will not damage the precision panel component (Figure 16.27, see page 339).

WELDING PROCESSES

It would be difficult to overstate the importance of welding in modern material joining processes. These processes make fabrication of complex metal assemblies possible. In welding, the materials to be joined are heated to the fusion temperature, at which point they are joined and allowed to cool. Filler material, generally of the same chemical composition as the parent metal, may be added to fill the weld joint during the process. Common indus-

FIGURE 16.18
Types of loads applied to adhesive bonded joints.

Peel-apart loads may be unsatisfactory for adhesive bonding

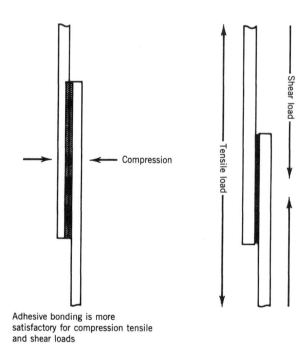

Adhesive bonding is more satisfactory for compression tensile and shear loads

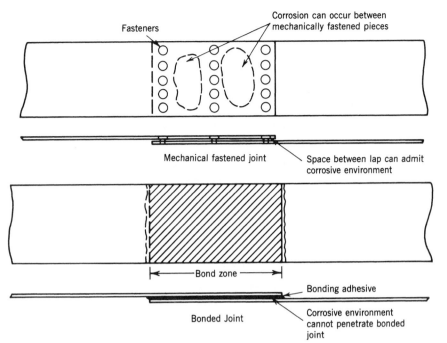

FIGURE 16.19
Bonded joints are resistant to corrosion.

trial welding processes include oxy-fuel welding, arc welding, resistance welding, solid state, and other welding processes.

Oxy-Fuel Welding and Cutting

The oxy-fuel welding process uses an oxygen-supported hydrocarbon (e.g., acetylene, methane, hydrogen, propane)

gas flame to heat the parts to welding temperature. The most commonly used gas is acetylene because of the resulting high flame temperature and heat content. The oxyacetylene process, often known as gas welding, is used mostly for repair welds on light to medium applications (Figure 16.28, see page 340), or where electric welding might not be available. Gas welding requires considerable technique skill on the part of the operator.

FIGURE 16.20.
Bonded joints distribute load better than mechanical joints.

FIGURE 16.21
Aircraft skin panels being cleaned prior to bonding
(Lockheed Co.).

In the gas welding process, gas and oxygen are separately delivered under regulated pressure to the welding torch. In the mixer within the torch the gas and oxygen are mixed. The fuel-oxygen mixture is ignited as it exits the torch nozzle and the operator adjusts the mix by valves on the torch assembly. The flame mixture may be adjusted to be neutral, oxidizing, or reducing. The neutral flame is a result of the complete combustion of acetylene, the oxi-

FIGURE 16.23
Bonded panels enter the large autoclave for curing
(Lockheed Co.).

dizing flame has an excess of oxygen, and the reducing or carburizing flame has an excess of fuel gas. The type of flame depends on the particular type of application. For example, carburizing flame can be used to inject carbon into the surface of the metal being heated, thus affecting the process of carburizing for surface heat treatment. However, the most commonly used type of flame is neutral.

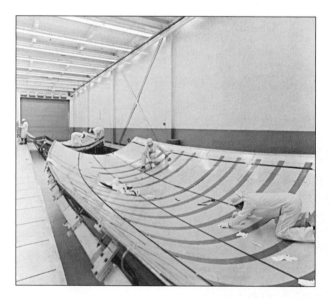

FIGURE 16.22
Panels are held in position with nylon tape prior to bonding
(Lockheed Co.).

FIGURE 16.24
Special transporters are used to handle the large panels
(Lockheed Co.).

FIGURE 16.25
Ultrasonic testing of panel bond joints (Lockheed Co.).

Oxyacetylene flame processes are also used extensively for cutting applications (Figure 16.29, see page 340). This process plays an important part in manufacturing. Oxyacetylene cutting torches can cut through metal several inches thick, making them indispensable for manufacturing of structural fabrications such as bridges, ships, and oil field and farming equipment.

FIGURE 16.26
Suction cups hold the panel while it is positioned for installation (Lockheed Co.).

FIGURE 16.27
Bonding technology requires special equipment handling as well (Lockheed Co.).

FIGURE 16.28
Oxyacetylene welding and cutting equipment.

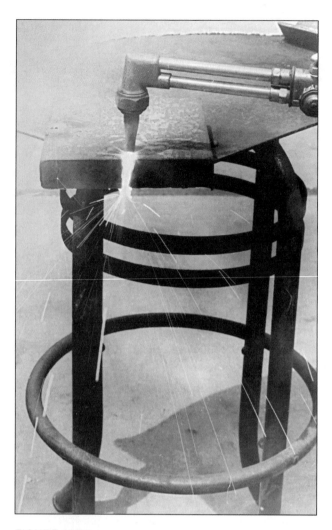

FIGURE 16.29
Flame cutting with an oxyacetylene cutting torch is an efficient, fast, versatile, and widely used process (RRK Machine and Manufacturing).

During this process, preheat jets (mixture of oxygen and acetylene) in the torch tip heat the metal to a red-hot state and maintain this temperature. When the proper cutting temperature is reached, a pure stream of oxygen is directed through a hole in the center of the torch tip. When the oxygen makes contact with the red-hot preheated metal, a high degree of rapid oxidation takes place almost instantly and the metal is cut at that point. An operator skilled in flame cutting can produce a very smooth torch-cut surface. Flame cutting has also been computer automated and is used where duplicate workpieces must be produced. Cutting torches that are solidly mounted and have their movement controlled by mechanical feeding mechanisms can produce very smooth flame-cut surfaces that may require little or no finishing machining operations.

Electric Arc Welding

Electric welding processes are mainstays in joining metals during manufacturing processes. The bulk of steel struc-tural fabrication and construction is done by electric arc welding processes (Figure 16.30, see page 341) or electric resistance welding processes.

Although the individual processes differ to some extent, they all convert the flow of electric current to heat so that the material to be welded may be heated to proper fusion temperatures. The welding system consists of a power supply capable of sustaining large currents at relatively low voltages and appropriate cables connected to the workpiece, thus completing the electrical circuit. Typical arc welding system configuration is shown in Figure 16.31, see page 341.

Several electric welding processes are in common use. They have wide capability and can be used on a variety of materials ranging from heavy sheet metal, structural metal, and plate metal to very thin sheet metal. Electric

FIGURE 16.30
Large structural steel fabrication joined by electric arc welding processes (Miller Electric Manufacturing Company).

FIGURE 16.31
Electric arc welding system components.

welding is also applied to many different metals. Common electric welding processes include shielded metal arc welding, inert shield gas arc welding, gas metal arc welding, gas tungsten arc welding, plasma arc welding, resistance welding, stud welding, and induction welding.

Shielded Metal Arc Welding   In the shielded metal arc welding (SMAW) process, current flows from the power supply through copper wire cables to the electrode holder, which is usually held in the welder's hand. The welding power supply may be powered from the utility line or it may be engine driven. Engine-driven power supplies permit electric welding to be accomplished at locations where no line power is available. This makes the welding system extremely useful for field repairs such as those often made on farm equipment.

A wire electrode is clamped in the electrode holder (Figure 16.32, see page 342). The electrode, often called a welding rod, is about 12 in. long and is usually coated with a hard dry coating material. The coating serves several functions during the welding process: (1) fluxing, (2) arc shielding, and (3) arc stabilization. The fluxing material cleanses and deoxidizes the molten metal. It also forms

slag, which slows down the freezing rate of the weld, and controls the shape and appearance of the weld bead. The coating also provides a protective atmosphere that serves to shield the weld zone from atmospheric oxygen and nitrogen. The arc stabilizing elements in the coating maintain a steady arc during operation.

The wire core of the welding rod acts as the filler metal in the joint between the materials being welded. As the operator touches the rod to the work, current flows across the junction and an electric arc is created. Heat develops from the large current being dissipated at the workpiece and in the electrode, which causes the filler rod to melt and deposit into the joint (Figure 16.33, see page 342).

Considerable skill is required on the part of the operator to properly maintain the arc length and to make correct movements so as to deposit filler metal neatly into the welded joint. Heat is controlled by adjusting the current flow from the power supply. Current flow is based on the type of rod used and the thickness of the metals being joined.

The operator (welder) must be protected from the intense heat and ultraviolet light coming from the welding arc by wearing appropriate clothing and a face hood containing a darkened-glass view port. This enables the welder to observe the welding being done while the eyes

FIGURE 16.32
The engine-driven power supply for electric welding is indispensable equipment for farm machinery repairs. The electrode holder and rod are seen in the welder's hand (Miller Electric Manufacturing Company).

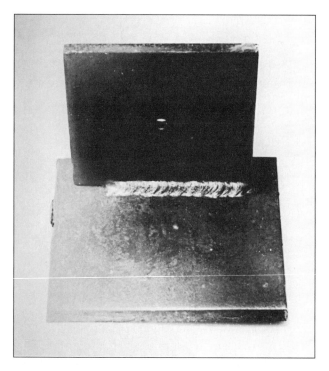

FIGURE 16.33
Electric arc weld.

are protected from the intense light emitted by the welding arc.

The shielded metal arc welding process is used widely in all types of manufacturing applications. It is especially popular in fabrication industries such as ship, bridge, pipeline, and tank manufacturing and in many other industries that fabricate products from structural metals, plates, and heavy-gage sheet steel materials.

Inert Gas Shielded Arc Welding   In the shielded metal arc welding process, the gases resulting from burning the coating on the welding rod provide a certain degree of protection from external air contamination of the weld zone. Since the weld takes place at elevated temperatures, contamination from the air surrounding the weld zone poses a significant problem. For example, during the welding of thin steel sheet metal, the heat from the welding process cannot be dissipated fast enough before oxidation of the welded material occurs. In addition, nonferrous metals with an even higher affinity for oxygen than steel will be consumed before they can be joined by the welding process.

An excellent method of controlling the gas contamination problem is inert gas shielded arc welding. In this type of welding, the entire weld zone is surrounded by an inert gas such as argon, helium, or carbon dioxide (Figure 16.34, see page 343) or by various mixtures of these gases. The inert gas displaces the air in the immediate vicinity of the

weld zone and welding takes place an uncontaminated environment. The major shielded gas welding processes include gas metal arc welding and gas tungsten arc welding. There are also several other arc welding processes.

Gas Metal Arc Welding   The **gas metal arc welding (GMAW)** process, also known as **MIG (metal inert gas)** welding, is a widely used and efficient process that can be applied over a wide range of metal thickness and types of materials. MIG is extremely well suited for production work when welding steel and aluminum.

The MIG system consists of a constant voltage power supply, wire feed unit, and a shield gas cylinder with pressure and flow regulators (Figure 16.35, see page 343). The MIG electrode is a continuous wire, preferably of the same material that is being welded. Electrode wire is fed continuously from a reel while the welding process is taking place (Figure 16.36, see page 344). At the same time shield gas flows from the welding gun (Figure 16.37, see page 344), surrounds the wire, and displaces air from the weld zone. Welding current, shielding gas flow, and electrode feed rate may be varied to meet the exact requirements of the welding job.

Variation of the MIG process includes the flux cored arc welding (FCAW) process, where the flux is contained inside a hollow electrode wire. Fluxed electrode welding may be performed with or without shielding gases.

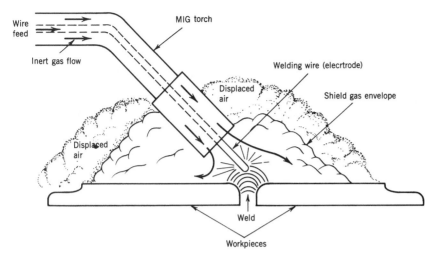

FIGURE 16.34
GMAW or MIG welding shields the weld zone with inert gas.

Because the process takes place in the inert gas environment, the weld quality of the MIG process is unsurpassed. The process has a number of applications, e.g., rehabilitation of worn parts by weld buildup (Figure 16.38, see page 344). In the illustrations, a shaft has been mounted in a lathe where it can be rotated. MIG bead can be deposited evenly and accurately on the workpiece (Figure 16.39, see page 345). After the welding is complete, the part may be machined to restore its original dimensions.

With the convenience of the continuously fed electrode and the inert gas shielding, the MIG process is highly suited for production welding of ferrous and nonferrous metals (Figure 16.40, see page 345). The MIG

process is highly suited to both semiautomation (Figure 16.41, see page 346) and full automation using industrial robots (Figure 16.42, see page 346). The production welding of the assembly represented in Figure 16.43 (see page 347) was accomplished in 30 min. using the robot, whereas the same job took 90 min. by the manual method.

Gas Tungsten Arc Welding   Gas tungsten arc welding **(GTAW),** commonly known as **TIG (tungsten inert gas)** welding, is a nonconsumable arc welding process that uses a tungsten electrode which does not melt during the welding process. Filler material is fed externally into the welded joint during welding. The process is somewhat

FIGURE 16.35
MIG welding equipment includes wire feeder and shield gas cylinder.

FIGURE 16.36
The MIG electrode and shield gas are fed to the weld in a continuous flow (Miller Electric Manufacturing Company).

FIGURE 16.37
The MIG torch.

like the oxyacetylene process in this respect, the difference being that the heat is supplied by an electric arc established between the tungsten electrode and the workpiece. TIG also uses a shielding gas to surround the weld zone and to displace atmospheric contaminants (hydrogen, oxygen, and nitrogen).

The TIG system (Figure 16.44, see page 347) consists of a power supply, cables, and a welding torch. In manual operation, the welder manually feeds the filler rod into the arc where it melts and is deposited into the welded joint (Figure 16.45, see page 348). Shielding gas flow and weld-

FIGURE 16.38
Lathe mounted workpiece undergoing weld buildup by the MIG process (Miller Electric Manufacturing Company).

FIGURE 16.39
Weld built up by the MIG process is virtually free of slag pockets and inclusions. After weld buildup is completed, the part may be remachined to restore original dimensions. (Miller Electric Manufacturing Company).

ing current are controlled by a foot pedal. Some TIG torches, which are used for higher welding currents, are water cooled. This helps dissipate heat, thus preserving the electrode and making the torch more comfortable to hold.

TIG process is extremely useful for small welding applications where detail welding is required. TIG can be used to weld metal a few thousandths of an inch thick. Since the process takes place in the inert gas envelope, it is possible to achieve superior quality welds. For this reason, TIG may be used where a weld of outstanding quality is required. An example would be the root pass in joining nuclear system piping. After a TIG root pass is made at the bottom of the weld joint, the remaining weld filling of the joint may be done by other processes such as MIG. TIG is also superior on nonferrous metals, especially aluminum.

Submerged Arc Welding Submerged arc welding (Figure 16.46, see page 348) is a shielded electric arc process. The welding process takes place buried under a fluxing agent that is being deposited in the joint ahead of the electrode. The fluxing compound acts in much the same way as a shielding gas in the MIG process by preventing atmospheric gases from getting in contact with molten metal. The fluxing compound also acts in much the same way as a coating in the SMAW process by cleansing impurities from molten metal and slowing down the cooling rate of the weld. In addition, flux prevents spatter, and thus results in a better quality weld (Figure 16.47, see page 349). Submerged arc welding

FIGURE 16.40
The operator can reach the work easily using an efficient portable MIG welder with overhead suspended wire feeder (Miller Electric Manufacturing Company).

FIGURE 16.41
Automated MIG welding equipment (Miller Electric
Manufacturing Company).

equipment consists of the wire feed mechanism, the flux
hopper, and the power supply. The flux tank (or hopper)
and wire feeder may be portable, and in the field they
can be connected to the power supply by cables (Figure
16.48, see page 349).

The flux hopper is air pressurized to provide the flux
feed. The process is suitable for automatic welding of pipe
and major structurals such as I beams, channel beams, and
H beams, but it is limited to a flat or horizontal welding
position so that the fluxing agent will remain in the weld
area.

Plasma Arc Welding   Heat for the plasma arc process
(Figure 16.49, see page 349) is derived by ionizing (creating an imbalanced electric charge) a plasma gas by an
electric arc. The resulting concentrated plasma arc can
deliver a large amount of heat at very high temperatures to
the weld zone. Plasma arcs can generate temperatures of
more than 40,000°F, much higher than can be achieved by
any fuel combustion process. Plasma arc welding is fast
and clean and its arc concentration properties permit high-
quality welds with minimal effect on the surrounding
welding area. The process is suitable for welding many
materials and is commonly used for cutting. Additional
uses of extremely high temperature plasma arcs are being
investigated and applied in many other areas of industrial
processing.

Resistance Welding

Most metals, particularly steels, have resistance to electrical current flow. This characteristic can be used to heat the
metal to welding temperatures, at which point the parts to
be joined are forced together, thus completing the weld.
No externally supplied filler material is required and the
welding process does not require any shielding gas. One
popular and versatile resistance process is spot welding
(Figure 16.50, see page 350). Workpieces are clamped
between two electrodes. The current flow heats the metal

FIGURE 16.42
The MIG welding process is
highly suited to industrial
robot automation (ESAB
Robotic Welding Division).

FIGURE 16.43
Robotic MIG welding can be done with great speed and quality (ESAB Robotic Welding Division).

to a fusion temperature and the mechanical pressure of the electrodes forces the material to fuse together. Since the welding process results in a relatively small weld size, usually several spot welds are necessary to ensure adequate strength in large fabrications.

The spot-welding process is very fast and is therefore highly suitable for volume production. Spot welding is widely used in the automobile industry and the nature of the

process is such that it was easily automated using industrial robots (Figure 16.51, see page 350). Microprocessor controls are used on spot welding equipment to control welding variables (e.g., current, contact time).

Resistance welding is also used to join the ends of band saw blades (Figure 16.52, see page 351). A band saw weld will require annealing (softening) after the welding process so that the band will have the flexibility required in service. Resistance welding can be applied to seams and long joints by using roller electrodes. The process called resistance seam welding is used in the manufacture of pipe and tube (Figure 16.53, see page 351).

Stud Welding   In the stud welding process, arc is established between the stud and the mounting surface. The stud is secured in a stud gun (Figure 16.54, see page 351) and a ceramic ferrule surrounds the base where the weld will occur. The ceramic ferrule serves two purposes: (1) it concentrates the heat, and (2) it serves as the mold that shapes the weld in the fillet form. The ferrule is broken off after the weld is completed. When the stud gun is energized, the solenoid in the gun withdraws the stud from the base metal, at which moment an arc is established. The arc heats and melts the base metal and stud. After a specified time interval, a mechanical spring is released, which forces the stud in the molten pool of metal, accomplishing the weld. The process takes place almost instantly.

FIGURE 16.44
GTAW or TIG welding system.

Line

Transformer

AC/DC power
supply

TIG torch

Work

Nonconsumable Tungsten
electrode

Arc is established between
electrode and work

Work
piece

Shield gas envelope

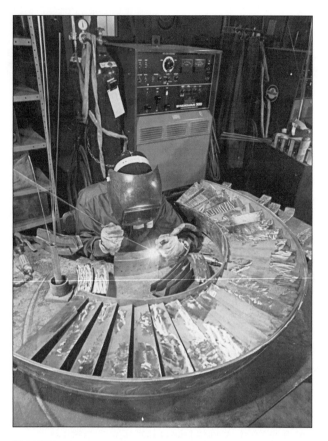

FIGURE 16.45
TIG welding on a turbine rotor. The TIG torch and filler rod are visible in the welder's hands (Miller Electric Manufacturing Company).

Studs of many different sizes and styles may be installed by this fast and efficient process. Stud welding is a very convenient method to attach a threaded fastener to a flat surface. Nonthreaded fasteners may also be installed by stud welding. These can be used as hooks or clips to which something may be attached.

Induction Welding   Heat generated for induction welding (Figure 16.55, see page 352) is generated by inducing a rapidly oscillating electromagnetic field in the parts to be welded. Parts may be placed in an induction coil field and induced current flowing in the workpiece as the field rises and falls in the induction coil causes rapid heating. Induction heating for welding is useful for production work. It also is used as a heating method for brazing.

Solid State Welding Processes

Forge Welding   In forge welding (Figure 16.56, see page 352), probably best known from the days of the village blacksmith, parts to be welded are heated in a forge fired with charcoal or coal. The fire is forced-air fed by a bellows or a motor-driven fan under the coal bed to increase the temperature. Parts to be welded are placed in the fire and are heated to a point slightly below melting. The parts are then removed from the fire and placed in contact. Pressure is applied by hammering, in the case of the blacksmith, or by mechanical pressing or rolling in more modern techniques. Assuming correct heating and scale-free surfaces, a weld occurs.

Ultrasonic Welding   Heat for ultrasonic welding (Figure 16.57, see page 353) is generated by subjecting the parts

FIGURE 16.46
Submerged arc welding.

FIGURE 16.47
Submerged arc results in a weld of exceptional quality (Miller Electric Manufacturing Company).

to be welded to high-frequency vibrations in the ultrasonic range. The mechanical energy causes the parts to heat as they are forced together by mechanical pressure. The process has applications for precision welding in a variety of materials, and where electromagnetic forces and intense heat can damage the product. The process can be used only for very thin material.

Other Welding Processes

Other welding processes include the ones which cannot be classified easily into one of the previous groups of processes. Some of these processes include thermit welding, electron beam welding, and laser welding.

Thermit Welding In the thermit process, parts to be joined are submerged in a bath of molten metal that is sufficiently hot to heat the parts to be welded to a fusion temperature. The thermit mixture consists of metallic oxide of the metal to be welded, reducing metal, and another metal

FIGURE 16.48
Submerged arc welding equipment includes the wire feeder and the flux tank (Miller Electric Manufacturing Company).

FIGURE 16.49
Plasma arc welding.

FIGURE 16.50
Spot welding.

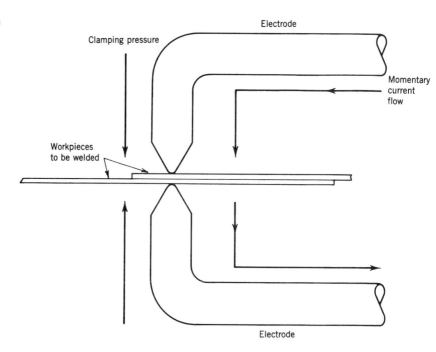

that will achieve a very high temperature when ignited. A typical mixture consists of iron oxide, aluminum, and magnesium. Magnesium is ignited, and as it burns it develops the necessary heat to start an exothermic chemical reaction between aluminum and iron oxide. As a result, aluminum oxide and pure iron are formed. Since the aluminum oxide is lighter, it will float to the surface where it is removed. Molten iron is now ready to be used for welding.

The thermit weld takes place in a mold so that the molten metal is retained as it solidifies. In this respect thermit welding is similar to casting (Figure 16.58, see page 353). Different refractory and high melting temperature materials can be used to make the mold. Parts to be welded must be cleaned and correctly aligned. Preheating is often required. An example of a thermit welding application is the joining of railway rail ends so that the rail becomes continuous, which makes for a much smoother roadbed.

FIGURE 16.51
Industrial robot spot welders on each side of the production line efficiently weld auto bodies with great speed and reliability (GM—Pontiac Division).

FIGURE 16.52
Resistance flash butt welding a band saw blade.

FIGURE 16.54
Stud welding.

Electron Beam Welding   Electron beam welding (Figure 16.59, see page 354) is a high-energy welding process that has found wide application in modern industry. Heat for the process is achieved by accelerating and focusing a stream of electrons in an electromagnetic field and allowing the beam to impinge onto a concentrated area on a workpiece. The kinetic energy of an electron is small. However, when that energy is multiplied by the number of electrons in the beam, it is sufficient to generate the heat that can melt the workpiece. To reduce or even eliminate the scatter of the electrons that occurs when an electron beam passes through air, the process is accomplished in a vacuum. The process is highly suitable for precision welding on parts that have already been machined. Since the weld penetration and location can be controlled precisely, this process has wide application in joining precision assemblies with a minimum of heat distortion. The elec-

tron beam process is also used for hardfacing teeth on band saw blades, thus permitting the manufacture of superior cutting bands for sawing difficult-to-cut materials.

Laser Welding   The laser, discussed in the previous chapter, is an intense beam of coherent light. Laser light is generated by particular agitation of atoms or molecules in gases, such as carbon dioxide and neon, or crystals, such as rubies. The light emitted from the gas or crystal laser is a single color and in an extremely tight or coherent beam. The coherent beam of a high-power laser can be extremely intense, which makes the laser quite useful for industrial applications. The laser has found many applications in modern industrial processes, including joining processes.

In laser welding (Figure 16.60, see page 354), heat is generated by directing the laser beam onto the parts to be welded. The beam is concentrated and also can be applied in areas of extremely small diameter. Precision welds (Figure 16.61, see page 355) of excellent penetration are possible on very thin materials (.001 in. thick).

Related in many ways to welding of metals, the ability of the laser to seal other materials such as human tissue has led to its many amazing applications in medicine.

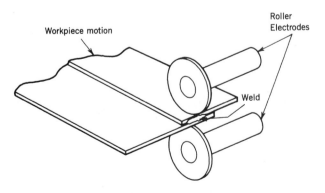

FIGURE 16.53
Resistance seam welding using roller electrodes.

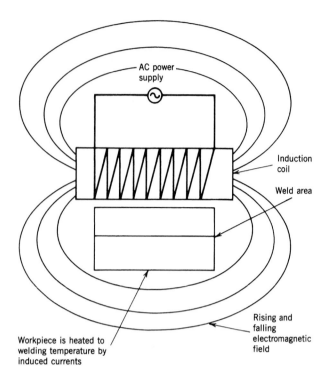

FIGURE 16.55
Induction welding.

The lasers can also be used to cut (Figure 16.62, see page 355), drill, plate, heat for selective hard facing, and etch markings on parts.

BRAZING AND SOLDERING

Welding requires heating parts above their melting temperature, where they will fuse into a common mass across the weld zone. If filler material is used, it too becomes part of the common mass. Alternately, in brazing and sol-

dering, the filler metal melts at a lower temperature than the melting temperature of the parts to be joined and the parts themselves are not melted. The filler metal does not actually become part of the common material in the weld. The joining of the material is purely mechanical, i.e., filler metal grips the parent metal by adhesion forces. The major difference between brazing and soldering is that the brazing processes use filler metals with higher melting temperatures than soldering processes.

Soldering and brazing do not provide as strong a mechanical joint as do the fusion welding processes. Nonetheless they have wide application as efficient material joining processes in manufacturing. It is important that these processes be applied only where soldering or brazing is a suitable joining method. They must not be used in applications where strength requirements necessitate the complete fusion bond accomplished in the welding processes.

Brazing

Brazing processes use filler material with a melting temperature above 840°F and below the melting temperature of the base metal. The process requires properly cleaned surfaces, which can be accomplished by mechanical (e.g., grinding, sand blasting, wire brushing, filing) or chemical means (e.g., solvents, alkaline baths, acid baths, salt bath pickling). In addition, the parts to be joined must be properly positioned with respect to each other. That is, the gap between the parts must be such that capillary forces can pull the filler material into a joint. A larger or smaller distance between the parts may adversely affect bonding. Heat sources commonly used in this process are the same as those used in welding: chemical, mechanical, and electrical.

Brazing filler metals are classified according to their chemical composition: aluminum-silicon, copper, copper-

FIGURE 16.56
Forge welding.

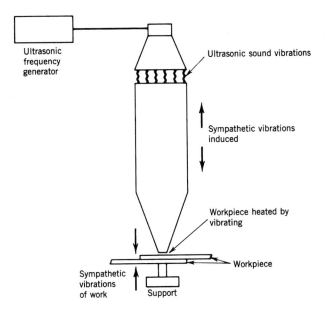

FIGURE 16.57
Ultrasonic welding.

zinc, copper-phosphorus, nickel-gold, heat-resisting materials, magnesium, and silver. The selection of braze material is based on the metal being brazed.

Some of the common brazing processes include: dip brazing, furnace brazing, induction brazing, infrared brazing, resistance brazing, and torch brazing. Joint strength is much higher with the brazing and braze welding process than with soft soldering processes.

Soldering

Soldering processes use filler material with a melting temperature below 840°F and below the melting temperature of the base metal. Typical soldering metals are alloys of nonferrous metals, e.g., lead and tin. The most commonly used solder is 50% lead - 50% tin alloy. Soldering is commonly used in joining sheet metal requiring low to medium strength, and in leak-proof joints. Solders are used extensively in the electronics industry for electronic-circuit card soldering (Figure 16.63, see page 355). The solder alloys have excellent electrical conductivity and lend themselves to automated soldering applications.

Soldering, similar to brazing, requires proper joint preparation and material cleaning to ensure properly soldered joints. Fluxes are almost always used to aid in cleaning during the soldering process. Soldering can be applied to many materials; however, there are some metals for which soldering is unsatisfactory.

The same processes and heating methods used for brazing are also used for soldering, e.g., dip soldering, furnace soldering, induction soldering, and infrared soldering.

THERMAL SPRAYING

Thermal spraying is a process of depositing a layer of metal or ceramic material onto the surface of a desired part. The process is used to provide surface coatings of

FIGURE 16.58
Thermit welding.

FIGURE 16.59
Electron beam welding.

different characteristics, such as coatings to improve abrasive wear or corrosion resistance, change thermal conductivity, or simply build up the surface. For example, thermal spraying can be used to apply a layer of aluminum on a steel structure (such as a bridge) to slow down or eliminate corrosion and to prolong the life of a structure. Also, disk teeth on farm equipment can be coated with wear-resistant material to reduce the wear caused by rock and dirt abrasion.

Several processes have been developed for thermal spraying: flame spraying, high-velocity oxyfuel (HVOF) spraying, arc spraying, plasma spraying.

In the processes of thermal spraying (Figure 16.64, see page 355), small droplets of molten metal or ceramics

FIGURE 16.60
Laser welding (Laserdyne).

are carried by the hot gases and are deposited on the workpiece, where they adhere. The strength of mechanical bonding at the interface between the part and the sprayed-material buildup is increased by making a rough surface on the substrate. Machining is normally used for round parts, while shot blasting is used for flat surfaces. Thermal spraying requires that the substrate be properly cleaned prior to depositing, and spraying should be performed right after the cleaning operation in order to avoid oxidation. The part to be sprayed should be preheated to approximately the same temperature that will be attained during spraying, in order to prevent cracking of the sprayed surface. The resulting spray deposit is porous. In the applications where porosity is undesirable, the part can undergo a fusing or recasting operation. In this operation the surface is heated to slightly above the melting temperature of the sprayed layer. This is most commonly performed with an oxyacetylene torch; however, other processes can be used also, such as laser heating.

WELDING PLASTICS

Welding of plastics is an increasingly important area since plastics are becoming widely used materials, especially where corrosion resistance and weight constraints are important. All plastics can be classified in two groups based on their chemical composition and their elevated temperature characteristics. The two groups are thermosetting and thermoplastic plastics. Thermosetting materials cannot be welded by any method that involves heating, while thermoplastic materials can be successfully welded.

Thermoplastics may be welded by several thermal or chemical joining processes. The various thermal joining processes are as follows: electromagnetic, friction, heated surface, high frequency, hot gas, implant, radiant, ultrasonic, and vibration.

Several steps are involved in thermal welding of thermoplastics—surface preparation, heating, application of pressure, diffusion, and cooling. Surface preparation is very important since most molded plastics are contaminated with mold release compounds. Heating can be accomplished by one of the previously mentioned methods. Pressure can be applied manually, in presses, or in automatic fixtures. Diffusion occurs when the interface between the parts becomes liquid, after which the solidification occurs during the cooling stage.

Solvents that dissolve the plastic material may be used in plastic joining. The plastic surface in the joint to be welded is rendered into a liquid state by the application of appropriate solvents. After the parts are joined and the solvents have evaporated, the joint becomes as homogeneous as the parent material.

Penetration zone of laser weld

FIGURE 16.62
Laser cutting (Photo courtesy of Coherent Inc., Palo Alto, California).

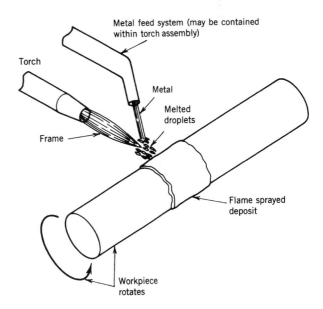

FIGURE 16.63
Electronic printed circuit card soldering using soft lead–tin alloy solders.

Metal feed system (may be contained within torch assembly)

Torch

Metal

Melted droplets

Frame

Flame sprayed deposit

Workpiece rotates

FIGURE 16.64
Thermal spraying.

355

Review Questions

1. Name three mechanical fastening methods.

2. Discuss the relative advantages and disadvantages of common mechanical fasteners.

3. An assembly must be taken apart often for maintenance. Which mechanical fastening system would be the best for this application?

4. Where would you most likely find stitching, tying, and snaps used as a material joining process?

5. What are the advantages of adhesive bonding?

6. What are the disadvantages of adhesive bonding?

7. In what industry would you likely find adhesive bonding and why?

8. List three plastic resins used in adhesive bonding.

9. Describe the inspection method used in the aircraft industry to check for defects in bonded joints.

10. List four common electric welding processes.

11. What is GMAW or MIG?

12. What are the advantages of GMAW?

13. What is GTAW or TIG?

14. How does GTAW differ from MIG?

15. Describe shielded metal arc welding and discuss its applications.

16. What is stud welding?

17. What is the function of the shield gas in shielded arc welding?

18. In the MIG process, how is the electrode delivered to the weld area?

19. What material is used for TIG electrodes?

20. If welding were needed where no access to utility company electric service was available, what welding processes could be used?

21. What is the resistance in resistance welding and what does it have to do with the process?

22. Describe laser welding and discuss its advantages.

23. Describe the thermit welding process.

24. What conditions are necessary for electron beam welding?

25. Describe the submerged arc process.

26. How do soldering and brazing differ from welding?

27. What are the most common soldering filler materials?

28. What are the most common brazing filler materials?

29. Discuss where thermal spraying, metallizing, and hard facing might be used.

30. How can plastic be welded?

Case Problems

Case 1: Welding Thin Materials

A new business begins to manufacture patio furniture made from thin-wall steel tubing. Shielded metal arc welding is initially employed, but metal burnthrough occurs immediately. The production supervisor, an experienced welder, recommends that shielded metal arc welding be replaced with TIG. The vice president of manufacturing suggests that this would be expensive and the problem could be just as easily solved by further training of the production welders in order to improve their welding skills. In your opinion, who is right and what would you recommend in this case?

Case 2: Collapsing Furniture

The Super Cut Rate Wood Furniture Manufacturing Co. is receiving customer complaints that some of its furniture is falling apart soon after purchase. Super Cut Rate sends its director of quality control to check on the problem. The QC director finds that on the furniture assembled by high-production stapling, the fasteners are pulling out when the furniture is subjected to normal-use loads. A new method of fastening must be used. Considering the primary factors of high production of a low-cost product, screws, bolts and nuts, and gluing are ruled out as being too expensive and labor intensive. What other methods might be employed to solve this production problem?

P roduct performance and manufacturing costs are major factors in motivating the designer to seek new product materials and new manufacturing methods to process them. For example, fuel efficiency in automobiles as well as aircraft has become an important consideration in recent years. To meet these requirements, the designer has turned to nonmetallic materials that demonstrate both light weight and strength. Numerous other advantages in using nonmetallic material such as plastics and advanced composites include corrosion resistance; resistance to denting, chipping, and changes in temperature; and ease of manufacturing.

The purpose of this chapter is to survey the major methods of processing these materials.

Plastics and Composites Processing

Objectives

This chapter will enable you to:

1. Identify common methods of processing plastics and composites.
2. Describe in general terms how the processes work.

THE WORLD OF PLASTIC AND COMPOSITE PRODUCTS

It would be difficult to envision the modern world without plastics and composites. The product line of these materials is vast and expanding every day. Although metals will always remain in widespread use, plastics and composites are fast replacing metals in a vast array of manufactured products. Why? Because of the often high cost of metals and the large energy requirements in metal processing, the ease by which plastics and composites may be processed compared to metals, and the high strength and light weight of plastic and composite materials, which now permits their use in applications where previous designs required metals.

In recent years, engineering development of plastic and composite materials has yielded products that are equal to or better than their metal counterparts in every way. One significant factor is weight. Because plastic and composite structures are generally lighter than their equivalent metal counterparts, they have become very popular in product applications where weight is a significant factor. Lighter weight vehicles both in the automotive and aircraft fields have generally increased fuel efficiency. Plastic and composite materials therefore have generated much interest as materials for these applications.

Composite engineering has yielded structures that meet design strength requirements, thus permitting extensive use of these materials in modern jet aircraft (Figure 17.1). The illustration shows an aircraft in which the fuselage and other major structural components are completely made from composite materials. When this aircraft is commercially produced, and powered by the efficient turbojet engine pusher propellers (Figure 17.2), it will have the speed and altitude capability of the pure jet,

FIGURE 17.1
Gates-Piaggio turboprop aircraft is an example of the use of composite material for structural aircraft components (Gates Learjet).

while at the same time it will also demonstrate very high fuel efficiency because of its light weight.

Advanced composite as well as plastic materials have also been used for major structural components on full-sized commercial and military jet aircraft (Figure 17.3). The advanced composite vertical stabilizer fin in the illustration weighs 25 percent less than its aluminum counterpart. Several aircraft manufacturers are building entire major structural components such as fuselages and wing structures from these advanced materials. The weight saving in aircraft and auto structures will play an important part in the fuel efficiency ratings of these products.

Another common example is the use of plastic material such as laminated products as a very durable substitute for wood. These materials are easily fabricated, easily cleaned and cared for, and can be made very thin and therefore very light while still retaining the beauty of wood grain designs. Figure 17.4 shows plastic laminate material as a wood substitute in aircraft cabinetry.

Since plastic and composites are generally in a liquid or softened state prior to their formation into products, these materials are highly suited to manufacturing one-piece complex shapes such as automotive body components. The truck and bus hood and fender components seen in Figure 17.5 (see page 360), are made from reinforced plastics and weigh less than 100 lb. They are also resistant to corrosion and impact damage and are unaffected by variations in temperature.

PLASTICS AND COMPOSITES PROCESSING METHODS

So far in this text, you have studied many processes in which metals are processed into useful products. Many of the same manufacturing process principles are applied in the processing of plastic and composite materials. These materials may be cast, molded, extruded, machined, fabri-

FIGURE 17.2
High strength but light weight plastic and composite materials greatly enhance fuel efficiency in aircraft (Gates Learjet).

FIGURE 17.3
The vertical tail fin of the Lockheed L1011, a full-sized commercial jet aircraft. This component is 25 percent lighter than its aluminum counterpart (Lockheed Co.).

cated, laminated, and rolled. The equipment used for processing plastic and composite material is specifically designed to process these materials.

Three major methods make up a large portion of plastic and composite processing techniques. These include blow molding, injection molding, and extrusion. Several other processes are also used.

Blow Molding

In blow molding, air is used to force a mass of molten plastic against the sides of a mold shaped in the form of the desired end product (Figure 17.6, see page 361). The sequence of illustrations (Figure 17.7 *a,b,c,* and *d,* see page 361) shows a common and versatile application of the blow molding process for the production of refrigerator liners. This process is particularly well suited to the forming of large fabrications, such as the example shown. After cooling, the mold is opened and the finished part is removed.

This method can also be used for sheet forming where a sheet of plastic is clamped between a form and die. Air pressure forces the heated plastic against the die where it assumes the die shape.

Blow molding may be used when the end product must have an enclosed hollow internal shape. An example of this application is containers such as the plastic jugs illustrated in Figure 17.8 (see page 362). Machinery used in this process will produce units at a high rate of production. The dies used to shape the container are seen in Figure 17.9 (see page 362).

Related to this process is vacuum forming, where a vacuum is drawn on one side of the material and air pressure on the opposite side forces the material against the mold or form.

Injection Molding

Injection molding is another versatile and widely used process. Examples of injection-molded products are found everywhere. In this process, molten plastic is forced into a metal die cavity that has been machined to the shape of the desired end product. When the plastic material has

FIGURE 17.4
Plastic laminates simulate the look of wood grain while taking advantage of the properties of plastics (Lockheed Co.).

FIGURE 17.5
Plastic and composite materials are highly suited to the manufacturing of one-piece complex-shaped products (Goodyear Tire and Rubber Company).

solidified sufficiently, the die is opened and the part removed. Figure 17.10 (see page 362) shows a cutaway view of a plastic injection molding machine (Figure 17.11, see page 363). Raw plastic material in small pellets (Figure 17.12, see page 363) is placed in the machine hopper. The raw material then enters the heater, where it is melted. Molten plastic is then pushed into the die cavity by hydraulic or direct mechanical pressure (Figure 17.13, see page 364).

Large-capacity injection molding machines may exert several hundred tons of pressure and can be used to fabricate large one-piece plastic parts. Examples include automotive body components, including fender/hood assemblies, bumpers, and grilles (Figures 17.14 and 17.15, see pages 364–365).

The injection molding process is extremely versatile for making parts with fine detail and complex shapes (Figure 17.16, see page 365). Examples include high production of small precision detailed parts such as those for plastic models.

Reaction Injection Molding Reaction injection molding (RIM) is the latest injection molding technique. In this process, two base resins (monomers) are mixed together just as they enter the mold (Figure 17.17, see page 366). A chemical reaction occurring at low heat takes place and the plastic material (polymer) of the end product is formed at that instant. The RIM process does not require the plastic materials to be heated before molding. Thus, the process is both energy efficient and very fast. Large parts weighing many pounds can be made in a few seconds and the process minimizes scrap.

Reaction injection molding machines (RIMM) may be of the platen design (Figure 17.18, see page 367). The machine illustrated is seen with essential safety guards removed so that details of construction may be seen. This machine is highly suited for production of large parts, such as auto bumpers and fenders.

Extrusion

Extrusion is another versatile and widely used plastic and composite processing method. In this process, plastic material is forced through an extrusion die that forms the end product (Figure 17.19, see page 367). The major component of a plastic extruder is the extruder screw (Figure 17.20, see page 368). The extruder may be of the twin screw design where two screws with variable lead sections are in mesh. Parallel screw extruders also are used. This system provides the mechanical force to push the plastic material through the extrusion die. Extrusion is the popular manufacturing process for products requiring long dimensions in one axis, such as plastic pipe and plastic molding.

The illustration (Figure 17.21, see page 368) shows a plastic extruder for manufacturing plastic pipe. The specific tooling for this product is attached to the machine on the left side.

Other Molding Processes

Other processes that make use of molding principles include compression molding, transfer molding, rotational molding, and solvent molding.

Compression Molding The compression molding process is similar to injection molding, but it is primarily used with thermosetting resins. In compression molding, the resin is in pellet or dry powder form and may be mixed

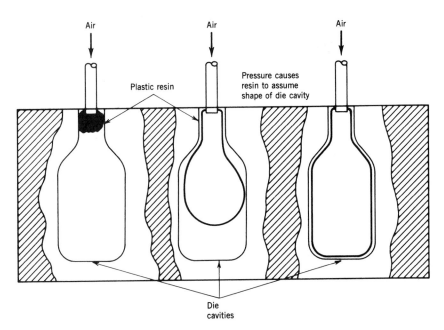

FIGURE 17.6
Blow molding process.

Air ↓   Air ↓   Air ↓

Plastic resin

Pressure causes
resin to assume
shape of die cavity

Die
cavities

FIGURE 17.7
(a) Partially blow-molded refrigerator liner. The blow molding die is shown in the open position. *(b)* The liner begins to assume the shape of the die. *(c)* The completed liner just prior to removal from the blow molding die. *(d)* Completed refrigerator liners after removal from the blow molding die (all photos courtesy of Whirlpool Corporation).

FIGURE 17.8
Continuous production blow molding machine for plastic containers (Plastic Machinery Incorporated).

with a binder agent to provide strength. In this respect it can become a reinforced plastic material. The dry resin is mechanically compacted into a mold shaped to produce the desired end product. Heat and pressure are then applied to melt the resin and cause it to flow throughout the mold. A thermosetting plastic material undergoes a fundamental chemical change in this process and a permanently hardened part is produced. The process is good for making handles and knobs for kitchen utensils since thermosetting resins can be subjected to heat without becoming soft.

FIGURE 17.9
Blow molding machine for plastic containers showing die cavities (Plastics Machinery Inc.).

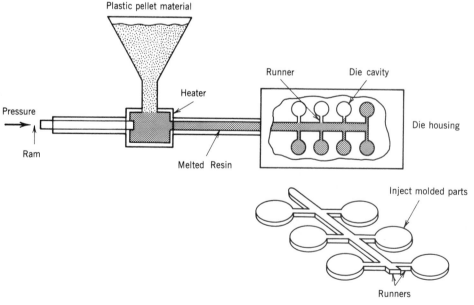

FIGURE 17.10
Injection molding process.

FIGURE 17.11
Cutaway view of an injection molding machine (Cincinnati Milacron Plastics Machinery Division).

Transfer Molding Transfer molding is similar to compression molding except that the resin is heated to a liquid state before being forced into the mold. This process may be used where intricate-shaped parts are required or where metal inserts are molded into plastic. Metal parts that are positioned in the mold will remain in location while the liquid resin flows around them. When the plastic has solidified, inserted metal parts are securely held in place.

Rotational Molding In rotational molding, the mold cavity is spun and centrifugal force causes the resin to flow into the mold. Heating is required to place the resin in a liquid state. The process is suitable for large parts and will yield products having little internal stress and very uniform wall thicknesses.

Solvent Molding In solvent molding, a mold of the desired end product shape is dipped into or filled with a liquid resin material. When the mold is withdrawn or emptied out, a thin layer or coating of resin adheres to the sides. When the solvent has evaporated, the formed part may be removed. Additional layers may be built up to increase the thickness of the end product and the process may also be used as a method to plastic-coat another part.

Other Plastic Processing Methods

Other plastic processing methods include casting, where liquid resins are poured into molds. The resin contains a catalyst that causes the resin to harden after a short period of time. Resin casting is popular in the homecraft area and will yield many decorative products. The process also can be used to encase an object in a transparent block of resin. Color may be added to clear casting resins. The process must be done quickly or the catalyst will harden the resin before it can be poured. Casting can be used to make sheets, rods, tubes, special shapes, film, and sheeting from both thermoplastic and thermosetting resins.

Rigid plastic and composite materials can be shaped effectively by most of the common machining processes. The material can be milled, turned, sanded and ground, broached, reamed, drilled, and tapped. A number of plastics are readily available in bar stock form for processing by machining methods. Plastics suitable for machined products include nylon, acetyl (delrin), fluorocarbon (Teflon), ABS, and PVC. Common machined plastic products include gears, bearings threaded fasteners, and bushings.

FIGURE 17.12
Plastic pellets, the raw material for injection molding and other plastic manufacturing processes (Whirlpool Corporation).

FIGURE 17.13
Injection molding machine, 250 ton capacity, with computer microprocessor controls (Cincinnati Milacron Plastics Machinery Division).

Processes used to join pieces of plastic in assemblies include welding by heat and ultrasound, gluing with solvents and other resin adhesives, sewing (sheet and film), and fastening with mechanical fasteners such as bolts and nuts, screws, and rivets.

Reinforced Plastic Materials

The mechanical properties of plastics can be enhanced greatly by adding a reinforcing agent to the resin. Plastic resins may be reinforced with cloth, paper, glass fibers, and other fibers such as graphite, thus forming advanced composite materials. Reinforced plastics differ from composites in the way in which the reinforcing fibers are established in the base resin material. Fibers used in reinforced plastic are generally in short pieces and are ran-domly oriented within the resin. Composite fibers, on the other hand, run unbroken through the resin and provide the major structural component for the material.

Although reinforced plastics may not have quite the same strength and load carrying capability as their true composite counterparts, they are nonetheless excellent structural material for a variety of products.

A popular process and one with wide application is fiberglassing, in which alternate layers of glass fiber fabric and resin are coated over a mold or form built in the shape of the desired end product. Short pieces of glass fiber may also be mixed with the resin while it is still in a liquid state. A popular application of this method is in manufacturing hulls for boats (Figure 17.22, see page 369) and other large hollow fabrications such as swimming pools.

FIGURE 17.14
One-piece plastic fender and cowel components made by injection molding (Goodyear Tire and Rubber Company).

FIGURE 17.15
Injection molding can produce intricate parts with extremely smooth and uniform surfaces such as these auto grille components (Goodyear Tire and Rubber Company).

The process is accomplished by building up layers of fiber strands or sheet fabric (Figure 17.23, see page 369) and resin. Pieces of glass fiber are laid over the desired form and the resin is applied by spray application or hand spreading (Figure 17.24, see page 369). The liquid resin and fiber mixture may be troweled and smoothed while it is still in a liquid state. After hardening, the fiber glass fabrication may be refinished by mechanical abrasive processes such as sanding and buffing (Figure 17.25, see page 369).

The end product of reinforced plastic is extremely strong, durable, and lightweight. It is also impact resistant and not subject to corrosion, and almost any shape can be fabricated given the proper tooling.

The process of **calendering** or rolling plastic material is used to form sheets and film. Plastic is considered to be film if its thickness is less than .010 in.; if it is thicker, the material is considered to be sheet. Sheet and film are formed by pressing the material between heated rollers. Thickness is determined by the spacing between the calendering rolls. Plastic coatings are also applied to cloth by this method, forming the many useful plastic-coated fabrics appearing in a large number of consumer and industrial products.

Plastic coatings may also be brushed or sprayed on metals or woods to protect them from corrosion. Clear plastic coatings are used as wood and metal finishing mediums. Plastics are also used in paints, where color as well as durability properties of the plastic can make a superior coating medium. Examples include epoxy- and polyurethane-based paints.

In the process of high-pressure lamination, which generally makes use of thermoplastic resins, sheets of the material are placed between heated steel platens, which are then compressed with hydraulic pressure. The process bonds the layers of the laminate into a rigid sheet.

Laminated sheets can also be formed around corners and then bonded to wood bases. The products resulting from this process are used extensively for kitchen countertops. Very attractive products that simulate woods, but have the advantages of plastics, are created by this process.

MANUFACTURING AND PROCESSING COMPOSITE MATERIALS

Composite materials are resins that have been reinforced with organic or inorganic materials. Composites differ from reinforced plastics in that the fiber structure is continuous throughout the structure. It is this particular structural design that gives the composite its superior mechanical properties. An example of a composite material is a plastic such as epoxy resin reinforced with graphite (carbon) fibers. Composite materials have many unique properties, especially their strength-to-weight characteristics, corrosion resistance, impact strength, and their ability

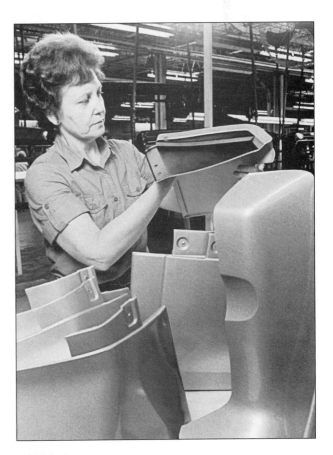

FIGURE 17.16
Complex shapes are easily produced by injection molding processes (Goodyear Tire and Rubber Company).

HT-RIMM
UNIVERSAL REACTION INJECTION MOLDING MACHINE

DRY GAS

DRY GAS

INJECTION

RECIRCULATION
METERING CYLINDER
FILLING

INSULATED
ENCLOSURE

HEAT
CONTROL
UNIT

HEAT
CONTROL
UNIT

MIXHEAD

MOLD

FIGURE 17.17
Reaction injection molding process (RIM) (Cincinnati Milacron Plastics Machinery Division).

FIGURE 17.18
Platen-type reaction injection molding machine (RIMM) uses the latest technique (RIM) in injection molding processes. This illustration shows the machine with essential safety guards removed so as to better show machine parts (Cincinnati Milacron Plastics Machinery Division).

to be formed into complex shapes by many of the previously discussed processes. Composite materials are undergoing extensive development. Their use in automotive and aerospace applications is already widespread and will be expanded in the future. Both thermoplastics and thermosetting plastics are used in advanced composite materials. Research has shown that the thermosetting plastic will probably be more popular for use in advanced composites. The reason for this is that many applications of advanced composites required the product to be subjected to considerable heat. The thermosetting plastics are more suitable to these applications.

Advanced composites consist of the plastic material reinforced with some type of stranded fiber material. The fiber reinforcement material usually makes up about half of the total material weight. Examples of the fiber materials used in composites are graphite and glass. The fiber functions as a structural component of the composite and is designed to take the load stresses applied to the composite structure. Were it not for the fiber portion of the composite material, only the resin portion of the material would be subjected to the applied loads. This structure alone would not be able to withstand the forces applied and would fail in service (Figure 17.26, see page 370).

Composite Manufacturing Methods

Three major techniques of manufacturing composite material are in common use. The first process is a reverse extrusion process or pulltrusion (Figure 17.27,

FIGURE 17.19
Extrusion process.

FIGURE 17.20
Twin-tapered extruder screws, uniquely engineered major components of a plastic extruder (Cincinnati Milacron Plastics Machinery Division).

see page 370) in which the fiber portion is pulled or drawn through the liquid resin and then through a heated die that forms the desired shape. The process is much like extrusion except for the pulling rather than pushing of the material through the extrusion die. Composite products manufactured by pulltrusion methods include structural members and tube.

In the technique of filament winding (Figure 17.28, see page 371), the fiber is wound back and forth on a cylindrical form. This method is used to produce cylindrically shaped products such as tanks or other pressure vessels. After curing, the form is removed, leaving the hollow composite product.

Laminating alternate layers of resin containing the structural fiber is the third major method of composite manufacturing (Figure 17.29, see page 371). This process is similar to the fiberglassing technique previously discussed. However, the composite fibers are continuous throughout the material, whereas in fiber glass short pieces of glass fiber are randomly situated in the resin structure.

COMPOSITE APPLICATIONS

As composite materials undergo further development, their uses will extend to a larger product line. In the beginning of this chapter, the applications in aircraft were mentioned briefly. In this application, the aircraft design engineer seeks materials with favorable strength-to-weight ratios. Strength coupled with light weight can result in products that are less heavy and thus require less fuel to propel them. This consideration is extremely

FIGURE 17.21
Plastic extrusion machine for plastic pipe. Tooling is seen attached to the machine on the left side (Cincinnati Malcron Plastics Machinery Division).

FIGURE 17.22
Reinforcing resin with glass fibers in the process of fiber glassing yields extremely durable structures such as boat hulls (Wellcraft Marine Corporation).

FIGURE 17.24
Smoothing fiber-glass reinforced resin in boat manufacturing (Wellcraft Marine Corporation).

important in the production of aircraft, spacecraft, and automobiles of today and tomorrow.

The honeycomb composite structure creates a product that demonstrates these highly desirable characteristics of light weight and strength. In Figure 17.30 (see page 372) aerospace technicians lay an aluminum honeycomb core on a graphite composite skin. The metal honeycomb, which is in itself a high-strength light-weight material, will be sandwiched between two layers of the graphite composite. The entire assembly will then be cured by heat and pressure, thus making a durable aircraft structural component.

Large components can be made more easily from composite materials than from metals. Figure 17.31 (see page 372) shows a large graphite composite component for a commercial jet liner. The use of honeycomb composite

FIGURE 17.23
Glass fiber application (Wellcraft Marine Corporation).

FIGURE 17.25
Finishing a fiber-glass reinforced boat hull (Wellcraft Marine Corporation).

FIGURE 17.26
Load applications to resins, reinforced plastics, and composites.

structure is seen in Figure 17.32 (see page 373). This large fabrication will become part of an aircraft jet fan reverser, once again an application where high strength and light weight are desirable engineering design specifications.

TOOL AND DIE MAKING FOR PLASTIC AND COMPOSITE PROCESSING

As is the case in processing most raw materials into useful products, special tooling is required in the plastic and composite processing industry. Here the raw material is most often in a liquid or flexible state prior to its formation into the end product. Therefore, considerable tooling is required to hold or form the material until steps are taken in the process that alter the state of the material to make it rigid.

In processing plastic and composites, much of the tooling appears in the form of molds or forms. Most of this tooling is made from metal and is manufactured to exact dimensional specifications. Figure 17.33 (see page 373) shows a large form tool and bonding fixture designed for shaping graphite composites.

Tool and die making is therefore an integral and indispensable part of any manufacturing industry. The tool

FIGURE 17.27
Manufacturing composites by the pulltrusion process.

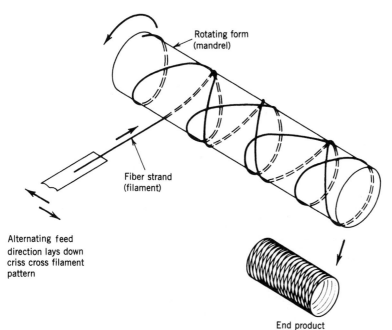

FIGURE 17.28
Manufacturing composite by
the filament winding process.

Rotating form
(mandrel)

Fiber strand
(filament)

Alternating feed
direction lays down
criss cross filament
pattern

End product

and die maker, who represents the upper end of the skilled machinist trade, will always be needed to build the tooling that makes mass production possible. The toolmakers of today and tomorrow have at their disposal all of the modern improvements that an advancing technology makes possible. Among these are computer aided design (CAD) and the versatility of computer numerical control machining (CNC) for making the high-precision tooling required by space-age manufacturing.

Modern methods of high-production plastic and composite part production will yield parts at very low cost. However, the tool and die cost must be included in the product price. Each plastic part produced includes a portion of the tooling price. The first part produced represents the full cost of tooling plus the material consumed in making the part. As each successive part is produced, the cost of tooling is spread further and further out. Eventually enough parts are produced to underwrite the tooling cost and from then on, the only cost is for material and production overhead expenses. If a large number of production parts are further produced, a good return is received on the tooling investment.

Production tooling, once manufactured, can be stored and reused at a later date. If a manufacturer receives an additional order for the same product, it is not necessary to remanufacture the tooling, and the customer may be able to purchase the manufactured product at a more reasonable price.

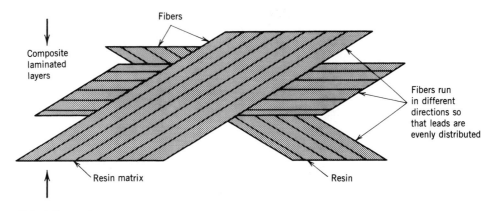

Fibers

Composite
laminated
layers

Fibers run
in different
directions so
that leads are
evenly distributed

Resin matrix

Resin

FIGURE 17.29
Manufacturing composites by the lamination process.

FIGURE 17.30
Aluminum honeycomb sandwiched and bonded between graphite composite skins makes an extremely strong and lightweight structure for the F-18 fighter aircraft (Northrop Corporation).

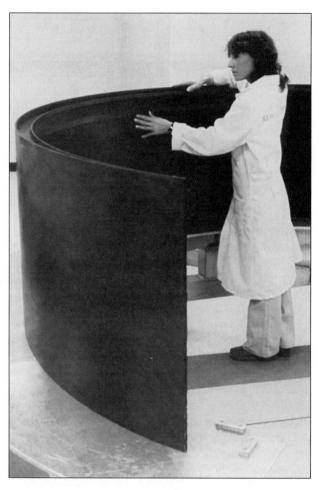

FIGURE 17.31
Commercial jet liner lightweight transcowling made from graphite composite material (Martin Marietta Corporation).

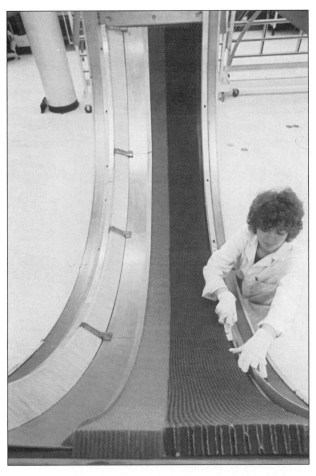

FIGURE 17.32
Aircraft fan reverser component showing composite fiber honeycomb structure (Martin Marietta Corporation).

FIGURE 17.33
Tooling is all important to the formation of plastic and composite products. A technician lays up graphite composite material on a large precision bonding fixture (Martin Marietta Corporation).

Review Questions

1. What factors are involved in selecting a plastic or composite material over metals?

2. What are the advantages of using these materials rather than metals?

3. List three major methods of processing plastics and describe how they work.

4. Give an example of a blow molded product.

5. Cite an example of an injected molded product.

6. Cite an example of an extruded product.

7. What is RIM and how does it differ from conventional injection molding?

8. Briefly describe molding processes other than RIM.

9. What types of plastic products can be made by machining?

10. Describe reinforced plastic material and cite a common example.

11. What is the purpose of reinforcing plastic and what materials may be used in this process?

12. Name a use of high-pressure laminates.

13. How does a composite material differ from a reinforced plastic?

14. What medium provides the structural component in a composite?

15. List and describe three major methods of composite manufacturing.

16. What process would be best for manufacturing a composite water tank? A composite structural beam? An aircraft composite deck walkway?

17. What is tool and die making and where does it fit into the plastics and composite manufacturing industry?

Case Problem

Case 1: Choosing Different Material

To conserve weight in an aircraft structure, an analysis is made of the aluminum structural beams that support the internal platform decks. Since these components are not flight-critical items, their strength-to-weight ratios can be maximized. In other words, they need only be strong enough to support required loads, but need not have the safety factors required of such flight-critical components as wing and tail spars.

One proposal advanced by structural engineering is to cut lightening holes in all deck beams to conserve weight. However, CAD analysis suggests that if enough material were cut from the beams to make a significant weight saving, the beams would be borderline as to their load-carrying capabilities. What would you suggest to solve this problem?

CHAPTER 18

Corrosion and Protection for Materials

Objectives

This chapter will enable you to:

1. Discuss the causes of corrosion.
2. Explain how corrosion may be slowed or prevented.

Because many materials, especially metals, react chemically with their service environments, it is often necessary to provide protective coatings and coverings for them to prevent or slow surface deterioration. Failure to provide surface protection can result in rapid destruction of many types of materials, thus requiring expensive replacement. Other material, such as wood and plastics, are subject to physical deterioration from weather, sunlight, and water must also be protected from the environment. The purpose of this chapter is to survey the various means by which materials may be protected from environmental degradation.

METAL CORROSION

Most common metals exist in nature chemically combined with other elements, thus forming the various oxides (ores). Examples include iron ores from which metallic iron is extracted and bauxite from which aluminum is derived. These ores are mined and refined by various processes to extract the metallic components that in turn become the common familiar metals. There is a natural tendency for many refined metals to revert to their natural state after the metallic component has been extracted. Because both air and water are common service environments for many metals and because oxygen is a large percentage of both, many metals will try to reunite with the oxygen present as they attempt to revert to their original oxide (ore) state.

Iron is a good example. Existing in nature as an oxide (iron ore), the metallic element iron is chemically combined with the oxygen in the air. When the metallic portion of the ore is separated from the oxygen during the iron ore refining process, the iron will tend to reunite with any oxygen present and revert to its original more stable state of iron oxide. This process, a slow oxidation of the iron, is known as rusting. If the temperature of the iron is raised to a high level, the oxidation process is much more rapid. However, not all metals oxidize in this manner. Gold, for example, does not readily unite with oxygen and is therefore found in nature in its pure form.

Metal corrosion can be classified according to the two major processes by which it occurs. One is the process just described, oxidation, and is called direct oxidation corrosion; the second process is galvanic corrosion, in which corrosion takes place at normal temperatures in a moist or electrolytic service environment.

The first process, direct oxidation corrosion, occurs slowly, as in normal rusting, or quickly, as in the case of steel, where heating to high temperatures for the purpose of hot rolling or heat treating leaves a layer of black oxide on the material. This is called mill scale and is noticeable as the black coloring seen on most structural steel.

Galvanic corrosion gets its name from the galvanic effect. This is the electrical phenomenon of current flow resulting from the presence of a potential (voltage) causing a current to flow through an electrolytic (conducting) medium. Galvanic corrosion is greatly enhanced if the metal being corroded is surrounded by an electrolyte. An electrolyte is any solution that can conduct an electric current. Fresh and saltwater are examples of electrolyte environments. Under these circumstances, the materials are subject to galvanic corrosion.

For galvanic corrosion to occur, the electrolyte must form a path for an electric current to flow through and a situation must exist where a potential (voltage) difference can exist. These conditions are easily met by placing two different metals in an electrolytic environment or by a situation in which two different areas on the same piece of metal act as the cathode and anode in the electric circuit. During the process of galvanic corrosion a small current flows in the circuit established between the cathode and anode. The electrolyte provides the path for current flow. In this corrosion process, the metal is actually deplated. Galvanic corrosion can be accelerated greatly if the electrolyte is efficient and the potential difference between cathode and anode is great. But the process can also take place where the electrolyte is barely identifiable. A fingerprint, for example, acting as the path for current flow, can cause galvanic corrosion to occur on a metal surface. For this reason some manufacturing of metal products subject to corrosion may be accomplished in clean-room environments where air moisture is carefully controlled. Parts are never touched directly.

PROTECTION METHODS

The particular method used to protect a given material from influences of the environment will depend on several factors, such as cost, how corrosive the environment is, and how directly the need to protect the material is tied to structural design requirements. Protection methods include:

1. Shielding the material from its service environment by coating it with another material. Examples include painting, plating, cladding, and treating with various oxides. As long as the integrity of the coating is maintained, the base material will be preserved.

2. Establishing the chemical makeup of the parent material such that service environments do not cause it to degrade. For example, the alloying of metals with noncorrosive alloys makes them less subject to corrosive forces. Although this is frequently done for many manufactured products, it is generally much more expen-

sive than coating a less expensive material with surface protection.

3. Allowing the corrosive activity to continue in a normal way, but to be concentrated on other nonstructural parts made from materials that are designed to accelerate corrosion and thus concentrate the corrosive activity. In other words, nonstructural parts of the assembly are sacrificed in order to concentrate and thus reduce the corrosive activity on essential structural parts.

An example of this third protection method is the **sacrificial anode** method, a technique used quite extensively where steel and water must coexist. Such a service environment is highly favorable to galvanic corrosion. To slow the corrosive process as it reacts against the base material, nonstructural sacrificial anodes are used. Such anodes are made from a different metal with a high susceptibility to galvanic corrosion. Thus, the corrosive process tends to concentrate on the anodes, which are sacrificed to preserve the base material. Although the method is effective, large fabrications may require many tons of sacrificial anodes—which ultimately must be replaced to maintain the concentration of electrolytic activity on the nonstructural members of a fabrication.

Cladding

Cladding is a thick layer of one metal applied to another metal to protect against corrosion to improve mechanical characteristics such as hard facing, or in some cases to improve appearance. Cladding may be done by weld buildup of a layer of metal, or by adhesive bonding one metal to another.

In the clad welding process, weld beads are laid down on the base metal side by side and layer by layer. The process is tedious if done by hand and usually requires a shield gas welding process to eliminate slag pockets and other inclusions in the cladding. After clad welding, parts may be machined to achieve desired dimensional characteristics.

Alloying and Oxidizing

Alloying   Alloying is an effective but often expensive method of material protection. Certain metals, such as chromium, have excellent anticorrosion properties. Plain carbon steel, on the other hand, has an affinity for the oxygen in air and will rust (oxidize) if exposed to a normal outdoor environment. It is also subject to galvanic corrosion. However, if the material is alloyed with chromium, creating stainless steel, corrosion by rusting may be reduced or totally eliminated. Figure 18.1 shows pieces of stainless and plain carbon steel that have been exposed to

FIGURE 18.1
Unprotected carbon steel (top) is heavily rusted. Stainless steel (bottom) shows little corrosion.

normal outdoor environments for the same period of time. The plain carbon steel shows considerable rust, whereas the stainless steel shows very little evidence of corrosion. However, even stainless steel will corrode in some environments, so it is important to select anticorrosive alloys when using this method.

Other metals that will provide anticorrosion characteristics include copper-nickel alloys (Monel), molybdenum, lead, nickel, and pewter. Lead, used in lead acid batteries for plates, serves in an extremely corrosive acid environment. Gold is also an excellent anticorrosive metal and is often used to plate other metals for corrosion protection.

Another possible advantage of alloying is that the anticorrosion properties exist throughout the thickness of the metal. Thus, even if the metal surface is damaged by abrasion, denting, or scraping, the anticorrosive properties will be preserved. This is not the case with plated, painted, or otherwise coated materials. Since the anticorrosion medium is only on the surface and most often is extremely thin, scratch or scrape damage can expose base material to the environment, permitting corrosion to begin. Figure 18.2 shows painted steel on which the integrity of the paint has been breached, thus exposing the bare metal to the environment. This exposes a place where corrosion can begin.

Oxidizing A metallic material that tends to react with oxygen and oxidize may be protected by accomplishing this very process only on an outside layer of the material. Common oxidizing processes include black oxidizing and anodizing.

In black oxidizing, parts are heated in a carbon or oxidizing salt environment and then quenched. Black oxidized parts take on a black color and are thus surface protected from rusting.

Anodizing is a popular surface protection process, especially for aluminum. The process is done electrically

and results in a thin layer of oxide on the surface of the part. This layer prevents further oxidation by the service environment (Figure 18.3). Once again, the anodizing layer must be maintained intact. This layer, which is less than .0005 in. thick, can be damaged to the point where the base metal is exposed to the corrosive environment.

Plating

The various plating processes make up a large portion of methods used to protect metals from their service environments. The theory behind plating is that if a thin layer of anticorrosive material can be plated to a base metal, then the material can be protected from corrosion. Figure 18.4 clearly demonstrates the application and advantage of plating as an anticorrosion measure. Note how the unprotected steel on the bracket is heavily rusted. However, the zinc-plated fastener shows little or no degradation from corrosion.

Plating processes are used widely and have numerous advantages. For example, sheet steel is used widely for a vast number of products. It is easily formed, can be effectively assembled by high-production methods such as spot welding, and is relatively inexpensive. Its primary drawback is that it is subject to rapid rusting in typical service environments. One solution is to use stainless sheet steel. However, this product is much more expensive and less easily fabricated.

Common low-carbon sheet steel is one product that can be protected effectively from rusting by plating it with an anticorrosive metal (zinc) in the process of **glavanizing.** Thus, the useful production properties of sheet steel

FIGURE 18.2
Damaged surface of painted steel exposes base metal and permits corrosion to occur.

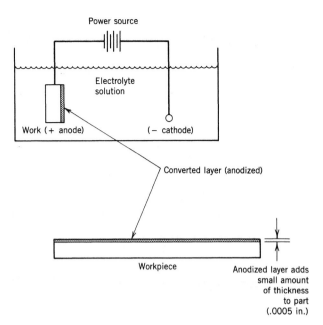

FIGURE 18.3
Anodizing process.

are maintained while preserving the material from corrosion from its service environment.

Electroplating Processes Many metals are electroplated onto other base metals by **electroplating** processes. Parts to be plated are placed in a tank containing a salt solution (electrolyte) of the plating metal. A direct current (DC) voltage is applied across the part to be plated (− cathode) and the source metal (+ anode). The current

flow causes plating metal ions to move from the source to the parts where are deposited as plated metal (Figure 18.5). Galvanizing is one common process where a zinc coating is electroplated onto steel (Figure 18.6).

Other metals that are electroplated include tin, cadmium, nickel, chrome, brass, silver, and copper. Gold has many industrial plating applications in electronics because of its superior electrical conductivity properties. Silver and brass plate add beauty and serve to resist corrosion (Figure 18.7).

Electroless Plating Processes Plating without electricity is also a widely applied process. Nickel is an example. The process of electroless nickel plating yields a more even thickness on complex parts than will electroplating processes.

Electropolishing

In the electropolishing process, a small amount of surface material is deplated from the part. The process is accomplished by placing the parts in an acidic electrolyte where a DC potential exists between parts (anode) and cathode (the other side of the power supply).

The process leaves an extremely smooth surface while not removing enough workpiece material to affect precision dimensional characteristics. It is also useful to smooth out part surfaces on irregularly shaped objects or where part surfaces are not visible or accessible.

FIGURE 18.4
Unprotected steel is heavily rusted while plated fastener shows no evidence of corrosion.

FIGURE 18.5
Electroplating process.

FIGURE 18.6
Examples of common galvanized products.

Painting

Painting is probably the most widely used process for surface protection of metals and other materials. Its primary advantages include ease of application, wide color variety, and reasonable cost. Its primary disadvantage is that because it is a surface treatment, damage can expose the base material to the corrosive environment. However, the process is in wide use and will probably remain a mainstay of surface protection.

There are many types of paints, and paint technology undergoes constant development. Lead-based oil paints have been used for many years but are giving way to water-based paints. Paints containing tin compounds are used as antifouling paints for marine applications such as ship hulls. These paints are toxic to the marine life that tends to attach itself to a ship's hull.

Many modern paints are plastic resin based, such as those from the acrylic group. Water is frequently the medium for carrying the pigment (solid paint particles with color). These paints are fast drying and do not cause the pollution problems that disposing of petroleum solvent fumes and liquids causes.

Paint Applications Paint is typically applied by dipping, brushing, rolling, and spraying. Spray technology is most popular in mass production applications and is the only way to achieve the truly smooth coat necessary in most manufactured product applications, such as on automobiles.

Spray painting can be done by air driven sprayers (Figure 18.8), but overspray must be captured and retained in order to meet most current air pollution requirements, especially if nonwater solvents are used. Personnel must also be protected during spray painting operations.

Electrostatic spray painting eliminates overspray and the resulting pollution. This process is extremely effective from two important standpoints—better coverage of product and less consumption of paint. In the electrostatic process, high DC voltage difference is maintained between the paint sprayer and the part to be painted. Paint particles flowing from the sprayer acquire the charge opposite to that of the part and are therefore attracted to it. The process is extremely fast, and since the part has an equal charge on all sides the paint is attracted to the back as well as the front. Thus, parts are literally painted all over at once (Figure 18.9).

FIGURE 18.7
Silver plating (left) and brass plating (right) add both beauty and resistance to corrosion.

FIGURE 18.8
Spray application of paint (Lockheed Co.).

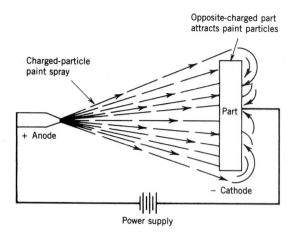

FIGURE 18.9
Electrostatic spray-painting process.

High-pressure paint pump sprayers have become very popular in recent years. In this equipment, the paint is atomized by pumping it through a small nozzle at very high pressure. High-pressure pumping is very fast in application and the technology is readily available.

Paint spraying, regardless of the application equipment, requires considerable skill on the part of the operator to achieve a smooth and even coat on the workpiece. Like many other industrial processes, spray painting can be accomplished effectively with the computer-driven industrial robot. Figure 18.10 shows some unique developments in this area. The auto bodies are positioned diagonally on the conveyer so that the spray-paint robots may access them for the automated production painting operation. Since painting must be accomplished on underside surfaces and overhead locations, the jointed-wrist type industrial robot is popular for this application.

Paints must dry after application. This can be a time-consuming operation and can be speeded up by the application of heat and circulating air. Parts are conveyed to a heated drying facility where heaters expedite the drying process.

PREPARING MATERIALS FOR SURFACE PROTECTION

Before material can be surface protected, it must be properly prepared. Failure to do this will result in a poor application of the particular surface protection process. Several methods are used to prepare materials.

Mechanical Methods

Commonly used mechanical methods physically abrade the material to remove mill scale, dirt, and corrosion. For example, in sand or shot blasting, high-velocity abrasive particles are propelled against the workpiece. Abrasives in the form of sandpapers, abrasive wheels, or wire wheels (brushes) may also be used.

Tumble cleaning and deburring is another popular method for both cleaning and removing burrs from machined workpieces. In this process, parts to be cleaned or deburred are placed in a vat or vibratory bowl with small pellets of abrasive material (Figure 18.11). The bowl is vibrated rapidly, causing the parts to be tumbled along with the abrasive material. Equipment used for this process ranges in size from small (Figure 18.12), to quite large (Figure 18.13, see page 382). Vibratory cleaning and deburring machinery also can be equipped with washing, drying, and material-handling conveyer systems (Figure 18.14, see page 382).

FIGURE 18.10
Production spray painting with industrial robots (Chrysler Corporation).

FIGURE 18.11
Vibratory deburring machine with a capacity of 160 ft³ showing compartments, workpieces, and abrasive pellets (Automated Finishing Inc.).

These methods are satisfactory but may be labor intensive, dirty, and sometimes dangerous. It is also difficult to clean all surfaces on a complex part by brushing or grinding. Sand blasting is a very abrasive process and will erode metal and wood surfaces rapidly. Using glass beads for the abrasive eliminates most of this effect, but this process must be done in a confined area where the fine dust can be collected and removed from circulation. Sand blasting must also be done under controlled conditions and the personnel must be protected from the flying sand.

Ultrasonic cleaning of parts is accomplished in a container with a cleaning medium. The bath is vibrated by ultrasonic sound waves. The process is a combination of both chemical and mechanical cleaning.

In the process of buffing, parts are brought into contact with a rapidly rotating wheel, usually made from layers of cloth sewn together. Various grades of abrasive materials in cake form are hand applied to the wheel. These materials include various metallic oxides and rouge. The buffing process provides highly polished mirror-like finishes on metals. However, it is labor intensive and often requires that the part be previously abraded with coarser materials so that final buffing will yield a desirable result. This process is popular in art metal and jewelry metal work.

Chemical Processes for Cleaning and Material Preparation

Chemical preparations are usually necessary prior to painting. In pickling, material is dipped in an acid bath to remove oxides. This process must be followed by a rinse to neutralize the pickling acid.

Solvents can be used to remove grease and oil. The parts can be placed in the vapor of the solvent. When vapor condenses it washes the parts. This process is called vapor degreasing. Alkaline solutions are also used as a washing agent.

FIGURE 18.12
Vibratory deburring machine with a capacity of 14 ft³ showing material-handling conveyers (Automated Finishing Inc.).

FIGURE 18.13
Large 160 ft³ capacity vibratory cleaning and deburring equipment (Automated Finishing Inc.).

To provide for a better bond between paint and material, two phosphate coating processes are used. In bonderizing the parts are dipped in a phosphating solution. The resultant coating permits paint to adhere more readily. The phosphate process of parkerizing also provides a corrosion-resistant primer coating under a painted surface.

Review Questions

1. Name the primary causes of corrosion.
2. Why not manufacture everything out of solid anticorrosive materials?
3. Describe plating processes and discuss their advantages and disadvantages.
4. What is anodizing?
5. What is galvanizing?
6. How can paint coatings be applied?
7. Are there any disadvantages to painting as a corrosion-prevention measure?
8. Describe electrostatic painting and discuss its advantages over air spraying.
9. How can materials be mechanically cleaned prior to painting or plating?
10. What chemical preparations are used to prepare materials for plating or painting?

Case Problems

Case 1: Abrasion Resistance

An escalator side plate is subjected to light but constant abrasive forces. In your opinion, what would be the best method listed below of protecting this material, with

FIGURE 18.14
Through-flow vibratory machines equipped with part washers, dryers, and material-handling conveyers. (Automated Finishing Inc.).

an emphasis toward minimum maintenance and long-term preservation of the product?

- Plating
- Painting
- Anodizing
- Making from solid material
- Galvanizing

Case 2: Rust Prevention

Large steel fabrications are designed to be used in a brackish estuary. The products are painted with the best available anticorrosive marine paints. What other methods may be used to further prevent corrosion of these items in their service environments?

CHAPTER 19

Processing Other Industrial Materials

Although metals and metalworking make up a large portion of materials and manufacturing processes, many other materials and processes are equally important in manufacturing. Ceramics and wood, for example, have long been in use. In the age of high technology, these materials are undergoing new developments, and many new applications are being found for them in the manufacture of industrial products.

Many of these other materials are used in conjunction with metals in all types of products. In many cases some of these materials, because they are cheaper, stronger, lighter, and easier to process, are replacing metals. Foremost among these are plastics and composites, as previously discussed. Important industrial materials discussed in this chapter include:

1. Glass
2. Ceramics
3. Wood, wood products, and paper
4. Fabrics
5. Rubber
6. Natural materials
7. Construction materials.

Objectives

This chapter will enable you to:

1. Be generally familiar with common industrial materials other than metal and plastic.
2. Be familiar with their manufacturing processes.

GLASS

Glass consists primarily of fused silica SiO_2, a major constituent of beach sand. The major uses of glass are widespread and include many common products such as windows, containers, and lenses. The glass window wall commercial building has become very popular in modern building construction (Figure 19.1).

The advantages of glass containers are unsurpassed. Glass will withstand chemical attack and can also withstand large variations in temperature. Lead crystal (Figure 19.2) is noted for its many beautiful designs in tableware. Lead is added to glass for windows that provide shielding from ionizing radiation while permitting radioactive equipment to be observed safely.

Another outstanding property of glass is its ability to be fashioned into precision optical lenses for eyeglasses, telescopes, microscopes, and many other optical instruments.

A relatively new and developing technology in the uses of glass is fiber optics. Here, the glass is drawn into a thin strand, much like a glass wire. The strands are grouped into a fiber-optic bundle and then sheathed in plastic for durability and protection.

Fiber-optic technology is a rapidly developing field for the transmission of information in telecommunications

FIGURE 19.1
Plate glass used in building construction turns entire walls into windows.

and computers. Figure 19.3 shows a fiber-optic bundle designed to transmit digitized data in an aircraft communications system. Note the difference in size between the fiber-optic bundle as compared to the standard aircraft cable in the lower right side of the illustration. The fiber bundle is much smaller and lighter and it can transmit more channels of data than its wire cable counterpart.

Sheet glass is manufactured on a production basis by the latest computer-controlled and automated production methods. The float method (Figure 19.4) produces a continuous ribbon of glass. This product is the material for sheet glass products such as windows, mirrors, and automotive items.

Glass for automotive windows is made by a lamination process in which a plastic interlayer is sandwiched between two layers of glass. This makes the material shatterproof. After assembly (Figure 19.5), the product is placed in an autoclave (Figure 19.6, see page 388), where a bonding between glass and plastic interlayer occurs.

Other glass product manufacturing processes include pressing, blowing, and combination press and blow operations (Figure 19.7, see page 388). These processes are done on production equipment in a highly automated manner (Figure 19.8, see page 389) and are used primarily in glass container production.

Hand glass blowing is a long-established process for producing containers and other glass art objects. Although the process is mostly automated for production applications, it is still done to some extent on an art and craft basis. In it a quantity of molten glass is retrieved from the furnace on the end of a blow pipe. Breath pressure is applied to the opposite end of the pipe causing the glass to inflate and assume a hollow shape. Rotating the blow pipe

preserves the desired round shape of the product (Figure 19.9, see page 389). Several types of tools and trimmers are used to further shape the glass before it cools (Figure 19.10, see page 390). The process lends itself well to the production of many beautiful and artistic glass products.

FIGURE 19.2
Lead crystal makes highly artistic and beautiful glass products.

Other glass manufacturing processes include cutting sheets into various shapes, edge beveling, grinding, and hole drilling so that glass pieces may be assembled by using mechanical fasteners.

Tempering turns standard glass into plate glass. The glass is heated in a large furnace to a semiplastic state. It is then removed from the furnace and quickly air quenched. This process sets up closely spaced swirl-shaped stress patterns in the glass. Tempering is done to provide glass with mechanical strength for such applications as refrigerator shelves, store fixtures, doors, and windows. Should plate glass be broken, many small relatively harmless pieces result, as opposed to large very hazardous slivers. Thus, plate glass is specified for storefront windows, doors, and floor fixtures that might be subject to breakage by weather or impacts. Tempered glass also exhibits superior mechanical strength, a requirement for many glass applications.

Glass in its multitude of product forms is an extremely useful and versatile industrial material. Annual glass production in the United States amounts to several million tons per year.

FIGURE 19.3
Glass fiber optics, a developing technology, uses strands of glass to transmit light signals in communications and computer systems. Note the difference in size of the glass fiber as compared to wire cabling that would carry equivalent electrical signals (Lockheed Co.).

FIGURE 19.4
Float glass manufacturing produces a continuous ribbon of glass (Libbey-Owens-Ford Company).

FIGURE 19.5
Assembly of laminated safety glass windshields involves placing an interlayer of plastic between two glass layers that have been cut and bent to the windshield shape (Libbey-Owens-Ford Company).

FIGURE 19.6
Curing automotive safety glass products in a large autoclave (Globe-Amerada Glass Company).

CERAMICS

High-technology product design has motivated great interest in and development of ceramic engineering materials in recent years. Ceramics are made from the metallic oxides of such metals as silicon, aluminum, magnesium, and others. Clay-based ceramic materials are also in wide use and are extremely popular in arts and crafts.

Materials used for industrial ceramic products include carbides, borides, nitrides, and silicides. Ceramic materials have some unique properties that make their applications for industrial products unique. These advantageous characteristics include having high melting points (up to 7,000°F), being chemically inert, being refractory (heat reflective), and withstanding large compressive (crush or pressure) loads. Certain ceramics such as those from the silicon nitride group are as hard as diamond and as light as aluminum. Possible disadvantages of ceramics material include brittleness and a tendency toward weakness when subjected to tension (pull-apart) loads.

FIGURE 19.7
Glass manufacturing process for containers (Wheaton Glass Division, Wheaton Industries).

FIGURE 19.8
Automated glass container manufacturing (Wheaton Glass Division, Wheaton Industries).

From the viewpoint of the industrial product designer, the ceramic property of resistance to high heat levels is one of the most interesting. Ceramic materials will withstand temperatures of several thousand degrees and still maintain their shape and strength. It is this characteristic that makes them useful for making parts that will be subjected to extreme temperatures. Examples include jet engine turbine components, reciprocating engine valves, and exhaust gas driven turbines in automobiles.

Manufacturing processes for ceramics include extrusion, pressing, casting, machining, injection molding, flame sprayed coatings, and sintering. After the product is shaped using raw ceramic materials, it must then be furnace fired to achieve its final strength and form characteristics. The plasma arc with its extremely high temperature characteristics has been used to produce high-quality ceramic powders, the raw material from which ceramic products are produced.

Uses of ceramics include cutting tools for machining, abrasives for grinding, and coatings on cutting tools to improve their performance characteristics. Industrial products include turbine rotors, engine valves, electrical insulators, bricks, tiles, dishes, and containers. Research is under way to develop mechanical ceramic components such as ball and roller bearings.

WOOD, WOOD PRODUCTS, AND PAPER

Throughout the history of manufacturing, wood has been an important and useful material with which to make a multitude of products. Since it is a natural and often readily available material, it is logical that its uses is widespread. Wood is also a renewable resource and modern forest conservation

FIGURE 19.9
Glass blowing and shaping by traditional methods (Wheaton Glass Division, Wheaton Industries).

FIGURE 19.10
Shaping blown glass using traditional hand tools (Wheaton Glass Division, Wheaton Industries).

practices should ensure that we have a continuing supply. With the increasing cost of metal and the energy expended in metal processing, the interest in wood as a structural building material is undergoing new interest.

It is interesting to note that wood was a fundamental fuel source well into this century. In fact, one factor that contributed to the Industrial Revolution, after which manufacturing shifted toward coal and petroleum-based fuel supplies, was a shortage of firewood accompanied by high wood prices. Today there is a renewed interest in wood as a fuel source for domestic heating and even for industrial and power generation purposes. However, the return to wood fuel on a large scale is unlikely because of cost, availability, air pollution control, and its primary value as a construction material.

Wood and Wood Product Processing

Many types of woods are used in constructing wood products. The primary use of wood is in structural lumber for homes and small building construction (Figure 19.11). Trees that grow straight and tall are most useful for lumber. These include pine, fir, spruce, and redwood. Major lumber supply areas include the western forests of North America. Trees are selected, felled, and transported to high-production sawmill operations where standard structural lumber is produced. The modern sawmill makes use of modern manufacturing automation and standard lumber is produced at a high rate.

Waste from sawing, which includes sawdust, slabs, and trimmings, was at one time burned at the mill site. Today, sawdust and wood chips are too valuable to be burned and are used to produce many other wood products, such as plywood and chip wood products. These materials play an important part in supplementing structural lumber as a building material.

Wood Products   In addition to structural lumber, wood products include plywoods in which thin layers of wood are glued into panels. The grain of each layer runs perpendicular to the next (Figure 19.12). Plywood is an indis-

FIGURE 19.11
Structural lumber is a versatile building material for house construction.

FIGURE 19.12
Plywood consisting of veneer layers glued together in sheets is an exceptional building material.

pensable building material in modern wood building construction. Plywood panels are extremely strong and are available in standard sized sheets in several thicknesses.

Other wood products include wood chip products such as flake board (Figure 19.13). This material allows wood waste to be used in a profitable manner. The material is quite suitable for a variety of products, such as cabinets and furniture.

Wood should not be discounted as a structural building material. Glue laminated beams (Figure 19.14) are well suited to significant structural building requirements. They will support large loads and are often lighter and cheaper than their metal counterparts.

Hardwoods The list of hardwoods used for furniture and flooring is long and distinguished. Common hardwoods used in furniture manufacturing include mahogany, rosewood, oak, teak, birch, maple, hickory, and walnut.

Several other woods from tropical areas are also used in furniture manufacturing.

Furniture manufacturing is a large industry employing many of the same manufacturing techniques and fastening systems used in metalworking. Wood must be cut and shaped and smoothed. Woodworking machine tools and processes resemble their metalworking counterparts in many ways. The most popular process is assembly with modern glues and adhesives as well as mechanical fasteners, including nails, staples, bolts, and screws. Typical wood manufacturing equipment includes lathes, shapers, sanding machines, and various types of saws. Great effort is placed on surface finishing and artistic design in the wood products (Figure 19.15). Woodworking is one area

FIGURE 19.13
Flake board, a wood chip product, makes an effective building material and makes use of wood chips that would otherwise become waste products.

FIGURE 19.14
Glue laminated wood beams are an excellent structural building material for buildings of large size. This material demonstrates high strength-to-weight ratios while minimizing cost of the structural components.

FIGURE 19.15
The value of hardwoods for fine furniture construction will never be replaced (Ethan Allen).

of manufacturing in which hand crafting (Figure 19.16) and hand carving still have an important place. Artistic specialty areas are unparalleled (Figure 19.17).

Paper

The importance of paper as a modern industrial material should not be underestimated. Paper is an indispensable material for the support of the modern life-style. Major uses of paper include printing and packaging. Modern

paper manufacturing makes available a large number of specialty paper types to meet the many specialized needs of the paper user. Heavy gage paper is used as an inexpensive construction material for many consumer products and paper has even been used for clothing. Rising energy costs and an emphasis on conservation of natural resources has prompted the growth of a large wastepaper recycling industry in recent years. Today, wastepaper is a valuable commodity and can be reused effectively rather than dumped or burned.

The primary raw material for paper is cellulose fibers from wood or other plants. Various forest trees are used for wood pulp and, at present, much paper is made from recycled wastepaper products. Various fabrics in rag form including felt, linen, and wool are also used in papermaking to achieve desirable paper characteristics, such as strength and texture.

The papermaking process is somewhat complex and, like many other continuous industrial processes, is highly automated, carefully monitored, and computer controlled. In the papermaking machine, a liquid mass of wood pulp and rag pulp is fed onto a traveling wire screen conveyer. As the wire screen moves, water drains from the wet pulp fiber mass transported by the screen. Suction systems remove more water, and the pulp mass begins to solidify as paper. Further operations remove additional water by pressing and rolling. Final finishing is accomplished by calendering rolls exerting many tons of pressure on the paper and providing the desired thickness and surface finish.

FABRICS

Fabrics are important for clothing and for upholstery both in the furniture and the auto industry. Fabrics may be divided into two major classifications, synthetic and nat-

FIGURE 19.16
The construction of fine furniture will always ensure employment of the artist and highly skilled craftsperson (Photo courtesy of Ethan Allen).

FIGURE 19.17
Construction of fine wood furniture often involves delicate hand work during the manufacturing process (Photo courtesy of Ethan Allen).

FIGURE 19.18
Air jet weaving machines produce high-quality woven fabrics at high production rates (Burlington Industries).

ural. Synthetic fabrics are primarily derived from plastic technology. Polyester is an example. Synthetic fabrics are in widespread use for clothing, carpeting, and upholstery applications. Natural fabrics made from natural fibers include cotton, wool, and silk.

Many types of fabrics, both synthetic and natural, are made by weaving threads of the material into cloth of sufficient size for manufacturing fabric products. Modern fabric mills make use of highly automated and computer-controlled water and air jet weaving machines (Figure 19.18).

Fabric processing involves cutting and joining primarily by stitching. All of the modern methods are applied, including computer control of cutting and

sewing. Mass production piece-part cutting of clothing pieces may be done by band sawing or with laser cutting systems. Modern fabric manufacturing is a continuous and highly automated process where computer process control is widely used (Figure 19.19). Fabric mills have made use of the latest factory automation such as the wire guided computer-controlled material-handling system shown in Figure 19.20. Fabric finishing and dyeing are maintained at a high level of consistency using computer-controlled processes (Figure 19.21).

FIGURE 19.19
State-of-the-art fabric production making use of modern CAM results in products of high quality (Burlington Industries).

FIGURE 19.20
Automation in fabric manufacturing material handling involves driverless computer-controlled transports (Burlington Industries).

FIGURE 19.21
Computer technology influences all phases of fabric manufacturing. Consistency, efficiency, and quality are computer monitored during the dyeing and finishing process (Burlington Industries Inc.).

processes for products such as tires. Extrusion is used for products such as tube, moldings, and hose. Calendering (rolling) is used to make sheet products and to coat rubber on fabrics and metal.

NATURAL MATERIALS

Natural materials include clay and adobe from which bricks are made. These materials have long and distinguished histories as fundamental building materials. Stone, including granite, marble, slate, and sandstone have also been used as material for buildings and decorative facings. Although these materials are giving way to concrete, steel, glass, plastic, and aluminum in modern buildings, they are still used to some degree as construction materials and are still very much in evidence in distinguished architecture throughout the world.

Leather, derived from animal hides, is another natural material with a long history of use and it is still much in use today. Shoe manufacturing makes use of leather. Although many modern shoes make use of plastic and other synthetic materials, leather still remains a superior material for this application.

Natural fabrics such as wool, cotton, linen, and silk still have wide appeal for clothing and other uses. Even though synthetic fabrics are in wide use, natural fabrics will always have a prominent place in fabric product manufacturing. They are processed by all the conventional manufacturing processes such as weaving, cutting, stitching, and bonding, and their product applications are widespread and well known.

RUBBER

Natural rubber or latex is derived from the natural gum of the rubber tree. Prior to World War II, this material was important in the manufacture of many products, and sources of supply were considered to be valuable. The geography of the war placed sources of natural rubber out of reach, and the needs of the allied military exceeded the available supply. The situation stimulated the development of synthetic rubber chemically derived from the basic chemical constituents of natural rubber. Once the chemistry of rubber was understood, a virtually unlimited supply of the material was then assured.

The applications of rubber are widespread. The material is used for coatings, tubing, hose, and tires. Rubber processing methods include many of the same processes used in plastics. Among these are various molding

CONSTRUCTION MATERIALS

Construction materials include all those from which structures are built, whereas engineering materials are those from which products are made. Construction materials include many natural materials such as stone and wood, or material made from natural components such as bricks and rammed earth (compacted dirt). All of these materials have long and distinguished histories and will probably always have some application for building certain types of structures.

In modern construction of large structures such as buildings, bridges, and highways, steel and concrete stand out as the most versatile and widely used materials. Structural steel is a totally essential material for modern large-scale construction. Modern steel production along with advanced metallurgical technology have developed high-strength structural steel that permits the safe construction of extremely tall buildings. Plans are presently being drawn for skyscrapers that will exceed 2000 ft in height (200 stories) but will occupy a minimum of space at their base. It is likely that future building construction in urban areas will favor high-rise design in order to save ground space. Structural steels will play an ever increasing part in this type of building construction.

When steel and concrete are used jointly, an unsurpassed construction material is created. Concrete (portland cement) is primarily a mixture of limestone and clay that has been pulverized to an extremely fine powder. This material is mixed with proper portions of sand, graded gravel (aggregate), and water, making a thick fluid mass. Large mixing machines that are often truck mounted are used to prepare the mix. Thus the concrete may be easily delivered to a job site.

In most cases where concrete is the primary construction material, it will be reinforced with steel reinforcing rod, often known as rebar. Steel reinforcing allows much more weight to be supported by concrete structures. Thus, the material is exceptional for dams, bridges, runways, highways, and buildings. Steel reinforcement rod and bar is often arranged in a lattice work and laid in the form before the concrete is poured. The lattice may be welded or wired together to hold it in position. After pouring the concrete, the steel lattice is permanently encased and can then act to increase the load-bearing capacity of a concrete structure.

Reinforced concrete has many advantages as a construction material. It can be cast into almost any shape using simply constructed wood forms. It can bear large loads. High-strength concretes containing polymers will withstand loads of over 10,000 psi.

Precasting of building wall components is a popular and versatile method of manufacturing building structural components (Figure 19.22). In this method, building walls are precast in molds, then removed and transported to the building site where they are bolted together. This method is very popular for commercial building and erection time is very fast. Cement walls for buildings can also be poured in place and when properly reinforced are suitable for high-rise buildings.

Review Questions

1. Name some advantages nonmetallic materials have over metallic materials.

2. What are ceramics and what are they used for?

3. What single property makes ceramic products so useful?

4. List some of the major manufacturing processes used to make ceramic products.

5. What are the major applications of wood and paper?

6. What processes are used in woodworking? Do these resemble machining processes in any way?

7. List some common natural materials.

8. What are the advantages, if any, to using natural material over synthetic materials? What are the disadvantages, if any?

9. What is concrete and what are its primary applications?

10. What can be done to concrete to enhance its strength?

FIGURE 19.22
Buildings constructed from precast concrete walls are erected quickly in a large variety of sizes and floor plans.

PART IV

MANUFACTURING SYSTEMS

P art IV discusses the design, tooling, and production aspects of manufacturing; modern automation methods; how the manufacturing industry functions; what job titles and general responsibilities are in the manufacturing industry; and how production manufacturing is accomplished.

Materials and specific processes are, of course, essential to the manufacture of any product. However, actual activities of manufacturing involve a great deal more than simply having material and processes available. In Part IV you will see how the materials of Part I and the specific processes of Part III are brought together to form a complete manufacturing capability. This capability produces the vast variety of products, ranging from individual parts to complex assemblies such as autos, aircraft, and consumer products, that establish the lifestyle in a modern industrialized society.

Accomplishing the miracles of modern production manufacturing requires the fully integrated effort of all the manufacturing team members. Products must be designed, prototyped, and tested. Production facilities and special tooling must be created. Manufacturing must be automated so that large numbers of exact duplicate product units may be made quickly, efficiently, and at competitive prices.

As the continuing research and development effort devises new materials and new methods, future manufacturing will be accomplished on an ever more automated and high-technology level. Individuals staffing manufacturing jobs in the age of high technology, will need to understand computers, industrial robotics, and fully automated factories. Training in new methods will play an ever increasing part in supplying the qualified manufacturing technicians and engineers of tomorrow.

CHAPTER **20**

Design, Tooling, and Production Lines

Objectives

This chapter will enable you to:

1. Explain the phases of the design process.

2. Describe the use of a CAD system and explain its function as a modern design tool.

3. Describe the functions of production tooling and describe common examples.

4. Describe the advantages and disadvantages of production lines.

P roducts seldom just materialize. Most are available only because a great deal of time has been invested behind the scenes in product design work. Design is the process of product creation on paper, or in the mind of the design engineer. The purpose of this chapter is to survey the major phases of design and to examine some of the modern tools that a designer has to work with. This chapter also examines the development of the support tools and equipment necessary to support product production and the production line system.

DESIGN

The first step in product development is design. Design is motivated by the following factors:

1. The need for a specific product

2. The level of available manufacturing technology

3. Mass production capability and production cost

4. Safety and reliability

5. Marketability.

Need

Most products are designed and produced because a need has developed for them. With the tremendous capability of modern industrial manufacturing, needs for products are easily developed and, in fact, they are often created through such techniques as the advertising of features and options on products, thus inducing a prospective customer to buy a new and better model. This is especially true in the consumer goods area. Therefore product designers are engaged in a constant process of designing and redesigning products to fulfill a continuing need.

Capability of Available Manufacturing Technology

The capability of the existing manufacturing technology is an extremely important consideration in product design and development. The term *state of the art* is frequently heard in product advertising. This means that the product is designed and built to the latest design specifications, and from the latest materials available. State of the art is probably more relevant to high-technology items, where products undergo significant design changes as new and better materials and techniques of manufacturing become available, than it is to continued production of standard products where the design is more or less fully developed. As technology develops new and better processes and materials, the designer naturally makes use of these to design and redesign products accordingly.

For example, during the vacuum tube age of electronics, hand-held calculators and small portable computers with low power requirements were not available. The advent of solid state electronics completely revolutionized this technology and permitted the electronic equipment designer to create an entirely new selection of products, heretofore unavailable.

The advent of integrated circuit electronics (ICs or microchips), coupled with solid state devices, has profoundly altered the technology once again, not only in the selection of products available, but in the way in which they are and will be produced. Microchips have opened many new avenues of design.

Production Capability and Costs

Most products that are made for the general public are successful only if they can be mass produced and marketed at affordable prices. The best product design is only as good as its ability to be competitively produced. Therefore a designer must consider these important questions when a product is being developed:

1. Is the production capability presently available?
2. If not, will sales of the product justify development of new production technology?
3. Can the product be manufactured and marketed at a cost that will return investment and profits?

These questions are complex and often occupy much of manufacturing and financial management's time. In some cases design of a product may also involve the parallel development of a new production capability necessary to its manufacture. This can be a costly and risky venture, but it also can be enormously profitable if successful. Manufacturing industries often dedicate large portions of their annual budgets to the research and development of new products and the manufacturing systems for them.

Safety and Reliability

Safety and reliability are critical factors in the design of many products, including aircraft, public works, ships, medical equipment, weaponry, automobiles, and consumer products. In recent years product safety considerations have attained greater importance. This has been because of increased public awareness resulting from improved communications and more intense interest in consumer affairs. The designer's responsibility is equal where personal and public safety are concerned.

Reliability and safety are closely tied. Indeed product safety can be measured by reliability. For example a backup electrical generator for emergency power must start and run reliably, as lives may depend on it. Aircraft control surface actuators must work reliably to ensure that the aircraft and passengers are safe, and automotive steering and braking systems must be extremely reliable. Safe products are generally reliable products, and reliable products can be designed for safety as well. These important considerations must govern the work of the designer in every design effort.

Marketability

If the manufacturing capability exists or can be developed, the next step involves marketing and distribution. It has often been stated that invention is one thing, but distribution is everything. This adage is very timely. If a product cannot be properly designed and manufactured in quantities to meet demand and then sold at an acceptable price, it will probably not be successful. The market environments of free enterprise economies are ruthless in selecting only those products that survive the tests of efficient cost-effective manufacturing.

Costs incurred in marketing include such things as advertising, national or international sales organizations, packaging, distributors and agents, distribution costs, and warranty services. The selling price to the end customer must reflect any or all of these expenses while being low enough to support sales and meet competition.

THE DESIGN PROCESS

The process of design generally begins with an idea that develops from a need determined in many cases by market research. Anyone can design; however, a manufacturer of technical products will generally employ individuals with specific engineering background as designers. Design engineers have training in the selection and applications of materials and manufacturing processes. They must work closely with the marketing and distribution departments in a manufacturing company; thus, product development must be an integrated and coordinated effort.

Design Phases

Product design involves the following phases:

1. Sketches
2. Prototype production and testing

3. Design review and evaluation

4. Production drawings.

Sketch Phase

The designer often begins work by freehand-drawing a sketch of the item to be manufactured. Various sketch formats are used, often in pictorial styles or artist's conceptions (Figure 20.1). These enable the designer to see what the finished product will look like.

Designing by **computer aided design and/or drafting (CAD)** can also be used to aid the engineer in the initial design phase. This developing technology is a spin-off from computer development and has rapidly become very popular. CAD systems are in fact revolutionizing traditional design processes. The computer in the CAD system can create two- and three-dimensional graphics (Figure 20.2) that can be manipulated by the designer in order to make quick adjustments and changes to the design. CAD graphics may be transferred to paper, if required, on a plotter (Figure 20.3). An advanced CAD system can subject a computer model of the design to many variables that it could encounter in its service environment, before an actual prototype or model is produced. This capability of CAD makes the designer's work much more reliable while saving large amounts of money and material that would otherwise be required to produce and test prototypes. Building an actual prototype can be a very expensive and time-consuming process. In the case of large and complex products such as buildings, ships, dams, or bridges, prototyping would be impossible. A computer model can be created, and subjected to the environmental variables that the design will encounter in service. All this can be done quickly and at low cost. CAD is also extremely versatile in testing a new design without the need to risk money, personnel, or expensive materials.

When using a CAD system, the designer can sketch ideas directly on the screen of a video display, use a digitizer pad (Figure 20.4), or enter data directly from the computer keyboard. The digitizer pad converts the fundamental elements of geometry such as points and lines into digital data (bits). The digitized data are understood by the system computer and the information appears simultaneously on the video display (CRT) as it is drawn on the digitizer pad. Some CAD systems employ a dual video screen and will produce screen images both in single color (monochrome) and in full colors.

Designing with a CAD system has many advantages. The computer can electronically generate three-dimensional drawings that may be rotated so as to view a part or assembly from different points. The operator may also zoom in and out on any portion of the drawing in order to reveal details. If the CAD system is used for schematic design, it may be done in layers of various color. Individual layers of the schematic may be electronically removed and replaced as required to study the interrelationship and position of overlapping systems in the design. This is particularly useful in integrated circuit (IC or chip) electronic design as well as in the layout of piping and electrical systems. Colors can also denote changes in stress or temperature in different areas of the design.

Another advantage of the CAD system is that many different part drawings may be stored permanently in the computer memory. These may be called up for review and revision at any time. Thus, the system speeds revised drawing information to the production shop floor.

The ultimate use of the computer as design tool is the ability to put designs to rapid test and evaluation prior to prototyping or manufacturing. This permits a design to be better evaluated prior to actual production. Considering the cost of modern production, the ability to evaluate

FIGURE 20.1
Freehand pictorial sketch formats.

Perspective

Isometric

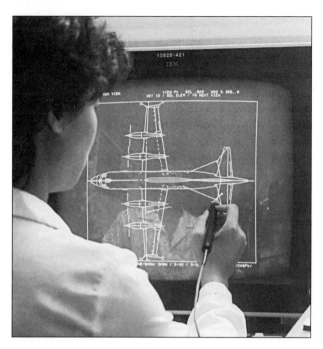

FIGURE 20.2
Computer aided drafting design or CAD (Lockheed Co.).

designs without having to produce actual prototypes or parts is a distinct advantage.

Examples of design evaluation by computer include such diverse activities as studying the air-fuel mixing patterns in an internal combustion engine (Figure 20.5).

Structural design may also be evaluated and weak points identified (Figure 20.6, see page 404). Redesign may then be accomplished immediately to ensure that the design meets specifications. Also a computer may be programmed with a large database representing a mathematical model of a prototype or production unit. By scanning the actual production model and recording measurements, an exact comparison of design dimensions and actual dimensions can be made by the computer (Figure 20.7, see page 404). Through this process, the engineer can determine how accurately production equipment is producing product units in conformity with design specifications.

When CAD systems are coupled with their manufacturing counterparts, or **computer aided manufacturing (CAM)** systems, a fully integrated **CAD/CAM, CADAM, (computer aided design and manufacturing)** or **CIM (computer integrated manufacturing)** system may be developed. Computer-generated information both for design and for production control may be fed directly to the production stations. In Figure 20.8 (see page 404), the computer locates and directs the robotic spot welding of the automobile body. Weld positions are shown on the computer display as the robot spot welder accomplishes its production welding task.

The integration of CAD and CAM, in all its forms, has become a major force in tying design to production in modern manufacturing. Integrated CAD/CAM systems and the steps toward complete factory automation will be discussed further in Chapter 21.

FIGURE 20.3
With the CAD plotter, hard copies of drawings are produced (Lockheed Co.).

FIGURE 20.4
Dual screen CAD terminal with digitizer (Intergraph Corporation).

Prototyping Phase

A prototype is a model from which a product may ultimately be mass produced. Prototypes may be the same size as the final product or they may be scaled down (Figure 20.9, see page 405). The purpose of prototyping is to build a working model of a newly designed product to determine if the design is acceptable. Although prototype designing can be done with the computer using a CAD system, sometimes there is no substitute for an actual working model, especially if physical product testing is to be done.

A manufacturing industry often employs people skilled in model making or prototyping; prototyping may also be jobbed out. However, costs can be high; especially if the design must be reworked. Nonetheless, some phase of prototyping is often an integral part of product design and development.

Prototyping in modern manufacturing can make use of some high-technology techniques. For example, a master model of an automobile where parts are made from plastic materials may be used to permit reliability engineers to examine how parts fit in a complex auto body assembly (Figure 20.10, see page 405). Each plastic part on the model can be removed and replaced with its production metal counterpart. The precision computer-linked scanning equipment shown in the illustration determines how each part fits compared to a perfect computer model of the desired design. Many thousands of mathematical data points representing design specifications are programmed into the scanner computer. This permits an extremely accurate comparison to be made between the prototype and its actual design specifications.

Design Review and Evaluation Phase

Test data from the prototype testing, or computer models if CAD is used, are carefully analyzed and evaluated (Figure 20.11, see page 406). If there are problems, the design will be modified accordingly. This may involve further prototyping. Sketches and/or drawings used in the prototyping will be revised as needed.

Production Drawing Phase

When a decision is made to take a proven design into production, finished manufacturing drawings are made and copies are distributed to all departments responsible for product manufacturing. In fully integrated CAD/CAM systems, electronic drawings are stored in the computer and are called to the computer display terminals or plotted on paper if required. In conventional practices where design is done on paper, rough sketches and drawings used for constructing prototypes are converted into finished manufacturing drawings in the drafting department.

Typical manufacturing drawings are usually produced in the standard orthographic format showing an object in any of several possible views (Figure 20.12, see page 406).

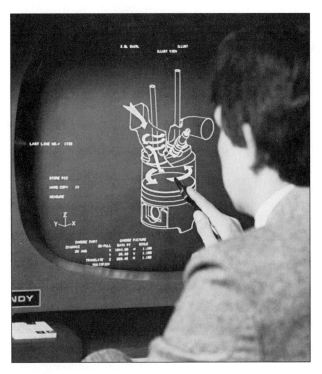

FIGURE 20.5
CAD used in evaluating fuel swirling in an internal combustion engine cylinder (Ford Motor Company).

FIGURE 20.6
CAD for evaluating a structural design (Ford Motor Company).

FIGURE 20.7
A coordinate measuring machine takes measurements on a production unit and compares them against a computer database containing precise design specifications (Ford Motor Company).

FIGURE 20.8
Integrating CAD and CAM in production spot welding. This process is also known as CIM (Chrysler Corporation).

FIGURE 20.9
Prototyping and modeling
(Lockheed Co.).

In addition, several types of auxiliary and section views may be used (Figure 20.13, see page 407). Occasionally a pictorial format such as isometric may also be used on a manufacturing drawing (Figure 20.14, see page 407). This is usually done to show relationships of parts in an assembly and is usually not dimensioned. Final-form drawings made from the designer's sketches are most often prepared in the company's engineering drafting department by professional drafters or generated by computer plotters.

Drawings are done to exacting standards, often on company-formatted drawing sheets. The completed drawings are then checked for accuracy and to see that no pertinent data have been omitted. They are then copied by various processes such as blueprinting and distributed to the various manufacturing departments or to jobbers that

are involved in supporting a company's production requirements. The original drawings are carefully preserved, and as production progresses they are revised as necessary to meet changing design requirements, improve production flow, or solve production difficulties.

Methods engineering is generally the responsibility of industrial engineers, who select and copy portions of drawings and generate operations sketches (op. sketches), manufacturing outlines (MOs), or assembly instructions. These documents accompany parts as they flow through a production process. A typical operations sketch shows only a small portion of a production drawing. It emphasizes dimensional data and other pertinent information necessary to a specific manufacturing operation on that portion of the part.

FIGURE 20.10
Plastic master cube used to check auto body parts against design specification (GM—Pontiac Division).

FIGURE 20.11
Wind tunnel testing of an
aircraft prototype (Lockheed
Co.).

Detail and Assembly Drawings Most product manufacturing uses two forms of drawings. First is the detail
drawing (Figure 20.15, see page 408), showing only one
part and containing all information required for its manufacture. Important information contained on a detail drawing includes:

1. Part name
2. Drawing number
3. Material specifications

4. Dimensions and locations of part and part features
5. Tolerances for size and geometry
6. Revision codes and dates of most current revisions
7. Any special notes regarding special handling during
manufacturing.

Second is the assembly drawing (Figure 20.16, see
page 409), a drawing showing several parts as they would
appear in a complete assembly or subassembly. Important
information contained on an assembly drawing includes:

FIGURE 20.12
The standard orthographic
format of manufacturing
drawings.

FIGURE 20.13
Auxiliary and section views on orthographic drawings.

Surface A Top view

Auxilliary projection of surface A eliminates distortion of hole feature in top, front, and side view

Surface A

Front view

Right side view

A

A

Section AA

Section views are used to reveal part's internal features

FIGURE 20.14
Isometric pictorial of a complex assembly (Lockheed Co.).

FIGURE 20.15
Detail drawing.

1. Assembly name
2. Assembly drawing number
3. Bill of materials, a breakdown of individual parts of which the assembly is composed
4. Dimensions and tolerances pertaining to the assembly
5. Revision codes and dates of last revisions
6. Any special notes regarding the assembly.

DEVELOPING PRODUCTION TOOLING

Once a product design is complete, a prototype is built and tested, and the decision is made to take the product into production, the next phase is production tooling. **Production tooling** includes all of the machinery and equipment necessary to support product manufacturing. The specific product to be manufactured dictates the size and capability of the production line required. The esti-

mated demand for the product dictates the required production line capacity.

Any mass-produced product will of course require applicable material processing equipment. This equipment may cover a wide range of standard machinery, such as machine tools. However, a given machine tool can usually accomplish many types of machining. A numerically controlled machining center, for example, can mill, drill, tap, ream, bore, index, position a workpiece, and perform other functions as well. If only hole drilling in a product were required then it would be more cost effective to use a machine that was limited to drilling only. On the other hand, a manufacturer involved with many different machined products may very well have use for the programmable multifunction use of the CNC machining center.

Even though a manufactured product may pass through many different processes during its manufacture, these individual processes are always repeated exactly on each part, making it necessary to hold and position the part while the specific process is being accomplished. Therefore, it is almost always necessary to develop spe-

408

FIGURE 20.16
Assembly drawing with bill of materials.

		BILL OF MATERIALS			
QTY	NAME	PART NO.	MAT'L.	DRAWING REF	
1	HOUSING	1	BRASS	62379-1	
1	SCREW	2	4-40 × ¼ STAINLESS	VENDOR	
1	CUTTER	3	CARBIDE	VENDOR	
2	SCREW	4	4-40 × ⁹⁄₁₆ STAINLESS	VENDOR	

	RRK MACHINE and MFG.	CUTTER ASSEMBLY		
		DWG	62379	REV

FIGURE 20.17
A drill jig is an example of simple tooling.

cial production tooling unique to the particular product being manufactured. One might say that this special tooling adapts the general multiuse material-processing equipment to specific manufacturing requirements. Included in special production tooling are the following:

1. Jigs
2. Fixtures
3. Dies
4. Molds
5. Patterns and templates
6. Cutters

Jigs   Jigs are special tools primarily designed to provide alignment of parts or tools during a manufacturing operation. A simple example of a jig is a drill bushing used in machining (Figure 20.17, see page 409). A drill bushing or drill jig consists of a hardened steel sleeve through which a drill can pass freely, but with minimum clearance. The drill jig aligns the drill and the resulting hole in the workpiece is then located in its proper position. Drill jigs may be complex tools designed to locate drills for drilling many holes in the same workpiece.

Jigs may also be used to hold large fabrications in position while welding or machining occurs. Alignment of such tooling or parts of an assembly may be done by the techniques of optical tooling. In Figure 20.18, a jig transit is used to align large portions of an aircraft structure during assembly. Optical tooling techniques have many applications in layout, alignment, and assembly of large fabrications such as ships, bridges, and aircraft.

Fixtures   Fixtures are special tools designed to hold, support, index, and align a workpiece while a particular

FIGURE 20.19
Large tooling fixtures for aircraft skin panels (Lockheed Co.).

manufacturing or assembly operation is performed. An example of a simple fixture is the common machine tool vise with soft jaws that have been machined to a special shape for the purpose of holding a workpiece in a particular position.

In Figures 20.19 and 20.20 special fixtures are used to hold aircraft skin panels and the leading edge rib for a

FIGURE 20.18
Using a jig transit to align aircraft sections (Lockheed Co.).

FIGURE 20.20
Large aircraft vertical stabilizer tooling fixtures being inspected (Lockheed Co.).

FIGURE 20.21
An entire car body is held in a
drilling fixture where body
panel holes are drilled in one
operation (General Motors
Corporation).

large aircraft's vertical stabilizer. Note the large size of this tooling.

Production tooling fixtures, especially in high-technology manufacturing, can be complex. In Figure 20.21 the entire auto body is held in position while numerous holes are drilled for body panel attachment (Figure 20.22, see page 412).

Dies   Dies include many types of special tools. They may be cavities into which materials are shaped by various processes. An example discussed in previous chapters is die casting, where molten metal is poured or pumped into a cavity shaped to the form of the desired product. Dies are also used in machining for thread cutting and in forging processes. In punch press work, dies are used in conjunction with punches to cut parts to exact shapes (Figure 20.23, see page 412). This type of tooling can produce a large variety of useful parts (Figure 20.24, see page 412).

Dies are also used in metal bending applications. Figure 20.25 (see page 413) shows a press brake type bending die and Figure 20.26 (see page 413) shows the type of product produced by it. Dies are also used extensively in the plastics industry.

Molds   Molds are most often used to form products from raw material in a fluid state. Products formed in molds assume the shape of the mold before they change form, such as by cooling or hardening. Examples are metal castings, fiberglass, and plastic resins. After the material solidifies by cooling or hardening, the mold is opened, removed,

or broken, leaving the finished product (Figure 20.27, see page 413).

Patterns and Templates   Patterns and templates cover a wide range of special production tooling. In Figure 20.28 (see page 414) the shape of the bend is checked against a template. In another example, a full-sized template of a hull plate for a ship's hull may be used to trace around in order to duplicate parts of the same size and shape.

In the fabric industries, paper patterns are used in cutting materials. These may be computer generated to maximize the number of pieces that can be obtained from a given bolt of material.

Cutters   Cutters are a large and important member of the production tooling family, especially in the machining area. Cutters generally remove workpiece material by shaving off small pieces or chips. They may be rotated against the workpiece, as in the operation of milling and gear hobbing or, as in turning, the workpiece may be rotated against a stationary cutter. In machining operations such as gear hobbing, a large number and variety of cutters are required (Figure 20.29, see page 414).

TOOL DESIGN AND TOOLMAKING

Tool design and toolmaking are among the most important activities supporting the manufacture of almost any product. Foremost in tool design is accuracy that will

FIGURE 20.22
Body panel drilling operation
sequence (General Motors
Corporation).

Sequence of Operations

1	Automatic on-line load	7	Drill 39 vertical holes
2	Clamp body at eight locations	8	Machine 39 horizontal surfaces
3	Position body per LVDT readings	9	Verify operations completed
4	Set back-ups at front, rear, top	10	Unclamp and release back-ups
5	Verify position for drilling	11	Automatic unload to conveyor line
6	Pierce four slots		

FIGURE 20.23
Punch and die tooling (Bracker Corporation/Leland J.
Balber Associates).

FIGURE 20.24
Punch and die tooling can produce a wide variety of end
products (Bracker Corporation/Leland J. Balber
Associates).

FIGURE 20.25
Press brake tooling for sheet metal bending (Bracker Corporation/Leland J. Balber Associates).

result in the end product dimensions being as standardized as possible. Therefore special manufactured tooling is built to exacting tolerances, often in the tenths of thousandths or smaller increments of inches. Building tools to critical tolerances requires the best quality machine tools available with experienced toolmaking personnel to operate them.

The toolmaker is usually an experienced machinist who has specialized in toolmaking activities. Toolmaking often is done within the manufacturing plant in a special machine shop area dedicated to this purpose. Toolmaking personnel are frequently paid premium wages for their expertise and experience.

A typical production tooling item must meet exacting dimensional standards, but must also be convenient to set up in the manufacturing process. The process operator

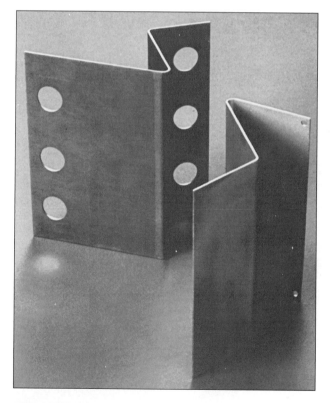

FIGURE 20.26
Example of press brake products (Bracker Corporation/Leland J. Balber Associates).

FIGURE 20.27
Mold tooling for manufacturing plastic pipe fittings (Crucible Specialty Metals Division of Colt Industries).

FIGURE 20.28
Inspecting an aircraft panel bend with a template
(Lockheed Co.).

should be able to easily load and unload the tooling with
production parts during manufacturing. If the tooling is to
be used in conventional machining or grinding operations,
provisions should be made to clear chips easily and
quickly before new parts are loaded in the fixture.

Production tooling often incorporates many heat-
treated parts so that accuracy and durability can be preserved despite much hard use. Most manufacturers identify and catalog special tooling so that it can be located
quickly for use whenever a need arises.

Design and production of tooling often involve as
much effort as design and production of a manufactured
product. In many cases the cost of tooling prior to manufacturing represents the major cost of production. Therefore it is often necessary to produce a minimum number of
production parts just to underwrite the tooling cost. Only
then does further production contribute to the earning
stream of the business. In large industries with complex
products, the payback process on tooling expenses can take
several years.

THE MODERN PRODUCTION LINE
AND MASS PRODUCTION

The culmination of design, prototyping, design review,
and development of production tooling, is the successful
mass production of the product. In the earliest examples
of mass production, it was soon discovered that work can
be accomplished most efficiently by setting up a series of
workstations, at which individual processing or assembly
operations are performed (Figure 20.30).

The product is then moved from one station to the
next station, where the next process is accomplished. This
is continued until the product arrives at the end of the production line with all manufacturing operations and/or
assembly operations completed. Production by this method

FIGURE 20.29
Many styles and sizes of
cutter tooling are required in
gear cutting (Illinois Gear, a
Household International
Company).

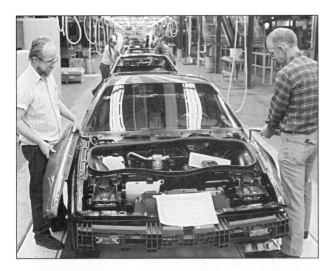

FIGURE 20.30
On the modern production line a portion of assembly occurs at each station (GM—Pontiac Division).

is effective for both small and large products (Figure 20.31). Products may be moved from workstation to workstation by hand (Figure 20.32) or on assorted mechanized conveying assemblies (Figure 20.33). A typical system used in the automotive industry uses floor conveyers and pallets to move the product along the production line (Figure 20.34).

There are many advantages to production by this scheme. Each workstation can be highly efficient and equipped with any specialized tools needed for the particular process occurring there. In most production accomplished by this method, only one or two processes are usually done at any one station. Hence, personnel staffing a specific station are specialists at their work. If proper supervision is applied, both efficiency and quality can result.

Machining Transfer Lines

One specialty type of production line is the machining transfer line where many repeat machining operations such as drilling, boring, and tapping must be accomplished, and it is not desirable to bring repeat product units to and from a machining center. Production under these circumstances would be inefficient as each unit would have to be mounted on the machine tool, machined, and then offloaded to make room for the next. A much improved system is the machining transfer line where continuous product flow may be maintained while at the same time the desired multiple machining operations can be accomplished. On the transfer line, a series of machine heads is located and tooled for the desired machining operation. The product is then moved along the line to each station where required machining is accomplished. In a sense, the machining transfer line is much like the larger factory production line, but it is used for specialized manufacturing operations. Transfer lines are very popular in the production of automobile engines and other high-production automotive mechanical systems.

Production Lines, Present and Future

Possible disadvantages to line production are that the work is often repetitive and therefore boring for personnel. Low-quality output can result. Line production of complex products and high-volume output can be very labor intensive and therefore very expensive. However, the production line, even with its possible disadvantages, is likely always to remain a major institution in manufacturing. Manufacturing industries are experimenting with many techniques to make the production line workers

FIGURE 20.31
Production lines are equally applicable to large products (Lockheed Co.).

FIGURE 20.32
Moving products along the line by hand is also efficient
(Del Manufacturing).

FIGURE 20.33
Product transportation by mechanized conveyer (Del
Manufacturing).

FIGURE 20.34
Palletized product
transportation on the
production line (General
Motors Corporation).

more comfortable at their job, and thus keep quality at a high level. There is also a major trend to further automate production lines using industrial robots to replace people for repetitive or hazardous jobs. With the wide capability of the programmable computer-driven robot, it is a certainty that these machines will become ever more present along the production lines of the world's manufacturing industries.

Review Questions

1. What factors motivate design?
2. What are the phases of design?
3. What is CAD and how is it used in design?
4. What are the advantages of CAD?
5. Why are design review and evaluation important?
6. What are the two major types of drawings found in manufacturing?
7. What information is contained on these drawings?
8. Describe the purpose of production tooling.
9. Who is responsible for building production tooling?
10. What major items are included in production tooling?
11. What are the advantages of production lines?
12. What are the disadvantages of production lines?

Case Problems

Case 1: Computers or Prototypes

A manufacturer of complex hydraulic equipment decides to reevaluate product design with the intent of branching into a new market requiring lightweight products (automotive and aircraft). One choice would be to build and test a large group of prototypes. This would take about one year and the cost for in-house design work and jobbed-out prototype development would exceed $500,000.

A CAD/CAM system could be purchased for about $750,000 that not only would evaluate designs by computer but also could be interfaced with the manufacturer's existing production capability, making it more efficient and productive. Complete phase in for the CAD/CAM system to fully support manufacturing would take three years, but the design review capability could be online almost immediately. The market potential for the new lightweight hydraulic equipment is excellent.

If the company must borrow the money for either project, which do you think would pay off in the long run from a standpoint of both efficiency and return on investment?

Case 2: Automate or Improve Working Conditions

A manufacturer is receiving poor quality work from production lines involving repetitive simple operations on simple parts. The production manager recommends that the company invest in industrial robots to replace all production line workers. The production line foreman suggests that the problem could be solved by rotating production line personnel so that their work is more varied and they take more interest in it. The company gainfully employs several hundred people, but at minimum or very low wages for production line work.

Considering all factors including costs of retooling the factory, the company's public relations image with the community, and the type of product produced, explain which choice you would make, purchase robots or adjust working conditions for production workers. Also explain why.

CHAPTER 21

Automation in Manufacturing

This chapter will enable you to:

1. Discuss reasons for automating manufacturing.

2. Describe the basic factors required for automation.

3. Discuss computer numerical control automation of manufacturing processes.

4. Discuss CAM, CIM, and flexible manufacturing systems.

5. Discuss the place of industrial robots in manufacturing automation.

Most modern manufacturing is automated. This means that the processes used are set up in such a way that they can be repeated over and over with little or no need for manual control except to start and stop the process or to change workpieces and tooling. Even the manual aspects of changing the workpiece and tooling also are fast becoming automated as new computerized manufacturing technology is applied in almost all phases of production. Automation trends, already at a high level in many manufacturing industries, will be refined and applied to an even higher degree in the future. The result of this trend will, in many cases, redefine the duties of today's typical production worker or machine operator. The purpose of this chapter is to explore the techniques used to automate modern manufacturing and to look at the types of automated equipment now being applied in the high-technology computer age.

It is necessary to produce most products in quantity so that the selling price can be low enough to make them attractive to a wide range of prospective buyers. There will always be some demand for one-of-a-kind, hand-crafted products. However, any such product that is built by hand crafting one unit at a time will, by the very nature of the production time involved, be much more expensive compared to its production machine-made counterpart.

There is no question that machine-made goods have a sameness about them, and the age of individual craftspeople is recent enough that we still think of manufacturing a product carefully by hand as being something special. It should be noted that even though the hand-crafted product's quality is high, it is not necessarily any better than its machine-made counterpart. However, hand-crafted products often do represent high quality, are frequently unique in design, and represent a certain dedication to manufacturing in a bygone era. In addition, as machine-made goods become more and more the standard of the industrialized world, the desire to possess something unique will always be a force for the preservation of the individualized craftspeople's dedication to their methods.

FACTORS IN AUTOMATION

Why automate manufacturing? In the opinion of some, automation is actually counterproductive as it would appear to eliminate jobs. The contrary view is that the manufacture, implementation, and maintenance of automated machinery creates many higher level jobs. The debate over jobs versus automation will always continue, but despite whatever negative factors it brings, automation is here to stay. Manufacturing, driven by practical economic reasons, will continue to exploit automation in an aggressive manner. Primary reasons for automating manufacturing include the following. Automation:

1. Allows maximum production at minimum costs and productivity factors.
2. Produces the volume required.
3. Allows standardization and interchangeability of parts.
4. Meets product demand and competition.

Costs and Productivity

Cost of manufacturing is a major if not the most important force behind automation. All manufacturing industries are controlled by the economic forces in play at any given time. Throughout history such factors as inflation, taxation, labor costs, material costs, and costs of marketing and distributing have driven the manufacturing industry to search constantly for better and cheaper ways to accomplish production.

There has been much debate in recent years as to the long-range results of automation and its effects on employment. Advances in manufacturing technology have affected employment levels for low-skilled workers and will have a more pronounced effect in the future. It is safe to say that in the longer term, there will be fewer low-skilled to semiskilled production jobs as automated manufacturing becomes more and more the standard of industry. If economic conditions place more demands on the individuals now staffing production jobs, these employees in turn will place more pay demands on their employers who will compensate by further automating industries, thus reducing labor intensity.

Even though automation might appear to be counterproductive, in the long run it probably will create more jobs. However, these jobs will be of a higher technical level, and production workers must be open-minded enough to accept retraining if they expect to remain gainfully employed.

Productivity generally can be defined as the ability to maximize the quantity and quality of production from all phases of manufacturing, including people and machinery. If plant equipment is antiquated, productivity will be low compared to modernized automated manufacturing and therefore the costs of the product may be higher. In labor-intensive manufacturing, productivity may be difficult to maintain because of the inconsistencies of the human work force. On the other hand, automatic machines do not become tired or bored, and are consistent in their output quality, assuming proper adjustment and maintenance.

Automatic manufacturing equipment can also operate under conditions that humans could not. For example,

robotic assembly can be done in total darkness, whereas such conditions would not be at all satisfactory for people.

Production Volume

It is simply not possible to manufacture certain high-volume products in a production environment where very many operations must be manually performed. Examples include nails, fasteners, lightbulbs, electronic components, and many other parts for assemblies of all types. These products are manufactured in quantities of billions each year, and the only way to meet volume requirements is to have highly automated production.

Standardization and Interchangeability

Few complex manufactured products are made totally by a single manufacturer. More likely a company will use many subcontractors to manufacture different parts for an assembly and then these will be brought together at a final assembly point. Examples are autos and aircraft. There are many advantages to manufacturing in this manner. Subcontractors become specialists in the production of individual parts or products. Their expertise develops keenly and their research and product development can be concentrated in a narrow area, resulting in a product of high quality with the ability to be mass produced.

Since a manufacturer of a general application product line such as electrical or hydraulic equipment may act as a supplier to many different prime contractors, **standardization** of the product is essential. This permits **interchangeability,** the key to any successful mass production of complex assemblies. The importance of standardization cannot be overemphasized. Lack of standardization in manufacturing impedes interchangeability in assembly, making manufacture of complex products time consuming or impossible.

Once again, automation plays an important role in achieving standardization because of the ability of the machine to duplicate products in large quantities with little or no variance in part dimensions and part feature locations.

Product Demand and Competition

Demand for a manufactured product created by need or want will often dictate the degree of automation required in production. For complex products such as ships, offshore oil platforms, and other large and complicated assemblies, demand will be low from a quantity standpoint. These products take time to complete, up to several months or years, and they are fairly labor intensive. Creating a highly automated production line for their production would be neither desirable nor cost effective.

On the other hand, such products as lightbulbs, autos, nuts and bolts, nails, home appliances, and electronics have high demand as well as wide market potential. In many cases, profit margins are very small on a per unit basis. Hence, very large production numbers are necessary, making automated production both desirable and necessary. Since products can be made quickly and efficiently using automated manufacturing, their costs can be kept low, making a vast line of products available at affordable prices.

Competition is a driving force in any free enterprise economy. Rarely do we find a product manufactured by only one maker. Competition tends to weed out producers with low quality, too high prices, or those who are unable to meet the production demand for their product. If the competition to industry A is automating its manufacturing facilities in order to bring a better, cheaper, and more available product to the customer, then industry B must do the same if it expects to compete in the marketplace. There are many examples where a manufacturer's inability or unwillingness to automate production resulted in business failure.

AUTOMATION SYSTEMS IN MANUFACTURING

Mass production of any product always involved repetitive operations. One of the primary functions of automation is to standardize repeat production operations so that they may be accomplished on a continuous unvarying basis. This permits design tolerances to be met on a large number of production parts and makes possible higher production rates. To do all this, automation techniques most often replace inconsistent manual functions with consistent machine functions. This is where automation can and does eliminate production jobs.

The process of automating a manufacturing process may take many forms. They range from rudimentary control, perhaps involving a simple mechanical device that operates part of a machine, to complex computer-driven feedback systems with considerable decision-making capacity over the process that they are controlling. In a true automatic system, the automation device has an ability to sense in some way factors about the process that is being accomplished and then initiate control functions.

Automation requires the following factors (Figure 21.1):

1. A process to automate

2. A sensing system for making decisions about the process

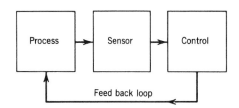

FIGURE 21.1
Factors for process automation.

3. A control action system that operates on the sensed information and then provides a process control function.

Almost any manufacturing process can be automated. This can be done at any phase of manufacturing, including the manufacturing processes themselves or systems that inspect and/or package a finished product.

Semiautomatic and Full Automatic

Consider a simple process such as the carriage feed on a lathe (Figure 21.2). The process must be initiated manually by the operator and disengaged manually at the end of the cut. This is not a fully automated process, but once engaged, it does serve to feed the tool automatically, without the requirement for external manual input. Thus, the process is semiautomatic.

By setting a feed stop, the operator can further automate the process. Initiation is manual, but feed disengagement is automatically accomplished. Two automation factors are present here, the process (automatic carriage feed) and the sensor (trip stop). This simple system does not have any feedback for further action after feed disengage. The operator must take over manual control at that point to withdraw and return the tool to the starting point. Therefore, the function is semiautomatic, requiring a manual initiation for each new cycle.

On a numerically controlled (NC) lathe or turning center, the feed might be engaged by computer control,

FIGURE 21.2
Semi-automatic control of the carriage feed on a lathe involves manual engage and manual disengage.

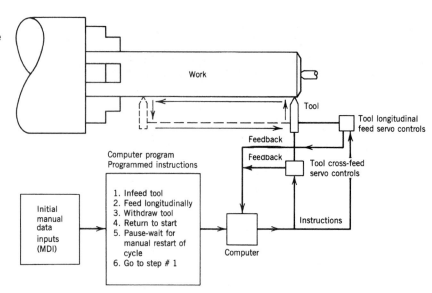

FIGURE 21.3
Full automatic programmable cycling.

disengaged, and the tool automatically withdrawn and returned to the beginning of the cut, or changed to a new tool. All three factors are present here for full automation, the process (tool feed), the sensor (computer control), and the action based on sensed information (tool withdrawal and return or tool change) (Figure 21.3).

Numerical Control Automation

It would be difficult to find a device that has had more influence on manufacturing automation than the computer. Computer control of production processes or computer aided manufacturing (CAM) as well as computer integrated manufacturing (CIM), and their design counterpart (CAD), have profoundly altered manufacturing in dramatic ways. One very important area that has been established for many years and continues to undergo development is that of **computer numerical control (CNC).**

CNC is the equipment and methods used to control a process by numeric instructions through computers driving electromechanical or hydraulic actuators. Although CNC has been applied to almost every manufacturing process, machining certainly is one area where its development is unparalleled. Perhaps the reason for this is that machining involves a wide variety of high-precision manufacturing requirements.

All of the previously discussed factors are present in an automated numerical control machining system. The processes to be controlled include tool and workpiece positioning, tool changing, feed and speed rates, and special functions unique to the specific process or machine tool. Electromechanical and electronic sensor systems are built into the computer control to make decisions and initiate appropriate actions.

NC Systems NC systems include open loop and closed loop. The open loop NC system does not make use of a feedback system to determine control action. This system was popular in the earlier days of NC before the development of more sophisticated computer servo controls.

Computer control technology permits sophisticated closed loop NC systems to be used, where sensed information is fed back to the control, thus initiating required control action. These computer numerical control (CNC) systems are widely used and very popular in automated manufacturing.

Another version of computer numerical control places several NC machine tools under the supervision of a single computer. This process, known as **distributed numerical control (DNC),** is popular in manufacturing automation.

NC Motion and Other Machine Functions

Major components of an NC machine tool move along fundamental machine axes. The centerline of the machine spindle is the Z axis. If the machine tool is a vertical machining center, the table and saddle components move at right angles to each other relative to the spindle position. Table motion is in the X axis and saddle motion is in the Y axis. If the machine tool is a horizontal spindle design, the spindle centerline is still Z axis whereas the X-Y plane turns into the vertical position (Figure 21.4).

FIGURE 21.4
Reference axes for CNC machine tools.

Turning center spindles are in the Z axis whereas cross slide motion is in X axis. Movement of the spindle, table, or saddle along each axis in either direction is defined in the program. Thus, each point within the entire volume of space in which the spindle may move or the workpiece may be positioned may be specified in the NC program. Modern CNC equipment has the capability to position the machine spindle and/or workpiece to specified locations within a few tenths of thousandths of inches. On large machine tools, this can be repeated accurately over long distances, making possible extremely accurate cuts and location of holes and other features in the workpiece. By simultaneously combining motions by the machine spindle and/or workpiece, the NC machine tool can accomplish angular and contour (continuous path) machining that could not be done by conventional hand operation methods. Thus, the NC machine tool can perform tasks that would be virtually impossible by conventional practices. The mathematics required for continuous path and angular machining is calculated by the control computer and the appropriate program codes are directly generated.

The NC Program   To accomplish machining tasks, the machine control unit (MCU) of the CNC machine must be programmed. This task may be done manually by the NC programmer, or it may be accomplished through the assistance of a computer. In manual programming, the programmer must be able to write the particular NC machine code used to direct each machine. In computer-assisted NC programming the programmer uses CAD data and/or simple English-like commands to generate a program, and the computer generates the NC machine code. NC programmers must have a knowledge of the machining process, the cutting tools required, and the workholding system, so in many cases NC programmers are experienced NC and CNC machine operators and machinists who have good math skills.

FIGURE 21.5
Part drawing for the CNC machining task.

OPERATION INSTRUCTION SHEET

PART NO. JEXAMP1.DWG	PART NAME: EXAMPLE	OPERATION: 10	SHEET 1 OF 3

NOTES: 1:1 SCALE
MATL. 6061 T6 AL.
REFER TO FINAL B/P FOR TOL., FINISHES, ETC NOT SHOWN

DRILL ⅜ DIA
(2) PLCS.

DRILL & TAP
½-13 UNC 28
(2) PLCS.

³⁄₁₆ RADIUS
(2) PLCS.

DRAWN BY KIBBE

The CNC program consists of a set of step-by-step instruction statements describing exactly what the machine is to do regarding positioning the cutting tool or workpiece, selecting the required tool, and accomplishing the required cutting operation at the correct speed and feed. Other statements in the program control machine functions such as coolant on/off, spindle stop/start, drill, tap and bore cycles, tool changes, feeds and speeds for machining, and workpiece indexing movements.

The NC program resides in the computer memory of the MCU. Programs can be transported to the MCU using computer disks, punched tape, or magnetic tape. Several programs may reside in the memory of the MCU.

Some types of CNC machine tools may be programmed by manually running them through the various operations of a machining task while the MCU memorizes these motions. (This is called teach programming.) Each manual operation becomes a computer program step and once the first run is accomplished, the machine tool has been programmed to repeat the job automatically.

Several documents may be required to assist the programming of a CNC machine tool. Generally the machine operator will need a part drawing (Figure 21.5) indicating dimensions, as well as an operation instruction sheet (Figure 21.6) indicating a list of machining operations, job setup instructions, and a list of gages used to measure the part. The CNC programmer may generate a program definition sketch defining various points, lines, and planes that will be specified in the program (Figure 21.7, see page 426). Figure 21.8 (see pages 427–428) contains an exam-

PART NO. JEXAMP3.DWG	PART NAME: EXAMPLE	OPERATION: 10		SHEET 3 OF 3	
SEQ	OPERATION DESCRIPTION	TOOL NO.	RPM	FEED
	SETUP INFO:			
	A) LOAD VISE TO NC MILL TABLE, JAWS PARALLEL TO Y AXIS.			
	B) INDICATE JAWS PARALLEL TO Y AXIS WITHIN .0005 T.I.R.			
	C) SET TOOLS IN APPROPRIATE HOLDERS PER TOOL LIST.			
	D) LOAD TAPE #001 INTO CONTROL MEMORY.			
	E) SET END LOCATING STOP ON VISE. LOAD A PART.			
	F) USING EDGEFINDER, LOCATE UPPER LEFT CORNER OF PART.			
	G) FROM THIS POINT, HANDLE OVER −10.0" IN X AXIS.			
	H) RESET AXIS READOUTS. THIS IS PROG. ORIGIN.			
	I) ZERO RETURN Z AXIS. PROGRAM BEGINS & ENDS HERE.			
	J) SET ALL OFFSET REGISTERS PER G.L.'S & T5 RADIUS.			
	K) USING STANDARD SETUP PROCEDURES (DRY RUN, MACH. LOCK, SING. BLK., ETC,) WORK ALL TOOLS IN TO CUT PART.			
	L) LAYOUT 1ST PART COMPLETE & PRESENT TO INSPECTION.			
	PROG. SEQ.:			
	CENTERDRILL (4) HOLES	1	2000	16IPM
	DRILL 3/8" DIA. HOLES (2) PLCS.	2	2000	10IPM
	DRILL 7/16" MINOR DIA. (2) PLCS.	3	1200	8IPM
	TAP 1/2-13 UNC 2B THREAD (2) PLCS.	4	315	CALC
	MILL 2" POCKET WITH 3/8" E.M.	5	3000	25IPM
	GAGES REQUIRED:			
	.370/.380 DIA. PLUG GAGE			
	.430/.440 DIA. " "			
	1/2-13 UNC 2B THREAD GAGE			
	3/16" RADIUS GAGE			
	S.M.I.			
			DRAWN BY:	

FIGURE 21.6
CNC operation instruction sheet defines job setup requirements, program sequence, and required inspection gaging tools.

FIGURE 21.7
Program definition sketch
defines part geometry that will
be used in the CNC program.

PROGRAM DEFINITION SKETCH

| PART NO. JEXAMP2.DWG | PART NAME: EXAMPLE | OPERATION: 10 | SHEET 2 OF 3 |

NOTES: 1:1 SCALE
MATL. 6061 T6 AL.

PROG. ORIGIN (−10.0 X FROM PART CORNER)

PT1   PT2

PT3   PT4

LN2

LN1   LN3

Z BASE

DRAWN BY: KIBBE

ple of a computer-assisted NC program and indicates what the various lines of the program mean. Figure 21.9 (see page 429) shows the machine control code generated from the computer-assisted program in the previous illustration.

Construction and Capability of NC Machine Tools

Because NC machine tools are designed for repetitive production of precision parts, they are constructed to the latest machine design available. Design effort is directed toward reliability of the mechanical and electrical/electronic components, as well as long-term maintenance of machine accuracy. This makes the NC machine tool more expensive than its manually operated counterpart. However, its higher initial cost is far outweighed by its truly amazing productivity and broad capability. Numerically controlled machine tools, as well as other types of NC material-processing equipment, have well established themselves as essential items in modern manufacturing automation.

With computer control, the NC machine tool has capabilities that could not be accomplished by manual operation. This capability has reduced or eliminated many constraints heretofore placed on the product designer. With the capability of new computer-controlled production, mass produced products are appearing that would have been totally unavailable only a few years ago. Technological advances in this type of computer aided

Programming language: COMPACT II

All INSERT STATEMENTS are REMARK and COMMENT statements used by the programmer. These are ignored by the machine computer.

Machine tool and link name

Punched taped identification

```
MACHIN, MAZVTCFAN1
IDENT, V-10 JEXAMP1.DWG S/U 1 TP 1 OP. 10
INIT, INCH/IN, INCH/OUT
```
Program in inch measurement or optional metric

```
SETUP, MOD10/5/20/1,CMOD3000/1/4/5/11,TPS,RPM3150,MAGPOS99,20X,10Y,OLX,OLY
```
Setup Statement

```
INSRT,PART:  PARTNAME    MACHINE PER OPER. INSTRU.
INSRT,PROGRAMMED BY:  KIBBE
INSRT,DATE:  /  /  .
INSRT,QQ
BASE,OXA,OYA,8.ZA
```
Establishes home position from the machine spindle

```
DPT1,11.XB,-1.YB,OZB
DPT2,PT1,2.X
DPT3,PT1,-1.5Y
DPT4,PT2,-1.5Y
DLN1,11.XB,OZB
DLN2,-3.YB,OZB
DLN3,13.XB,OZB
```
Geometry definitions describe parts and operation used to direct cutter path

Tool description and number

```
INSRT,
INSRT,QQ
INSRT,QQ     TOOL# 1:  CENTERDRILL (4) HOLES.
INSRT,QQ
ATCHG,TOOL1, TLCMPO,6IPM,850RPMRH,6GL
```
Length of tool and holder to spindle end

```
MOVE,PT1,2Z
DRL,PT1,.1Z,.2A
DRL,PT2,.1Z,.2A
DRL,PT3,.1Z.,.2A
```
RPM of spindle and direction of rotation

Offset used

Automatic tool change

Drill starts at 1.000 in. over part surface. Drills 2 in. deep at point 4

```
DRL,PT4,.1Z,.2A
MOVE,2Z
HOME
```
Full rapid traverse positioning move to next location

```
INSRT,QQ
INSRT,QQ     TOOL# 2:  DRILL 3/8" DIA. (2) PLCS.
INSRT, QQ
ATCHG,TOOL2,TLCMP2,10IPM,2000RPMRH,6GL
MOVE,PT1,2Z
DRL,PT1,.1Z,.7A
DRL,PT2,.1Z,.7A
MOVE,2Z
HOME
```
Returns to home (start) position, turns spindle off

```
INSRT,QQ
```

FIGURE 21.8 *(SHEET 1 OF 2)*
The CNC program contains all the codes necessary to instruct the machine tool to perform its assigned task.

```
INSRT,QQ       TOOL #3:  DRILL 7/16" MINOR DIA. (2) PLCS
INSRT,QQ
ATCHG,TOOL3,TLCMP3,8IPM,1200RPMRH,7GL
MOVE,PT3,2Z
DRL,PT3,.1Z,.7A
DRL,PT3,.1Z,.7A
MOVE,2Z,CUF ─────────────── Generates a canned (used over and over) drilling cycle
HOME
INSRT,QQ
INSRT,QQ       TOOL #4:  TAP 1/2-13 UNC 2B THREAD (2) PLCS.
INSRT,QQ
ATCHG,TOOL4,TLCMP4,315RPMRH,8GL
MOVE,PT3,2Z
FLT,PT3,.5Z,.1A,13PITCH ────────── Given pitch of thread in tapped holes, the computer
FLT,PT3,.5Z,.1A                     calculates and generates correct feed rate for the
MOVE,2Z,COF                         tapping operation
HOME
INSRT,QQ
INSRT,QQ       TOOL #5:  MILL 2" POCKET ── Generates a canned tapping cycle based on the
INSRT,QQ                                    previous data
ATCHG,TOOL5,TLCMP5,3000RPMRH,.0937TLR,25IPM,6GL
MOVE,OFFLN1/XL,-4.5YB,-2.Z,.010STK

CUT,-.6ZB ─────────────── Leaves .010 in. of material on part for finish machining
CUT,OFFLN2/YS,PARLN1
CUT,OFFLN2/XS,PARLN2
CUT,-1.5Y
MOVE,OFFLN1/XL,-4.5YB,OSTK,100IPM
CUT,OFFLN2/YS,PARLN1
CUT,OFFLN3/XS,PARLN2
CUT,-1.5Y
MOVE,2Z
HOME
END
```

FIGURE 21.8, *Continued—SHEET 2 OF 2*

manufacturing permit new and innovative product design profoundly affect us all.

Types of NC Machine Tools   Generally speaking, NC machine tools can be classified into two major groups. These include machining centers and turning centers. Machining centers are equivalent to milling machines in the conventional shop, and turning centers perform lathe operations.

Machining Centers   CNC machining centers are one of the most popular types of automated machine tools. They can have vertical or horizontal spindles and they can be equipped with a variety of features that can increase their versatility. These include tool changers where a number of cutting tools are mounted on a belt or rotary drum. Figure 21.10 (see page 430) shows a CNC machining center with twin tool holders. When the machine tool is in operation, the NC computer program selects the correct tool from the belt and mounts it in the machine spindle (Figure 21.11, see page 430). When the cut is completed, the program causes the tools to be removed from the spindle and returned to the tool holder.

Multiple workpieces may be mounted on several pallets on the same machine tool (Figure 21.12, see page 431). Pallets may be indexed to present different parts of the workpiece to the cutter (Figure 21.13, see page 431). In many cases, the machining center will be used as a component in a work cell as part of a **flexible manufacturing system (FMS).** In this case, a provision is made to completely off-load a work-holding pallet from the machining center (Figure 21.14, see page 431) and transport it to another machine or to a new cell or station in the manufacturing system. CNC machining centers can accommodate large workpieces in duplex arrangement

```
        TOOL #1 : CENTERDRILL (4) HOLES

N001G91G28Z0T01
N002S63M03
N003T00M03
N004G92Z280000
N005G90G00X110000Y-10000Z160000
N006G81G99Z13900R141000F600
N007X130000
N008X110000Y-25000
N009X130000
N10G80Z161000
N011X0Y0Z280000

        TOOL #2 : DRILL 3/8" DIA. (2) PLCS

N012G91G28Z0T02
N013T01M06
N014G27Z80000
N015G90G00X110000Y-10000S73M03
N016G45Z160000H02
N017G81G99Z134000R141000F1000
N018X130000
N019G80Z161000
N02G45Z280000H02
N021X0Y0

        TOOL #3 : DRILL 7/16" MINOR DIA. (2) PLCS

N022G91G28Z0T03
N023T02M06
N024G92Z280000
N025G900G00X110000Y-25000S66M03
N026G45Z170000H03
N027G81G99Z144000R151000F800
N028X110000
N029G80Z171000
N030X0Y0

        TOOL #4 : TAP 1/2-13 UNC 2B THREAD (2) PLCS.

N032G91G28Z0T04
N033T03M06N034G92Z280000
N035G90G00X110000Y-25000S53M03
N036G45Z180000H04
N037G84G99Z164000R165000F2423
N038X110000
N039G80Z185000
N040G45Z280000H04
N041X0Y0

        TOOL #5 : MILL 2" POCKET

N042G91G28Z0T05
N043T04M06
N044G92Z2800000
N0455G90G00X111037Y-45000S76M03
N046G45Z260000H05
N047G01Z134000F2500
N048Y-31037
N049X128963
N050Y-46037
N051X110937Y-45000F1000
N052Y-30937F2500
N053X129063
N054Y-45937
N055G00Z154000
N056G45Z28000H05
N057X0Y0
N058G91G28Z0
N059T00
N060M11
N061M16
N062T05
N063M12
N064M30
=&S
```

FIGURE 21.9
The computer output for the CNC program.

(Figures 21.15 and 21.16, see page 432), permitting both sides of the work to be machined at the same time.

Turning Centers CNC turning centers (Figure 21.17, see page 432) are also an extremely popular and versatile member of the family of equipment used for automated machining. As on the machining center, the turning center's capability lies in its ability to bring a variety of tools into use to machine the workpiece.

Turning center tooling is often turret mounted (Figures 21.18 and 21.19, see pages 432–433). Since the machine spindle holds the workpiece, all the tooling for machining on the outside diameter (OD) as well as the inside diameter (ID) of a workpiece may be installed and remain on the machine during a production run. In many cases, both ID and OD turning operations may be accomplished simultaneously. An example includes thread cutting inside and outside at the same time (Figure 21.20, see page 433).

The CNC turning center is also suitable for shaft turning operations (Figure 21.21, see page 433) and can be equipped with a part-handling robot (Figure 21.22, see page 434) that can load and unload parts from the machine tool automatically.

NC Multispindle Profilers The NC multispindle profiler is another important member of the NC family of automated machine tools. The profiler is a multiple-spindle version of the vertical or horizontal machining center. These machine tools are often quite large. Workholding tables can be 70 or more feet long. The spindles on this machine tool are carried on a gantry supported on each side of the table on a vertical spindle model. On the horizontal spindle model, the work-holding table is vertical and the spindle gantry is supported on a floor rail.

A common design of profiler is the vertical spindle model with three or more spindles (Figure 21.23, see page 434). The horizontal spindle model may also have three or more spindles (Figure 21.24, see page 435). Automatic tool changes are also used on multispindle profilers. These are similar to automatic tool changes commonly found on many NC machining centers. The NC profiler has numerous advantages. These include the capability to machine complex parts in multiple batches. With computer numerical control, the NC profiler is a favorite in the aircraft industry where complex contour machining of duplicate parts is routinely accomplished.

Trends and Other Applications in Numerical Control CNC is by no means limited to machining applications. It can be applied in almost any manufacturing process, including fabric cutting, typesetting, flame cutting, punch press operations (Figures 21.25 and 21.26, see page 435), pipe bending, welding (Figure 21.27, see page 436), tool and cutter grinding, and nontraditional machining

FIGURE 21.10
CNC machining center with twin work-holding pallets and automatic tool changing (Cincinnati Milacron).

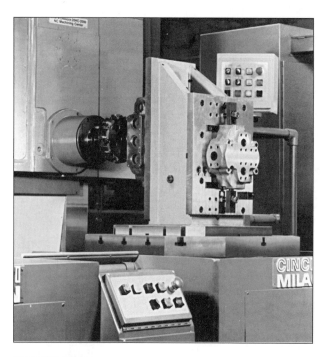

FIGURE 21.11
A face milling cutter is delivered from tool holder to spindle for a face milling task (Cincinnati Milacron).

processes such as EDM, as well as measurement, inspection, and industrial robotics processes.

Numerical control now is and will continue to be a mainstay in manufacturing automation for the foreseeable future. As computer technology advances, NC will advance as well, continuing its important place in manufacturing automation. As computer technology becomes more sophisticated, it will be applied in many new and interesting automation trends in manufacturing. One such area now undergoing development is computer acoustic emission analysis of machining processes. In each machining operation, characteristic acoustic emissions (sounds) are emitted by the machine and tooling during production. These sounds are excellent measures of cutting tool performance. In fact, during conventional manual machining operations, the experienced machinist or machine operator will listen to the sound of the cut as well as observe the visual action. A change in the characteristic sound often will indicate a problem such as a dull or broken tool, almost before the visual aspects of the situation appear. In electronic acoustic analysis, sounds emitted from the production processes are instantly compared to a database in the computer representing the correct characteristic sounds. If the comparison is true, production continues. If the sounds are not comparable, the computer control will stop the machine or take other appropriate corrective action. This is another example of the high degree of sophisticated computer process control that is fast becoming the industry standard.

CAM, CADAM, CIM, AND FLEXIBLE MANUFACTURING SYSTEMS FOR THE INTEGRATED FACTORY

CAM, CADAM, and CIM

In the previous chapter, CAD or computer aided design/drafting was discussed. The counterpart of CAD in the man-

FIGURE 21.12
Multiple workpiece pallets on a CNC machining center with a variety of work-holding methods (Cincinnati Milacron).

ufacturing side is CAM or computer aided manufacturing. These two developing technologies go hand in hand to form CADAM (computer aided design and manufacturing) systems. CIM refers to computer integrated manufacturing. In integrated manufacturing the design and manufacturing engineering functions work together to rapidly create manufacturable products and agile, efficient manufacturing processes. Automated manufacturing equipment is integrated into a system whose individual machines and devices communicate together as a cohesive unit.

In fully integrated CADAM systems, once design data have been generated it can be used to generate the programs needed to operate manufacturing equipment such as CNC machining centers (Figure 21.28, see page 436), flexible manufacturing systems, or industrial robots. Steps in between such as time-consuming and costly

FIGURE 21.13
With pallet indexing capability, different sides of a workpiece or different workpieces can be positioned for machining tasks (Cincinnati Milacron).

FIGURE 21.14
A workpiece pallet may be off-loaded or on-loaded when the machining center is a cell in a flexible manufacturing system (Cincinnati Milacron).

FIGURE 21.15
Duplex CNC machining centers can machine both sides of the workpiece at the same time (Cincinnati Milacron).

FIGURE 21.16
One side of the duplex machining center showing the large workpiece pallet (Cincinnati Milacron).

drafting can be eliminated. Unfortunately, jobs for drafting people may be eliminated as well, bringing up once again the matter of employment under changing technology. The draftspersons of today and the future will have to be computer operators and programmers and be deeply involved with manufacturing engineering as well.

FIGURE 21.17
CNC turning center (Cincinnati Milacron).

Flexible Manufacturing Systems

The ultimate in integration is to completely integrate a fully automated manufacturing plant. This concept is already in use in modern manufacturing and the trend toward automation will continue. However, many jobs are difficult to automate, and some amount of manual labor is necessary. It should be noted that manual labor tasks in future highly automated manufacturing most likely will be on a low-skill, low-pay level.

Major steps toward full factory automation include the flexible manufacturing system with work cells. The

FIGURE 21.18
Turning center turret mounted tooling for ID and OD machining (Cincinnati Milacron)

FIGURE 21.19
Turning center turret tooling for OD shaft turning
(Cincinnati Milacron).

FIGURE 21.20
Turning center used for simultaneous OD and ID thread
cutting (Cincinnati Milacron).

flexible manufacturing system (FMS) consists of manufacturing cells that are equipped with machining centers
or other automated manufacturing equipment. Figure
21.29 (see page 437) shows a 10-station FMS with an automated material-handling system. Figure 21.30 (see page
437) shows a manufacturing cell equipped for vertical
turning operations. Note how the cell is equipped with
pallets for several workpieces. As one part is machined,
another is ready for loading. This increases efficiency
greatly, as the machining centers in the cell are constantly
kept engaged in production (Figure 21.31, see page 438).
Part loading and off-loading also can be accomplished by
industrial robots (Figure 21.32, see page 438).

The entire system of machining centers and cells may
be tied together with an automated material-handling system designed to move parts between cells and cell workstations (Figure 21.33, see page 439). Part transfers are
accomplished automatically by automatic guided vehicles
(AGVs), which are guided from station to station along a
track installed in the factory floor (Figure 21.34, see page
439). Part inspections are also accomplished within the
FMS system at the inspection cell (Figure 21.35, see page

FIGURE 21.21
OD shaft turning with turret-mounted tooling (Cincinnati
Milacron)

FIGURE 21.22
CNC turning center with robot arm for loading and off-loading parts. The automated material-handling system is seen at the left (Cincinnati Milacron).

FIGURE 21.23
Two views of the vertical spindle CNC profiler. The machine is equipped with three spindles and is engaged in machining complex contour aircraft parts (Lockheed Co.).

440). The entire FMS is under the control of a central computer that commands machining center activity as well as material transport and transfer systems.

The FMS has many advantages. Among these are the obvious advantages of increased productivity by automating the process in general. The FMS, because it is built up by selecting only those modules required for specific operations, is extremely versatile (Figure 21.36, see page 440). A further advantage is the flexibility of each cell and machining center to do many different jobs. Thus, the production of the system can be reprogrammed easily to handle a different part without the need for expensive and extensive retooling or the purchase of new manufacturing equipment.

INDUSTRIAL ROBOTS

The industrial technologist, manufacturing technologist, and manufacturing engineer have always sought better, faster, and less expensive ways to accomplish production. In times past when human labor was easily procured at low cost, industries employed many production workers in all types of often repetitive work. In more modern times it has become more difficult and certainly more expensive to hire people to do much of the mundane work of production manufacturing. Modern production workers not only demand wages that are in line with the costs of living, they also demand safe and comfortable working conditions and fringe benefits, all of which add to overall labor costs and must be reflected in the final price of manufactured goods.

Although it is unlikely that automation will displace the production worker completely, it is certain that more and more of the less skilled production work will be taken over by automated equipment. Therefore, the trends toward replacing people with machines will continue. One important contribution toward this is the industrial robot (Figure 21.37, see page 440).

What is a robot? A robot is a fully automatic device that can accomplish a specific task such as assembling or

FIGURE 21.25
CNC punch press (Bracker Corporation/Leland J. Balber Associates).

FIGURE 21.24
Two views of the horizontal spindle CNC profiler. Workpieces are mounted on the vertical face of the large table (Lockheed Co.).

FIGURE 21.26
CNC sheet metal punching is both versatile and productive (Bracker Corporation/Leland J. Balber Associates).

435

FIGURE 21.27
Robotic welding (ESAB Robotic Welding Division).

loading or unloading parts in a machine tool. By adding computer control, the robot may be programmed to do many different functions, thus becoming an extremely versatile, untiring aid to manufacturing automation.

Basic Industrial Robotic Systems

A simple industrial robot may be of the pick-and-place design. These machines, also called polar robots, have limited capability and are primarily designed to pick a part from one location, turn, and place the part in another location or on an assembly.

Since industrial robots are built by people to accomplish tasks heretofore done by people, it is logical that they be endowed with human characteristics. A robot of the jointed-wrist design is sophisticated and can be programmed with motions in multiple axes. The jointed-wrist capability (Figures 21.38 and 21.39, see pages 440–441) permits needed flexibility for complex movement necessary in many industrial applications. Such robots have various types of grippers (fingers) to permit picking up of objects (Figure 21.40, see page 441). The jointed-wrist robot has the capability to turn up and under an assembly, and is thus very useful for such production tasks as painting, welding, and other tasks requiring the turn, bend, and multiposition capability of wrist motion. Depending on the application, mobility may be another important robot feature, permitting the robot to reach the work that it is intended to accomplish (Figure 21.41, see page 441).

FIGURE 21.28
Integrating CAD and CAM in machining (Lockheed Co.).

FIGURE 21.29
Ten-station flexible manufacturing system or FMS with automated parts transfer between cells (Kearney & Trecker).

The true robotic system has sensors and the ability to feed back information to its computer control. Through this, decisions can be made by the robot regarding its assigned task. This is to be differentiated from the abilities of a simple positioning device with only one or two functions and no real programmable capacity. It is also different from a fixed-cycle automated operation where a simple task is carried out repeatedly without any variance and there is no easy way to change or reprogram the function. Sensory capability of industrial robots may include sight, accomplished through use of a video camera. Robots can recognize parts in a bin and then select the correct one to be placed in an assembly (Figure 21.42, see page 442).

The Teachable Robot Programming the teachable industrial robot is accomplished by running the machine through each of its desired positions and functions during an initial manually controlled cycle. The computer remembers each position and cycle that it is taught. Once this is accomplished, further cycles are repeated automatically. Reprogramming is accomplished by clearing the computer's memory of previous instructions and teaching the robot new instructions. If programming is to be saved for future repeat applications, disk drives may be employed.

FIGURE 21.30
Vertical turning work cell, a component of the FMS (Giddings & Lewis).

FIGURE 21.31
CNC work cell machining center with workpiece pallets (Kearney & Trecker).

FIGURE 21.32
Loading and off-loading parts using an industrial robot (Kearney & Trecker).

Industrial Robot Applications

Applications for industrial robots are numerous. Robots can be used to position workpieces in a process or on a machine. They can load and unload parts from a machine (Figure 21.43, see page 442). Many types of assembly operations are well suited to robotics (Figure 21.44, see page 442). Robots can move parts between or to workstations in a factory. In Figure 21.45 (see page 442) AGVs follow a wire embedded in the floor for guidance as they carry engine assemblies to the auto assembly line.

Since the auto industry has shifted to the unibody construction where the car body has become the primary structural unit, robotic spot welding has come into wide use in auto production. Robot welders located on both sides of the assembly line (Figure 21.46, see page 443) permit production spot welding of auto bodies with great accuracy and at a high production rate (Figure 21.47, see page 443). Robots are also used for many specialty applications such as automatic belt tensioning on auto engines (Figure 21.48, see page 444).

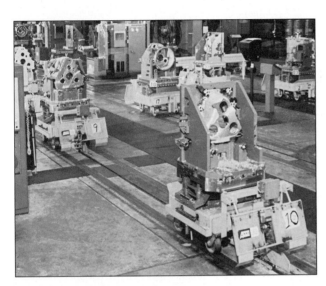

FIGURE 21.33
Transporting workpieces mounted on machining pallets between cells in the FMS (Kearney & Trecker).

FIGURE 21.34
Workpiece transfer in a modern fully automated FMS (Kearney & Trecker).

FIGURE 21.35
The high-precision multiaxis workpiece inspection cell in an FMS (Kearney & Trecker).

FIGURE 21.37
Programmable teachable industrial robot (Positech Corporation).

FIGURE 21.36
Specialty FMS cell machining differential housings for the auto industry (Kearney & Trecker).

FIGURE 21.38
Industrial robot hydraulic jointed-wrist component (Bird-Johnson Fluid Power Division).

FIGURE 21.40
Styles of robotic finger grippers (Barrington Automation Ltd.).

FIGURE 21.39
Motions of the jointed robotic wrist permit great flexibility in applications (Bird-Johnson Fluid Power Division).

FIGURE 21.41
Mobility is often essential to robotic welding applications (ESAB Robotic Welding Division).

FIGURE 21.42
Sight for industrial robots, a most desirable capability
(Lockheed Co.).

FIGURE 21.43
Using a robot to load and unload parts on a CNC turning
center (Cincinnati Milacron).

FIGURE 21.44
Using a robot for an assembly operation (Boeing photo)

ENGINE MERGE
TO BODY

FIGURE 21.45
Robot carriers move engine assemblies to car bodies
(GM—Oldsmobile Division).

FIGURE 21.46
Robot spot welders on both sides of the production line weld auto bodies with great speed and accuracy (Giddings & Lewis).

FIGURE 21.47
CNC robot welders apply 3000 spot welds to each auto body (Chrysler Corporation).

FIGURE 21.48
Belt tension adjusting, a
specialty application for a
specialty type industrial robot
(GM—Oldsmobile Division).

**AUTOMATIC BELT TENSIONING
AND SECURING**

Review Questions

1. What factors prompt manufacturing automation?

2. Are there any negative aspects to automation?

3. What factors are necessary to automation?

4. What is the difference between semiautomatic and full automatic?

5. Explain in general terms what CNC automation is.

6. What are the advantages of CNC?

7. What are the common types of CNC machine tools?

8. What are CAM and CIM and how do they work?

9. What is a flexible manufacturing system?

10. What components comprise an FMS?

11. What are the advantages of an FMS?

12. What is an industrial robot?

13. What sensory capability might an industrial robot have?

14. What is a teachable robot?

15. List some applications for robotics in manufacturing.

Case Problems

Case 1: The Level of Automation

A manufacturer of precision mechanical equipment has many varied precision machining requirements including precision hole drilling, milling, and turning. The products produced consist of several precision subassemblies. The production requirements often are varied depending on the particular models being produced and the quantities required. Automation in machining is desirable because of its ability to meet required close tolerances. The manufacturer may equip the factory with several types of automated machine tools. These include machines that have fixed automation cycles, numerical control (NC) machine tools that are easily programmed for variable requirements, and flexible manufacturing systems that not only can be programmed for different tasks but also will provide a greater degree of automation by shifting parts from station to station.

Considering the primary factor of the need to machine many different close tolerance parts, which automation system would be most applicable—fixed cycle automation, numerical control, or numerical control flexible manufacturing systems? Discuss advantages and disadvantages of each.

Case 2: Robotic Applications

Given the following situations:

A. One automated assembly operation requires merely that certain parts be selected from a supply and then placed in a location for fastening.

B. A second automated assembly operation requires the same parts to be selected from a supply and then placed in a location for fastening. However, in this case, a visual evaluation also must be made to properly reference the part position prior to completing the assembly.

Which operation is suited to robotics: both, only A, only B, neither?

With common readily available robotic technology, which process would be most suited to robotics?

If a common industrial robot was used, what added equipment might make this machine more suited to handling situation B?

CHAPTER 22

The Manufacturing Industry

Objectives

This chapter will enable you to:

1. Describe the structure of a typical manufacturing industry.

2. Identify general job titles in manufacturing and describe responsibilities of each job.

3. Describe the flow of a product through the production manufacturing process.

The manufacturing industry is the place where the products that make an industrialized society function are made. Manufacturing industries make up a large portion of total business and are the places of employment for a large segment of the labor force. The purpose of this chapter is to describe the structure of a typical manufacturing industry, to describe the various jobs and responsibilities of the people who fill them, and to see how a product flows through the manufacturing process.

There are many types of manufacturing industries. They range in size from one- or two-person businesses to major corporations employing several hundred thousand employees. The operations of major manufacturing corporations are accomplished on a worldwide scale, and the dollar value of sales can range from a few thousand dollars in a small business to upward of several billion dollars for a large multinational corporation. Many large companies are parts of larger groups or conglomerates. Companies engaged in the same or similar product manufacturing may form technology groups. In this manner, duplication of product lines is eliminated and technical expertise may be pooled for more efficient business operations, product design, and production. In other cases large manufacturers of complex products such as automobiles may own subsidiaries that supply them with component parts. This ensures that a supply of required parts is always available. The end product manufacturer may also tap much diverse expertise for all phases of business if the secondary industries are owned by the parent corporation. Regardless of the size of the business, all have a basic management structure where job responsibilities are delegated in much the same way.

BUSINESS STRUCTURE

A typical manufacturing industry of average size and having most of the typical departments and job titles will be organized along the lines shown in the organizational chart of Figure 22.1.

Upper Management

Upper management includes the company president, chief executive officer (CEO), chairman of the board of directors, and one or more vice presidents. The responsibilities of the CEO, president, and/or chairman of the board of directors are to oversee the entire operation of the business. The head of the company is responsible to the stockholders and, in many cases, upper management is elected by the stockholders to run and manage the business.

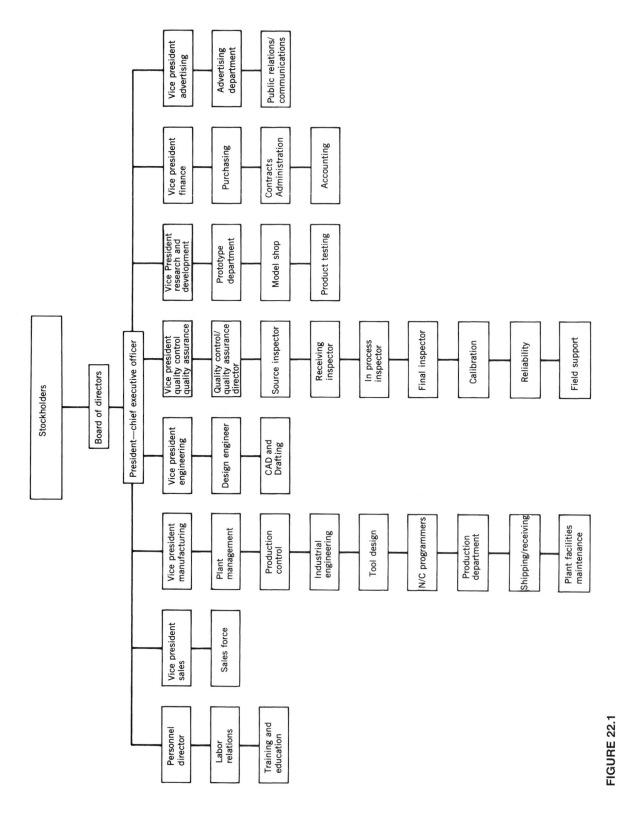

FIGURE 22.1
Basic organizational framework of a typical manufacturing industry.

Vice presidents (VPs) are likely to be specialists in several areas. Examples include manufacturing, finance, advertising, research and development, engineering, quality assurance, and marketing. The VP of manufacturing is responsible for all aspects of the manufacturing operation. The VP for finance is responsible for company loans, stockholder dividends, stock and bond issues, and accounts payable and receivable. The advertising VP is responsible for advertising the company's products if this is done in house. In many cases, advertising is jobbed out to advertising agencies that specialize in product promotion.

If the company does research and development of new products or is in the business of product development for outside customers, the VP for research and development oversees this aspect of the business. Engineering involving product design is the responsibility of the VP for engineering.

Middle Management

Below upper management is middle management. In this area of management many job titles and far-flung responsibilities are found. Middle management includes such job titles as plant managers, department heads, and various other titles. It is here that much of the specific management relative to production of the product is accomplished.

One important job responsibility in middle management is industrial engineering (IE), sometimes called manufacturing engineering or industrial technology (IT). Personnel who staff these jobs provide an invaluable service to the smooth running of a manufacturing industry.

The industrial or manufacturing engineer (IE) or industrial technologist is usually trained in broad areas of manufacturing. This individual knows both the business end of manufacturing as well as the processes used in making the product. Thus, an IE is able to communicate effectively with both upper and lower management.

Some of the responsibilities of industrial engineering include jig, fixture, and tool design, plant equipment and facilities planning, CNC programming, and many aspects of quality assurance and control. Although product design is the responsibility of the design engineer, industrial engineers are often the individuals who make production of the product ultimately possible.

Production Support Management

Many other individuals also contribute to the production process. Although their contributions may be indirect, they are nonetheless vital to the successful function of the business. Job titles in this area include purchasing agents, who are responsible for purchasing tools, equipment, and raw material required for the production process, and production schedulers, who make sure that the proper material is ordered and arrives at the production facilities in a timely manner. This permits production to continue in an orderly fashion with no delays due to a lack of material. The company personnel department is responsible for hiring necessary individuals to meet the staffing needs of the company.

Another vital function in any manufacturing business is filled by individuals involved in plant facilities and production equipment maintenance. Their job is to keep the production line moving. Production manufacturing is at best a costly business—and if the production line stops, pay for the employees does not. The line must be restored to service as soon as possible. Maintenance personnel are therefore always on call in many high-production manufacturing industries. In modern computer-automated manufacturing, the electromechanical and the computer numerical control electronics technicians are often directly responsible for uptime on expensive production equipment. Uptime refers to the time that a piece of manufacturing equipment is actually producing product units. Downtime refers to times when a machine is not producing. Machine downtime can occur after breakdown, during required maintenance, or when the machine is waiting for material to arrive and be loaded for processing. It is naturally desirable to have as little downtime as possible. Thus, the mechanical and electronic maintenance technicians are often very busy in fast-paced production plants. They often are called to work during night shifts or weekend periods in order to keep production equipment operating at peak efficiency. As computer automation trends continue to predominate in modern manufacturing, the maintenance technician will play an ever increasing part in keeping manufacturing equipment in an uptime mode.

Quality assurance and control personnel perform many vital tasks in all phases of manufacturing. Their responsibilities range from source and receiving inspections, first article inspections, production process inspections, final inspections, product reliability, and after-production warranties.

Lower Management

Lower management is directly responsible for production worker supervision and the hour-by-hour production of the product. Job titles at this level include line supervisors and fore, lead, and setup persons.

The Production Worker

No discussion of business structure would be complete without mentioning those who do the actual work of production and production support. Job titles are many and varied in production work and include machine operators, machinists, toolmakers, toolroom attendants, plumbers, sheet metal workers, mechanics, inspectors, electronics assemblers, helpers, electricians, and electronic and mechanical system maintenance technicians.

PREPARING FOR PRODUCTION MANUFACTURING

The business of manufacturing begins when an order is received by the manufacturer or when the company's own designers create a new product that has sales potential. In many cases the order is solicited by the company's sales force as it works in the marketplace to generate business. The sales force is supported by the company's advertising department or an outside advertising agency. The sales force will attend trade shows to demonstrate and promote the company's products and they will make calls on prospective customers.

When an order is received, upper manufacturing management and financial management make the necessary decisions as to how production will be financed and accomplished. If necessary and justifiable, new production facilities may be built or purchased after careful study to determine the profitability of the proposed production. There is always some risk. It is likely that the manufacturer will have to borrow much of the money needed to finance the production. The sales of the product must return this investment so that the company can repay its loans while at the same time meeting its daily financial requirements. These include meeting payroll and purchasing material for the product being manufactured.

Estimating Costs

When the order (Figure 22.2) is received at the manufacturer's facility, it is processed and a price and time of delivery are quoted to the customer. In many cases a manufacturer is asked to bid on a job. Under this system, the customer shops around for the required products and asks several competing companies what their best price is for the manufacturing required. Bids are generally provided at no cost to the customer and the manufacturer must carefully prepare the bid so that it reflects actual production costs plus acceptable profits. Underbidding to get a contract may result in the manufacturer losing money on the job or being unable to complete the work. Both circumstances can result in a business failure. Cost estimating involves a careful analysis of all the costs involved in production of a product. Major factors influencing costs include:

1. Employee wages, salaries, and benefits
2. Interest on borrowed money
3. Price of material
4. Tooling and production machinery
5. Packaging and shipping
6. Quality control and assurance
7. Plant and equipment maintenance and security
8. Utilities
9. Product liability insurance if required
10. Stockholder dividends and other returns to investors
11. Design expense
12. Rent, lease, or construction of manufacturing facilities
13. Guarantees and services included in purchase price
14. Company profits.

All these factors represent a portion of the end product price. By process of careful cost accounting, each expense can be represented as a percentage of the product price. For example, if employee benefits represent 10 percent of the product price, then $10.00 of a $100.00 product represents that portion of the cost attributed to employee benefits. Product costs are also influenced by other complex business cycle factors that all have their direct or indirect influences on the price of a product. Included in these are taxes, depreciation, investment tax credits, and projected sales, often known as sales forecasting. If a manufacturer has a strong sales position in a given market and the national economy is in an upturn, the company may be willing to risk venture capital in plant and/or product expansion, while at the same time holding its prices in line or maybe even reducing prices, if continued strong sales justify this. However, in times of economic recession, or resulting lowered product demand, high rates of interest on borrowed funds, or foreign as well as domestic competition, a company may take a more conservative approach in new plant and equipment investments, since a large debt in a declining economy and/or market can cause a business failure and the unemployment of many workers in the labor force. Accurate cost estimating thus becomes a critical issue in any business and the manufacturer continually strives to estimate costs efficiently while reducing operating costs at all levels.

One method of cost estimating reduces all expenses to a common denominator of production hour rates or, even more simply, the price per unit of production. Production hour rates are often used by service industries or manufacturing concerns that act as incidental vendors. Job machine shops are an example. When a customer asks for a bid on a job, the price is quoted at so much per hour

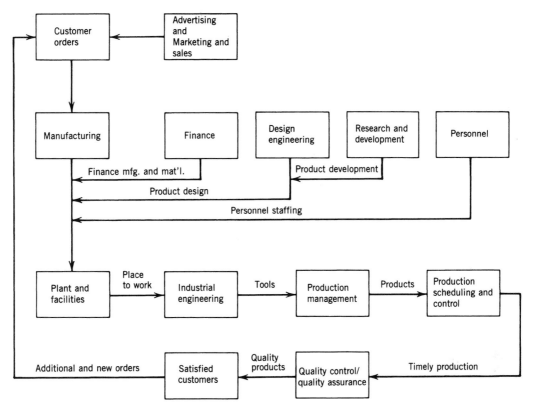

FIGURE 22.2
Manufacturing requires the coordinated efforts of all the team players.

of production time plus any materials that are purchased for the job. The production hour rate reflects all the costs of production with the exception of the material. The estimation of production hours requires that the estimator be generally informed of the processes involved and then be able to estimate accurately the time needed to accomplish the task at average rates of productivity.

Example

production hour rate = $40.00 per hr

time estimate = 10 hr

cost = (time in hours)
(production hour rate)

= (10 hr) ($40.00 per hr)

= $400.00

When products, particularly those that become parts of more complex assemblies, are manufactured on a continuous basis, it is simpler to estimate and quote prices simply on a cost per unit basis. This method is particularly suited to price quotations on various quantities. We have all heard the adage "cheaper by the dozen." In production manufacturing, this bit of philosophy usually turns out to be true.

Once the production process is established, costs are divided by the number of products produced. The more products produced, the less the unit costs for each one.

Example

production unit cost = $1.35 per unit

production rate = 10 units per hr

cost of 20 units = ($1.35 per unit) (20)

= $27.00

hour cost = (units per hr)
(cost per unit)

= (10 per hr)
($1.35 per unit)

= $13.50 per hr

Production hour rates can be converted to cost per unit and vice versa. It is necessary to know how many product units can be produced in a production hour or how many production hours are required to produce a given number of product units. It is important that all costs be reflected in a quotation of any hour rates or costs per unit. If they are not, underbidding may occur and a cost overrun above the estimate given may result. Certain contractual arrangements between customer and vendors do not permit restructuring of estimated costs once prices have been agreed on.

To establish a production hour rate and then a cost per unit estimate, it is necessary to evaluate carefully all the variables that will affect the final price. These expenses can be expressed in terms of percentages of total operating budgets in a business to determine what a customer will have to be charged per worker hour or per unit of production.

Examples

Cost per unit, knowing production rate and production hour rate:

cost per production hour = $45.00

production rate per = 9 units
production hour

$$\text{unit cost} = \frac{\text{production hour cost}}{\text{production rate of units}}$$

$$= \frac{\$45.00}{9} = \frac{\$5.00}{\text{unit}}$$

Production hour cost, knowing unit price and rate of production:

unit price = $12.50

production rate = 15 per hour

$$\text{production hour cost} = (\text{unit price}) \frac{(\text{production rate hour})}{}$$

$$= (\$12.50) \frac{(15 \text{ units}}{\text{production hour})}$$

$$= \frac{\$187.50}{\text{production hour}}$$

Production hour cost breakouts as percentage of expenses. Cost of production hours = $200.00 representing:

Labor 67 percent	(.67)($200.00) =	$134.00
Benefits 12 percent	(.12)($200.00) =	$24.00
Utilities 2 percent	(.02)($200.00) =	$4.00
Material 19 percent	(.19)($200.00) =	$38.00
Total 100 percent		$200.00

Production hour cost as a percentage of unit cost. Unit cost = $1250.00:

Labor 67 percent	(.67)($1250.00) =	$837.50
Benefits 12 percent	(.12)($1250.00) =	$150.00
Utilities 2 percent	(.02)($1250.00) =	$25.00
Material 19 percent	(.19)($1250.00) =	$237.50
Total 100 percent		$1250.00

Facilities, Design, and Purchasing

A production facility is necessary that will accommodate the required equipment and personnel; depending on the product, a new factory may have to be built and personnel hired and/or relocated to staff it. Plant facilities and industrial engineering are responsible for the layout of the factory floor and setup of the production and/or assembly lines in such a way that product flows smoothly through the production process.

If product design is required, it is the responsibility of the engineering design department. Engineering design may be accomplished with the aid of the computer (CAD) in modern manufacturing; production tooling, jigs, and fixtures are developed by the industrial and manufacturing engineering staff. Construction of tooling is done in the tool and die shop or is jobbed out to a vendor.

The purchasing department, staffed by purchasing agents or buyers, is responsible for procurement of materials, goods, or services necessary to support the manufacturing operation.

PRODUCTION MANUFACTURING

When design is completed (Figure 22.3), manufacturing facilities are established and staffed, material is procured, and tooling is in place, production manufacturing may begin. As the major assembly components are completed, they are fitted together to make up the final product (Figures 22.4, 22.5, and 22.6). The final product is possible if and only if all of its many parts are completed and assembled. Sequence of assembly must be properly han-

FIGURE 22.3
Designing in the high-technology age will be done with CAD (Lockheed Co.).

FIGURE 22.4
Major wing subassemblies are separately constructed (Lockheed Co.).

FIGURE 22.5
Major fuselage subassemblies are lowered into place (Lockheed Co.).

FIGURE 22.6
Fuselage sections take shape
in a large production facility
(Lockheed Co.).

dled so that all the components will fit where they belong
in the final assembly.

As the large components go together, work on internal parts that can be done at the same time continues
(Figures 22.7, 22.8, and 22.9). Industrial engineering is
constantly busy solving production problems while designing new tools, jigs, and fixtures. Plant and production
maintenance keeps plant and production equipment functioning properly.

Production control and production scheduling are
responsible for the timely arrival of materials and subassemblies so that the final product may be manufactured
with no delays. Quality assurance and control personnel
carry out their responsibilities at every stage in the production process. Today, control personnel are assisted with
sophisticated instruments (Figure 22.10) in performing
daily tasks.

When major subassemblies of a large and complex
product are finally ready for joining (Figures 22.11 and
22.12, see pages 453–454), the end product begins to take
on a familiar shape. Behind the scenes, in the meantime,
all of the many parts, components, and subassemblies are
being manufactured by thousands of subcontractors much
in a similar manner. Other complete complex subassemblies are shipped from their respective manufacturers for
installation or assembly in the final product (Figures 22.13
and 22.14, see pages 454–455).

The product reaches its final assembly stages while at
the same time new product units are only beginning their
manufacture. The entire process may take place under one
roof (Figure 22.15, see page 455). End of production sees
(cont. p. 456)

FIGURE 22.7
All the details must be completed in proper sequence
(Lockheed Co.).

FIGURE 22.8
Millions of small parts such as rivets must each be installed with the same care and attention (Lockheed Co.).

FIGURE 22.9
Internal work on major subassemblies takes place simultaneously (Lockheed Co.).

FIGURE 22.10
High-technology quality control. Robot-controlled water jets inspect epoxy graphite composite bonded joints on a jet engine fan reverser (Martin Marietta Corporation).

FIGURE 22.11
Completed fuselage sections ready to join wing assemblies (Lockheed Co.).

FIGURE 22.12
The final product takes on a
familiar look (Lockheed Co.).

FIGURE 22.13
Major complex
subassemblies such as
aircraft engines are readied
for installation on the airframe
(Lockheed Co.).

FIGURE 22.14
Gas turbine engines are hoisted for installation on the airframe (Lockheed Co.).

FIGURE 22.15
As finished product nears completion, new products begin their trip through the manufacturing process under the same roof (Lockheed Co.).

FIGURE 22.16
Final detailing, testing, and delivery are the culmination of the manufacturing process (Lockheed Co.).

the final product off the production line ready for final detailing, testing, and delivery to the customer (Figure 22.16). The production effort is a team effort. Thousands of individuals contribute a small part in the making of an incredibly complex product. Their effort is rewarded by the production of a magnificent example of modern industrial manufacturing capability.

Review Questions

1. Outline the structure of a typical manufacturing industry.

2. Discuss some of the responsibilities of the industrial engineer.

3. Who is most likely to be responsible for installing a new electrical service to a piece of production equipment?

4. The company needs 10 million rivets. Who places the order for them?

5. Who would you approach if you were seeking employment with the company?

6. Who is finally responsible for company financial affairs?

7. Who designs production tooling?

8. Who designs products?

9. Describe the flow of production in a typical manufacturing industry starting with order placement.

10. Who do you think is the most important person in the company? Why?

Case Problems

Case 1: Estimating the Cost

Manufacturer X produces a product at a price of $.0234/production unit. The production rate for this company is 325/hr. Manufacturer Y, a competitor, sells an identical item for $.0205/production unit that can be produced at a rate of 375/hr. Calculate the cost of an order for 15,000 parts from each company, and calculate the time to produce each order. Calculate the total cost of each order and calculate the cost per hour of each company's production.

Case 2: Cost Breakouts

A manufacturer builds and sells a product costing $4500 per production unit. Represented in the product cost are:

Wages	60 percent
Employee benefits	11 percent
Utilities	2 percent
Material	20 percent
Other expenses	7 percent

Calculate each dollar cost as represented by these percentages. If the time involved is 48 worker hours, calculate the worker hour cost in this product. Manufacturing 100 units of the product will result in a 16 percent saving. Calculate the worker hour cost and product price under this condition.

Glossary

Acicular Needle-shaped particles or structures as found in martensite.

Acoustic emission Machines and cutting tools emit characteristic sounds when operating at optimum efficiency. These sounds are digitized by computer and the data are compared against a database already in the computer. The information is used to determine that the tool is operating as it should. If not, the computer may take action to stop the machine or indicate a problem to the operator.

Adhesives Materials or compositions that enable two surfaces to join together. An adhesive is not necessarily a glue, which is considered to be a sticky substance, since many adhesives are not sticky.

Aggregate Small particles such as powders that are used for powder metallurgy that are loosely combined to form a whole; also sand and rock as used in concrete.

Aging The process of holding metals at room temperature or at a predetermined temperature for the purpose of increasing their hardness and strength by precipitation; aging is also used to increase dimensional stability in metals such as castings.

AISI Abbreviation for American Iron and Steel Institute.

Allotropy The ability of a material to exist in several crystalline forms.

Alloy A substance that has metallic properties and is composed of two or more chemical elements of which at least one is a metal.

Amorphous Noncrystalline, a random orientation of the atomic structure.

Anisotropy A material that has specific physical properties in different directions. Rolled steel is strongest in the direction of rolling.

Annealing A heat treatment in which metals are heated and then cooled very slowly for the purpose of decreasing hardness. Annealing is used to improve machinability and to remove stresses from weldments, forgings, and castings. It is also used to remove stresses resulting from cold working and to refine and make uniform the microscopic internal structures of metals. Also called *anneal*.

Anodizing Subjecting a metal to electrolytic action, as takes place at the anode of a cell, in order to coat it with a protective or decorative film; used for nonferrous metals.

Assembly drawing One of the two major types of typical manufacturing drawings, showing several parts in an assembly and containing pertinent information regarding the assembly, such as a bill of materials. See *Detail drawing.*

ASTM Abbreviation for American Society for Testing and Materials.

Austempering A heat-treating process consisting of quenching a ferrous alloy at a temperature above the transformation range in a medium such as molten lead; the temperature of the quenching medium is maintained below that of pearlite and above that of martensite formation to produce a tough, hard microstructure.

Austenite A solid solution of iron and carbon and sometimes other elements in which gamma iron, characterized by a face-centered crystal structure, is the solvent.

Austenitizing The process of forming austenite by heating a ferrous alloy above the transformation range.

Autoclave A sealed chamber used to heat and cure parts assembled by adhesive bonding techniques.

Automation Processes and equipment used in mass production manufacturing and designed to accomplish repeat operations with little or no external control or inputs by people.

Bainite An austenitic transformation found in some steels and cast irons. The microstructure consists of ferrite and fine dispersion of cementite.

Bill of materials A list generally appearing on an assembly drawing indicating the name of the parts in the assembly as well as their detail part drawing numbers and possibly the vendor source of the parts.

Binary data Information expressed in terms of zeros and ones called bits. Binary encoded information can be represented as ON (1) or OFF (0) positions of electronic switches within a computer's central processor.

Blanking The operation of cutting a shape with a die from sheet metal stock. The hole material is saved and used for further operations.

Breaking point The final rupture of a material being pulled in tension, after it has reached its ultimate strength.

Bright anneal The softening process for ductile metals that have been work-hardened. The metal is enclosed in a container with inert gas to avoid scaling and then heated to its recrystallizing temperature. Also called *process anneal.*

Brinell hardness The hardness of a metal or alloy measured by pressing a hard ball (usually 10 mm diameter) with a standard load into the specimen. A number is derived by measuring the indentation with a special microscope. The abbreviation *HB* refers to the Brinell hardness number.

Briquette A compacted mass of usually fine material such as metallic powder used in powder metallurgy. Also called a *green compact.*

Brittleness The property of materials to not deform under load but to break suddenly; for example, cast iron and glass are brittle. Brittleness is opposite to plasticity.

BUE, Built-Up Edge A condition in which a metal being cut with a tool adheres to the point or cutting face of the tool, becoming in effect a new rough cutting edge of the tool. This situation causes tearing and a rough finish in the metal being cut.

CAD Computer aided design or computer aided drafting. Drawings are made by computer instead of pencil and paper in this system.

CADAM Computer aided design and manufacturing.

CAD/CAM Computer aided design/computer aided manufacturing. Systems where design is done by computer (CAD) and then actual machine control programs are generated directly (CAM).

Calendering A process that involves rolling of the product into sheets to achieve desired surface finishes and thickness.

Calibration The processes of comparing measuring instruments and gages to known measurement standards and then adjusting them to conform with the standards.

Carburizing A process that introduces carbon into a heated solid ferrous alloy by having it in contact with a carbonaceous material. The metal is held at a temperature above the transformation range for a period of time. This is generally followed by quenching to produce a hardened case.

Case hardening A process in which a ferrous alloy is hardened so that the surface layer or case is made considerably harder than the interior or core. Some case-hardening processes are carburizing and quenching, cyaniding, carbonitriding, nitriding, flame hardening, and induction hardening.

Cast iron Iron containing 2 to $4\frac{1}{2}$ percent carbon, silicon, and other trace elements. It is used for casting into molds. Cast iron is somewhat brittle.

Casting A process of producing a metal object by pouring molten metal into a mold.

Catalyst An agent that induces catalysis, which is a phenomenon in which the reaction between two or more substances is influenced by the presence of a third substance (the catalyst) that usually remains unchanged throughout the reaction.

Cellulose A polysaccharide of glucose units that constitutes the chief part of the cell walls of plants. For example, cotton fiber is over 90 percent cellulose and is the raw material of many manufactured goods such as paper, rayon, and cellophane. In many plant cells, the cellulose wall is strengthened by the addition of lignin, forming lignocellulose.

Cementite Fe_3C. Also known as iron carbide, a compound of iron and carbon.

Centrifuging Casting of molten metals by using centrifugal forces instead of gravity. The mold (or molds) is rotated about a center where molten metal is poured and allowed to follow sprues outward and into the mold cavity.

Ceramic materials These materials demonstrate great hardness and resistance to heat and are used to make cutting tools, coatings on tools, parts subjected to very hot conditions, abrasives, and mechanical parts.

Ceramics Metallic oxides of metals such as silicon and aluminum.

CF Cold finished.

CIM Computer integrated manufacturing. The system that ties design directly to production. A variation of CAD/CAM.

CIP Cold isostatic pressing.

Cladding The joining of one metal (usually sheet or plate) to another by using heat and pressure or by an explosive force. With this method, a thin sheet of more expensive metal or one less likely to corrode may be applied to a less expensive metal or one more likely to corrode.

CNC Computer numerical control. The computer is an integral part of the machine control unit in CNC.

Coining (embossing) Shaping a piece of metal in a mold or die, often creating raised figures or numbers.

Cold drawing Reducing the cross section of a metal bar or rod by drawing it through a die at a temperature below the recrystallization range, usually room temperature.

Cold finish Refers to the surface finish obtained on metal by any of several means of cold working, such as rolling or drawing.

Cold rolling Reducing the cross section of a metal bar in a rolling mill below the recrystallization temperature, usually room temperature.

Cold working Deforming a metal plastically at a temperature below its lowest recrystallization temperature. Strain hardening occurs as a result of this permanent deformation.

Colloid Thick viscous substances, like table jelly or gelatin, which will not pass through a semipermeable membrane.

Compacting The process of making a solid shape in a die by using punches to press together loose powder or a blend of powders; also called molding.

Comparison measurement Measurement accomplished by comparing a known standard to an unknown dimension.

Composite fibers The strands of material used as reinforcement extending through a resin or other matrix in a composite material. An example is carbon fibers in an epoxy matrix. Loads applied to the structure are carried by the fibers.

Composite material Materials exhibiting a much higher strength than the matrix or base material because of reinforcement fibers.

Compressive strength (ultimate) The maximum stress that can be applied to a brittle material in compression without fracture.

Compressive strength (yield) The maximum stress that can be applied to a metal in compression without permanent deformation.

Computer graphics Drawings, pictures, graphs, and so forth generated by a computer.

Cores In metal casting, the hollow parts that are made by using formed sand shapes that are strengthened by baking or by using epoxy.

CR Cold rolled.

Crazing Minute surface cracks on the surface of materials often caused by thermal shock.

Creep Slow plastic deformation in steel and most structural metals caused by prolonged stress under the yield point at elevated temperatures.

Crystal unit structure or **unit cell** The simplest polyhedron that embodies all the structural characteristics of a crystal and makes up the lattice of a crystal by indefinite repetition.

Crystalloid A substance that forms a true solution and is capable of being crystallized.

Cutting fluid A term referring to any of several materials used in cutting metals, such as cutting oils, soluble or emulsified oils (water based), and sulfurized oils.

Decarburization The loss of carbon from the surface of a ferrous alloy as a result of heating it in the presence of a medium such as oxygen that reacts with the carbon.

Deformation Alteration of the form or shape as a result of the plastic behavior of a metal under stress.

Dendrite A crystal characterized by a treelike pattern that is usually formed by the solidification of a metal. Dendrites generally grow inward from the surface of a mold.

Density The density of a body is the ratio of its mass to its volume.

Deoxidizer A substance that is used to remove oxygen from molten metals; for example, ferrosilicon in steel making.

Detail drawing One of the two major types of production drawings showing an individual part. See *Assembly drawing*.

Die casting Casting metal into a mold by using pressure instead of gravity or centrifugal force.

Diffusion The process of atoms or other particles intermingling within a solution. In solids, it is a slow movement of atoms from areas of high concentration toward areas of low concentration. The process may be *(a)* migration of interstitial atoms such as carbon, *(b)* movement of vacancies, or *(c)* direct exchange of atoms to neighboring sites.

Dimensional measurement Equipment and process used to evaluate size of a part or form and location of part features.

Dimensional tolerance Acceptable range of part size or form and positions of part features that will accommodate design specifications.

DNC **1** Direct numerical control. A system in which several machines are controlled by a central computer. **2** Distributed numerical control. The modern form of DNC in which a system of several CNC machines utilizes a central computer for data handling and storage.

Drawing **1** Reducing the diameter of wire or bar stock by pulling through successively smaller dies. **2** Reducing the wall thickness of tubing by drawing it through dies. **3** Forming sheet stock into cuplike parts by pushing it through a die. **4** Tempering of hardened tool steel.

Ductility The property of a material to deform permanently, or to exhibit plasticity without rupture while under tension.

EBM Electron beam machining. The process by which material is removed by an intense beam of electrons impinging on the workpiece.

Eccentricity A rotating member whose axis of rotation is different or offset from the primary axis of the part or mechanism. Thus when one turned section of a shaft centers on an axis different from that of the shaft, it is said to be eccentric or to have "runout." For example, the throws or cranks on an engine crankshaft are eccentric to the main bearing axis.

ECDB Electrochemical deburring. The process in which electron flow and chemical action are used to remove burrs from a workpiece.

ECM Electrochemical machining.

EDM Electrodischarge machining.

Elastic deformation The movement or deflection of a material when an external load is applied that is less than the elastic limit.

Elastic limit The extent to which a material can be deformed and still return to the original shape when the load is released. Plastic deformation occurs beyond the elastic limit.

Elasticity The ability of a material to return to its original form after a load has been removed.

Elastomer Any of various elastic substances resembling rubber.

Electrochemical deburring A manufacturing process wherein an electric current flowing through an electrolytic fluid removes (deplates) material from a workpiece, rounding corners and removing sharp edges.

Electrolyte A nonmetallic conductor, usually a fluid, in which electric current is carried by the movement of ions.

Electromechanical Combining electrical or electronics and mechanical principles, such as a solenoid-operated valve in a hydraulic system.

Electron beam welding The fusion of material by energy imparted from an intense beam of electrons.

Electroplating Coating an object with a thin layer of a metal through electrolytic deposition.

ELG Electrolytic grinding. The process where material is removed by chemical and abrasive processes.

Embossing The raising of a pattern in relief on a metal by means of a high pressure on a die plate.

Equilibrium A condition of balance in which all the forces or processes that are present are counterbalanced by equal and opposite forces. Processes where the condition appears to be one of rest rather than of change.

Eutectic The alloy composition that freezes at the lowest constant temperature, causing a discrete mixture to form in definite proportions.

Eutectoid The alloy composition that transforms from a high temperature solid into new phases at the lowest constant temperature. In binary (double) alloy systems, it is a mechanical mixture of two phases forming simultaneously from a solid solution as it cools through the eutectoid (A_1 in steels) temperature.

Extrusion Creating structural materials in length by forcing a solid metal piece (often heated) through a shaped die by using an extremely high force (similar to squeezing toothpaste from a tube).

Fatigue in metals The tendency of a metal to fail by breaking or cracking under conditions of repeated cyclical stressing that takes place well below the ultimate tensile strength.

Fatigue strength The amount of stress that can be applied to a metal without failure while it is subjected to ten million or more cycles of load reversals. In mild steels, the fatigue strength is about 50 percent of the tensile strength.

Feedback A signal that is returned to the controller indicating what action has or has not occurred, creating closed-loop control.

Ferrite A magnetic form of iron. A solid solution in which alpha iron is the solvent, characterized by a body-centered cubic crystal structure.

Ferrous From the Latin word *ferrum,* meaning iron. Describes an alloy containing a significant amount of iron.

Fiber **1** The directional property of wrought metals that is revealed by a woodlike appearance when fractured. **2** A preferred orientation of metal crystals after a deformation process such as rolling or drawing. **3** Cellulosic plant cells that are used for manufacturing paper and other products. **4** Strands of materials used as reinforcement in plastic products and other materials.

Fiberglass A resin matrix reinforced with glass fibers for strength. A reinforced plastic manufacturing material with many applications.

Filament winding A composite manufacturing process where the end product is to have a hollow internal shape. A filament of the fiber is wound around a form, then bonded in place with the resin matrix.

Final inspection The last inspection done on parts at the completion of manufacturing processes.

First article inspection Inspection performed at the beginning of a manufacturing process to determine whether the first parts produced are correct.

Fixture Production tooling designed to hold and align parts during manufacturing processes.

Flame hardening A means of surface hardening steel or cast iron by applying heat, followed by a quench.

Flash The extrusion of extra material in die casting dies, forging dies, or plastic injection molding dies along the parting line.

Float glass A glass manufacturing process that produces a continuous sheet or ribbon of glass.

Flux A solid, liquid, or gaseous material that is applied to solid or molten metal in order to clean and remove the oxides.

FMS Flexible manufacturing system. A system of material processing stations or work cells connected by automatic material handling and transfer equipment, all under computer control.

Forging A method of metal working in which the metal is hammered into the desired shape, or is forced into a mold by pressure or hammering, usually after being heated to a plastic state. Hot forging requires less force to form a given part than does cold forging, which is usually done at room temperature.

fpm Feet per minute. An English system unit used to measure cutting speed in a machining operation. Also *sfm* and *sfpm.*

Fracture A ruptured surface of metal that shows a typical crystalline pattern. Fatigue fractures, however, often display a smooth, clamshell appearance.

Full automatic Processes in which all phases, once started, are accomplished without the need of further manual input.

Fusion The merging of two materials while in a molten state.

Galvanic corrosion A common type of corrosion process in which a potential difference through an electrolyte causes a deplating (corroding) of the material.

Galvanizing The application of a layer of zinc to the surface of iron and steel for protection from corrosion.

Geometric dimension Dimensions that pertain to the form of a part and the position of part features rather than to size.

Geometric tolerance Tolerances pertaining to geometric dimensions of form and position.

Glue laminated beam A structural wood beam made by gluing thinner boards together until a desired dimension for beam thickness is reached. Glue laminated beams will support large loads and can span long distances with only end support.

GMAW Gas metal arc welding. A shield gas welding process also known as *MIG* welding.

Grain Individual crystals in metals.

Grain boundary The outer perimeter of a single grain where it is in contact with adjacent grains.

Grain growth Called recrystallization. Metal grains begin to reform to larger and more regular size and shape at certain temperatures, depending to some extent on the amount of prior cold working.

Graphics In this text, a reference to pictures generated by computers.

Graphite fiber Strands of carbon in graphite form used in composite materials as the main load-bearing constituent.

GTAW Gas tungsten arc welding. A shield gas welding process also known as *TIG* welding.

Hardenability The property that determines the depth and distribution of hardness in a ferrous alloy induced by heating and quenching.

Hardening The process of increasing the hardness of a ferrous alloy by austenitizing and quenching; also, the process of increasing the hardness of some stainless steels and nonferrous alloys by solution heat treatment and precipitation.

Hard-facing When a hard surface is desired on a soft metal, a hard material (usually another metal) is applied to the surface. Some methods employed to do this are arc welding, spray welding, and electroplating.

Hardness The property of a metal to resist being permanently deformed. This is divided into three categories: the resistance to penetration, the resistance to abrasion, and elastic hardness.

Heading Enlarging the end of a rod or bar of metal by hot or cold upset forging.

Heat sink A large mass of metal that has a high thermal conductivity used to stabilize the temperature of a part held in contact with it. In welding, the mass of the base metal often acts as a heat sink to the weld metal.

HIP Hot isostatic pressing.

Hot pressing Forming or forging tough metals such as alloy steel at high temperatures.

Hot rolling A process of forming metals between rolls in which the metals are heated to temperatures above the transformation range.

Hot-short Brittleness in hot metal. The presence of excess amounts of sulfur in steel causes hot-shortness.

HSLA High strength, low alloy (steels).

Hydraulics **1** The dynamics of liquids (hydrodynamics), specifically, the study of the flow of water in pipes. **2** Pertaining to fluid power where mechanical actions and force are developed by the pumping of fluids (hydraulic oil) at high pressure. Hydraulics may be used to develop large mechanical advantages in many types of mechanisms, such as hydraulic presses and machine tools.

ICAD Integrated computer aided manufacturing. A version of CAD/CAM in which computers are used to control manufacturing processes.

Impact test A test in which small notched specimens are broken in an Izod-Charpy machine. This test determines the notch toughness of a metal.

Inclusions Particles of impurities that are usually found during solidification and are usually in the form of silicates, sulfides, and oxides.

Induction hardening Heating the surface of cast iron or tool steel by means of electromagnetic currents followed by a quench.

Industrial engineer An engineer who places emphasis on manufacturing methods, tooling, and facilities.

Ingot A large block of metal that usually is cast in a metal mold and forms the basic material for further rolling and processing.

In-process inspections Part of quality control and assurance in which inspections occur when required during a manufacturing process.

Interchangeability The concept that parts manufactured by many different manufacturers to the same dimensional specifications may be interchanged in assemblies with no difficulty.

Interface **1** A surface that forms a boundary between two phases or systems. **2** That part of manufacturing equipment where different systems or types of equipment connect to each other to make complete functioning systems, such as computer numerical control and flexible manufacturing systems.

Involute Geometry found in modern gears that permits mating gear teeth to engage each other with rolling rather than sliding friction.

Ion An atom or molecule electrostatically charged by losing or gaining one or more valence electrons.

IPM Inches per minute.

IPR Inches per revolution.

Iron The term *iron* always refers to the element Fe and not to cast iron, steel, or any other alloy of iron.

Isomerism **1** Atoms are said to be isomeric when their nuclei contain the same number of protons and neutrons but differ in energetics and behavior. **2** Compounds are said to be isomeric when they have the same elementary composition (i.e., their molecules contain the same numbers and kinds of atoms) but different structures, and hence, properties. It is believed that differences are due to the arrangement of atoms in each molecule.

Isothermal transformation (I-T) Transformation in some metals that takes place at a constant temperature.

Jig Production tooling designed to guide a cutting tool during production manufacturing processes.

Kaolin A fine white clay that is used in ceramics and refractories composed mostly of kaolinite, a hydrous silicate of aluminum. Impurities may cause various colors and tints.

Killed steel Steel that has been deoxidized with agents such as silicon or aluminum to reduce the oxygen content. This prevents gases from evolving during the solidification period.

Lamellar An alternating platelike structure in metals (as in pearlite).

Laminates Composed of multiple layers of the same or different materials.

Laser Light amplification by stimulated emission of radiation. A device in which heat is derived from the intense coherent beam of laser light energy. This intense, narrow beam of light is used in some welding and machining operations.

Lattice A term that is used to denote a regular array of points in space. For example, the sites of atoms in a crystal. The points of the three-dimensional space lattice are constructed by the repeated application of the basic translations that carry a unit cell into its neighbor.

Lignin A substance that is related to cellulose that with cellulose forms the woody cell walls of plants and the material that cements them together. Methyl alcohol is derived from lignin in the destructive distillation of wood.

Liquidus The temperature at which freezing begins during cooling and ends during heating under equilibrium conditions, represented by a line on a two-phase diagram.

Machinability The relative ease of machining that is related to the hardness of the material to be cut.

Macroscopic Structural details on an object that are large enough to be observed by the naked eye or with low magnification (about 10 ×).

Macrostructure The structure of metals as revealed by macroscopic examination.

Malleability   The ability of a metal to deform permanently without rupture when loaded in compression.

Manufacturing outline (MO)   A document that accompanies parts through the manufacturing process and indicates the steps required, types of machines, and gaging equipment required.

Martempering   Heat treatment of martensitic material, in which martensite transforms to tempered martensite, composed of stable ferrite and cementite phases.

Martensite   Iron phase supersaturated in carbon that is the product of austenite transformation.

Mass production   Continuous production of duplicate product units such as appliances, cars, or any parts of these products.

Measurement standards   Known standards to which production gages and other measurement tools are periodically compared to ensure their conformity and accuracy. Gage blocks are an example.

Metalloid   A nonmetal that exhibits some, but not all, of the properties of a metal. Examples are sulfur, silicon, carbon, phosphorus, and arsenic.

Metallurgy   The science and study of the behaviors and properties of metals and their extraction from their ores.

Methanol (methyl alcohol, wood alcohol)   An alcohol produced by the destructive distillation of wood or made synthetically.

Microscopy   The use of, or investigation with, the microscope.

Microstructure   The structure of polished and etched metal specimens as seen enlarged through a microscope.

MIG   Metal inert gas welding. Also known as *GMAW*. A shield gas welding process that uses a continuous wire electrode fed from a reel.

Modulus of elasticity   The ratio of the unit stress to the unit deformation (strain) of a structural material is a constant as long as the unit stress is below the elastic limit. Shearing modulus of elasticity is often called the modulus of rigidity.

Mold   A general classification of tooling used to form or shape a product while the material is in a molten or liquid state. Mold tooling is found in many manufacturing processes including metal and plastic casting, injection molding, fiberglassing, and composite part manufacturing.

Monomer   A single molecule or a substance consisting of single molecules. The basic unit in a polymer.

M$_s$ Temperature at which *m*artensite *s*tarts to form; in other words, *m*artensite transformation *s*tarting temperature.

Muffled furnace   A gas- or oil-fired furnace in which the work is separated from the flame by an inner lining or "muffle."

NC   Numerical control. Equipment and processes that control manufacturing processes by digital data, often involving the use of computers.

NC axes   On NC manufacturing equipment, the primary paths along which major components of the machine or workpiece move.

NC program   A step-by-step set of instructions written in a format understood by an NC machine and defining all of the information necessary to accomplish the manufacturing task required.

NIST   National Institute of Standards and Technology. A U.S. government organization that, among other functions, maintains measurement standards for U.S. industry.

Nitriding   A process of case hardening in which a special ferrous alloy is heated in an atmosphere of ammonia or is in contact with another nitrogenous material. In this process, surface hardening is achieved by the absorption of nitrogen without quenching.

Nondestructive testing   Application of physical principles for detection of flaws or discontinuities in materials without impairing their usefulness.

Nonferrous   Metals other than iron and iron alloys.

Normalizing   To homogenize and produce a uniform structure in alloy steels by heating above the transformation range and cooling in air.

Notch toughness   The resistance to fracture of a metal specimen having a notch or groove when subjected to a sudden load, usually tested on an Izod-Charpy testing machine.

Open-die drop forging   Also called smith forging. In open-die forging a drop hammer delivers blows of great force to a heated metal that is shaped by manipulating it under the hammer.

Operations sketch   A portion of a part drawing used in a specific manufacturing step emphasizing certain dimensional information and indicating the steps to be accomplished.

Oxidation   The slow or rapid reaction of oxygen with other elements; burning. In metals, overoxidation during heating under oxidizing conditions often results in permanent damage to metals.

Oxidation-reduction   A chemical reaction in which one or more electrons are transferred from one atom or molecule to another.

Part drawing   A drawing of an individual part; a detail drawing.

Patterns   A general class of production tooling representing the shape and/or size of the parts to be made.

Patterns can be used to form molds in casting, and as measurement and gaging tools to check or to guide material cutting and preparation activities.

Pearlite The lamellar mixture of ferrite and cementite in slowly cooled iron-carbon alloys as found in steel and cast iron.

Peel load In metal, plastics, or composites, a force that acts to peel apart joined pieces.

Peening Work hardening the surface of metal by hammering or blasting with shot (small steel balls). Peening introduces compressive stresses on weld surfaces that tend to counteract unwanted tensile stresses.

Perforating Piercing many small holes close together.

Performance inspection An inspection designed to evaluate the performance of a part or product. Examples include the burn-in test for electronics and automobile engine tests.

Periphery The perimeter or external boundary of a surface or body.

Permeability In casting of metals, the term is used to define the porosity of foundry sands in molds and the ability of trapped gases to escape through the sand.

Petrochemicals Chemicals derived from petroleum; substances or materials manufactured from a component of crude oil or natural gas.

Phase A portion of an alloy, physically homogeneous throughout, that is separated from the rest of the alloy by distinct boundary surfaces. The following phases occur in the iron-carbon alloy: molten alloy, austenite, ferrite, cementite, and graphite.

Pickling A process in which metal parts are dipped in acid for the purpose of cleaning and etching prior to plating, painting, or further cold working.

Piercing Cutting (usually small) holes in sheet metal.

Pig iron The product of a blast furnace. It is a raw iron that usually contains about 4.5 percent carbon and impurities such as phosphorus, sulfur, and silicon.

Pitch diameter For threads, the pitch diameter is an imaginary circle, which on a perfect thread occurs at the point where the widths of the thread and groove are equal. On gears, it is the diameter of the pitch circle.

Plasma An ionized gas of extremely high temperature achieved by passing an inert gas through an electric arc. Plasma arcs are used in welding, cutting, and machining processes.

Plastic deformation Deformation that occurs when so much stress is applied to a solid that it does not return to its original condition.

Plasticity The quality of material such that it can be deformed without breaking. Clay is a completely plastic material. Metals exhibit plasticity in varying amounts.

Plating The process of depositing a layer of one metal on another, often done electrically, for the purpose of corrosion protection, appearance, improved electrical conductivity, and other engineering requirements.

Plotter A computer-driven drawing machine used to convert computer drawings to paper drawings.

P/M Powder metallurgy. A process of producing a metal object from metal powders by compaction and sintering.

Poisson's ratio When a rod of elastic material is elongated by stretching (strain), the lateral (crosswise) dimensions will contract; Poisson's ratio is the ratio between the strain and the amount of lateral contraction.

Polymer A chemical compound or mixture of compounds formed by polymerization and consisting essentially of repeating structural units.

Polymerization A chemical reaction in which two or more small molecules combine to form larger molecules that contain repeating structural units of the original molecules.

Precast concrete Concrete products that have been precast in molds and then removed and transported to construction sites. Precast concrete walls for buildings are examples.

Precipitation hardening A process of hardening an alloy by heat treatment in which a constituent precipitates from a supersaturated solid solution while at room temperature or at some slightly elevated temperature.

Preform To bring to the approximate shape and size before the final forming takes place. Used in forging and powder metallurgy.

Process anneal See *Bright anneal.*

Production line The set of stations where various steps in the assembly or making of parts and products is accomplished in a manufacturing industry.

Production tooling Tools, equipment, and special devices necessary for the manufacture of a product.

Productivity A measure of manufacturing efficiency.

Programming The preparation of machine control instructions for numerically controlled manufacturing equipment. Programming may be done manually or by computer (CAM).

Prototype A working model of a product to be manufactured and used for testing and design evaluation, often full sized.

PSI Pounds per square inch.

Pulltrusion A process that is the opposite of extrusion and used in composite part manufacturing.

Punching The operation of cutting a hole in sheet metal using a die. The hole material is scrapped.

Quality assurance All activities in manufacturing directed toward ensuring production of a high-quality product.

Quality control One segment of quality assurance often responsible for dimensional inspections during production.

Quenching The process of rapid cooling of metal alloys for the purpose of hardening. *Quenching media* include air, oil, water, molten metals, and fused salts.

Reciprocate A movement back and forth along a given axis.

Recovery The relaxation and reduction of locked-in internal stresses in cold-worked metals.

Recrystallization A process in which the distorted grain structure of metals that are subjected to mechanical deformation is replaced by a new strain-free grain structure during annealing.

Reduction Reaction involving oxidation-reduction.

Refractory Materials that will resist change of shape, weight, or physical properties at high temperatures. These materials are usually silica, fire clay, diaspore, alumina, and kaolin. They are used for furnace linings.

Residual stress Stresses induced within the structure of a material by cold working, machining, and heat treatments.

RIM Reaction injection molding. A chemical reaction between two resins generates the heat required to make the mix a liquid.

Rimmed steel A low carbon steel (insufficiently deoxidized) that during solidification releases considerable quantities of gases (mainly carbon monoxide). When the mold top is not capped, a side and bottom rim of several inches forms. The solidified ingot has scattered blow holes and porosity in the center but a relatively thick skin free from blow holes.

Robotics Systems and equipment used to operate industrial robots.

Rockwell Hardness A hardness test that uses a penetrator and known weights. Several scales are used to cover the very soft to the very hard materials. The Rockwell "C" scale is used mostly for steel.

RPM Revolutions per minute.

Runout An eccentricity of rotation as that of a cylindrical part held in a lathe chuck being off center as it rotates. The amount of runout of a rotating member is often checked with a dial indicator.

Sacrificial anode A metal slug, usually magnesium and zinc, designed to concentrate galvanic corrosion upon itself and thus save a more important structure on which the anode is attached, such as a ship's hull or a buried pipe line.

SAE Abbreviation for the Society of Automotive Engineers.

Scrap Materials or metals that have lost their usefulness and are collected for reprocessing.

Sealant A sealing agent that has some adhesive qualities: it is used to prevent leakage.

Semiautomatic A process in manufacturing that requires some degree of manual input, but acts without this input for at least part of the cycle.

Sensor A device used to measure some quantity and then send the information back to the control equipment.

Servo A motor or valve that can be controlled by an electrical signal, thus converting electrical inputs to mechanical action outputs.

sfm or **sfpm** Surface feet per minute. Also *fpm* (feet per minute).

Shear load A load that tends to force materials apart by application of side slip action.

Shearing A concentration of forces in which the bending moment is virtually zero and the metal tends to tear or be cut along a transversal axis at the point of applied pressure.

Shell molding A form of gravity casting of metal (usually a high melting temperature metal) in which the mold is made of a thin shell of refractory material.

Shield gas Usually an inert gas used to displace air from around a weld zone, thus keeping the weld uncontaminated.

Shot peening A cold working process in which the surface of a finished part is pelted with finely ground steel shot or glass beads to form a compression layer.

SI Système Internationale. The metric system of weights and measures.

Sintering A process of fusing compacted material such as metal powders into a solid piece by applying heat sufficient to bond, but not melt, the particles.

Skelp The name of the semifinished steel of which butt-welded pipe is made.

Sketch A simple drawing made without the use of drawing instruments.

Slag (dross) A fused product that occurs in the melting of metals and is composed of oxidized impurities of a metal and a fluxing substance such as limestone. The slag protects the metal from oxidation by the atmosphere since it floats on the surface of the molten metal.

Slip planes Also called slip bands. These are lines that appear on the polished surface of a plastically deformed metal. The slip bands are the result of crystal displacement, defining planes in which shear has taken place.

Slurry A watery mixture of insoluble material, such as mud, lime, or plaster of paris.

Smelting The process of heating ores to a high temperature in the presence of a reducing agent such as carbon (coke) and of a fluxing agent to remove the gangue.

Smith forging See *Open-die drop forging.*

Soaking A prolonged heating of a metal at a predetermined temperature to create a uniform temperature throughout its mass.

Sodium silicate Na_2SiO_3. Also called water glass.

Solid solution Found in metals at temperatures below the solidus. Some of the types of solid solutions are continuous, intermediate, interstitial, substitutional, and terminal.

Solidus Seen as a line on a two-phase diagram, it represents the temperatures at which freezing ends when cooling, or melting begins when heating under equilibrium conditions.

Solubility The degree to which one substance will dissolve in another.

Solute A substance that is dissolved in a solution and is present in minor amounts.

Solution heat treatment A process in which an alloy is heated to a predetermined temperature for a length of time that is suitable to allow a certain constituent to enter into solid solution. The alloy is then cooled quickly to hold the constituent in solution, causing the metal to be in an unstable supersaturated condition. This condition is often followed by age hardening.

Solvent A substance that is capable of dissolving another substance and is the major constituent in a solution.

Spalling Breaking small pieces from a surface, often caused by a thermal shock.

Specific gravity A numerical value that represents the weight of a given substance with the weight of an equal volume of water. The specific gravity for pure water is taken as 1.000.

Spheroidizing Consists of holding carbon steel for a period of time at just under the transformation temperature. An aggregate of globular carbide is formed from other microstructures such as pearlite.

Spinning Cold drawing ductile sheet metals into cylinders and other shapes having rotational symmetry in a spinning lathe.

Springback The tendency of a formed metal part to return to some extent to its former shape because of the elasticity of the metal.

Sputtering To dislodge atoms from the surface of a material by collision with high-energy particles for the purpose of depositing a metallic film on a part.

Stainless steel An alloy of iron containing at least 11 percent chromium and sometimes nickel that resists almost all forms of rusting and corrosion.

Standardization The practice of designing machinery components with as many like features as practical.

Statistics Mathematically generated data derived from small samples of large batches on which inferences may be made about the specifications of a large batch without the need to inspect every part.

Steel An alloy of iron and less than 2 percent carbon plus some impurities and small amounts of alloying elements is known as plain carbon steel. Alloy steels contain substantial amounts of alloying elements, such as chromium or nickel, besides carbon.

Strain The unit deformation of a metal when stress is applied.

Strain hardening An increase in hardness and strength of a metal that has been deformed by cold working or at temperatures lower than the recrystallization range.

Strength The ability of a metal to resist external forces. This is called tensile, compressive, or shear strength, depending on the load. See *Stress.*

Stress The load per unit of area on a stress-strain diagram. *Tensile stress* refers to an object loaded in tension, denoting the longitudinal force that causes the fibers of a material to elongate. *Compressive stress* refers to a member loaded in compression, which either gives rise to a given reduction in volume or a transverse displacement of material. *Shear stress* refers to a force that lies in a parallel plane. The force tends to cause the plane of the area involved to slide on the adjacent planes. *Torsional stress* is a shearing stress that occurs at any point in a body as the result of an applied torque or torsional load.

Stress raiser Can be a notch, nick, weld undercut, sharp change in section, or machining grooves or hairline cracks that provide a concentration of stresses when the metal is under tensile stress. Stress raisers pose a particular problem and can cause early failure in members that are subjected to many cycles of stress reversals.

Stress relief anneal The reduction of residual stress in a metal part by heating it to a given temperature and holding it there for a suitable length of time. This treatment is used to relieve stresses caused by welding, cold working, machining, casting, and quenching.

Swaging Compacting or necking down metal bars or tubes by hammering or rotary forming.

Temper In ferrous metals, the stress relief of steels that are hardened by quenching for the purpose of toughening them and reducing their brittleness. In nonferrous metals, temper is a condition produced by mechanical treatment such as cold working. An alloy may be cold worked to the hard temper, fully softened to the annealed temper, or worked to intermediate tempers.

Tension load   A load applied to joined parts that attempts to separate them by a pulling or stretching action.

Thermal conductivity   The quantity of heat that is transmitted per unit time, per unit cross section, per unit temperature gradient through a given substance. All materials are in some measure conductors of heat.

Thermal expansion   The increase of the dimension of material that results from the increased movement of atoms caused by increased temperature.

Thermal shock   A stress induced on the surface of a material such as carbide tools or fire brick caused by a rapid rate of heating and surface expansion.

Thermal stress   Shear stress that is induced in a material due to unequal heating or cooling rates. The difference of expansion and contraction between the interior and exterior surfaces of a metal that is being heated or cooled is an example.

Thermoplastic   Capable of softening or fusing when heated and of hardening again when cooled.

Thermosetting   Capable of becoming permanently rigid when cured by heating; will not soften by reheating.

TIG   Tungsten inert gas welding. See *GTAW*.

Tolerance   The allowance of acceptable error within which the mechanism will still fit together and be totally functional.

Tool and die making   The processes of building specialty production tooling to support manufacture of a product.

Tool design   The process of designing cutting tools, jigs, and fixtures that will support product manufacturing.

Tool geometry   The proper shape of a cutting tool that makes it work effectively for a particular application.

Tool steel   A special group of steels that is designed for specific uses, such as heat-resistant steels that can be heat treated to produce certain properties, mainly hardness and wear resistance.

Tooling   Generally, any machine tool accessory separate from the machine itself. Tooling includes cutting tools, holders, workholding accessories, jigs, and fixtures.

Toughness   Generally measured in terms of notch toughness, which is the ability of a metal to resist rupture from impact loading when a notch is present. A standard test specimen containing a prepared notch is inserted into the vise of a testing machine. This device, called the Izod-Charpy testing machine, consists of a weight on a swinging arm. The arm or pendulum is released, strikes the specimen, and continues to swing forward. The amount of energy absorbed by the breaking of the specimen is measured by how far the pendulum continues to swing.

Transducer   A device by means of which energy can flow from one or more transmission systems to other transmission systems, such as a mechanical force being converted into electrical energy (or the opposite effect) by means of a piezoelectrical crystal.

Transformation temperature   The temperature(s) at which one phase transforms into another phase; for example, where ferrite or alpha iron transforms into austenite or gamma iron.

Transition temperature   The temperature at which normally ductile metals become brittle.

Ultimate strength   Also tensile strength. The highest strength that a metal exhibits after it begins to deform plastically under load. Rupture of the material occurs either at the peak of its ultimate strength or at a point of further elongation and at a drop in stress load.

Ultrasonic   Sound frequency above that able to be heard by the human ear.

UM   Universal milled (steel plate).

Uncertainty principle   A principle in quantum mechanics. It is impossible to assert in terms of the ordinary conventions of geometrical position and of motion that a particle (as an electron) is at the same time at a specified point and moving with a specified velocity.

Upset forging   The process of increasing the cross section of stock at the expense of its length.

Valence   The capacity of an atom to combine with other atoms to form a molecule. The inert gases have zero valence. Valence considers positive and negative properties of atoms as determined by the gaining or losing of valence electrons.

Video display   The television display screen of a computer. Also known as a CRT (cathode ray tube).

Viscosity   The property in fluids, either liquid or gaseous, that may be described as a resistance to flow; also, the capability of continuous yielding under stress.

Visual inspection   An inspection made during manufacturing based on simply looking at a product to determine if any deficiencies are apparently visible.

Void   A cavity or hole in a substance.

Voltage   Electromotive force or a surplus of electrons at a point in an electric circuit. Voltage is equivalent to pressure in a fluid power system.

Vulcanization   The process of treating crude or synthetic rubber or similar plastic material chemically to give it useful properties, such as elasticity, strength, and stability.

Water jet   A manufacturing process in which a material is cut by a high-pressure jet of water often containing an abrasive material to enhance cutting action.

Weldment A unit formed by welding together an assembly of pieces.

Work hardening Also called strain hardening, in which the grains become distorted and elongated in the direction of working (rolling). This process hardens and strengthens metals but reduces their ductility. Excessive work hardening can cause ultimate brittle failure of the part. Stress relief is often used between periods of working of metals to restore their ductility.

Wrought iron Contains 1 or 2 percent slag, which is distributed through the iron as threads and fibers imparting a tough fibrous structure. Usually contains less than 0.1 percent carbon. It is tough, malleable, and relatively soft.

Yield point The stress at which a marked increase in deformation occurs without an increase in load stress as seen in mild or medium carbon steel. This phenomenon is not seen in nonferrous metals and other alloy steels.

Yield strength Observed at the proportional limit of metals and is the stress at which a material deviates from that proportionality of stress to strain to a specified amount.

Index